Nanostructured Materials for Advanced Technological Applications

NATO Science for Peace and Security Series

This Series presents the results of scientific meetings supported under the NATO Programme: Science for Peace and Security (SPS).

The NATO SPS Programme supports meetings in the following Key Priority areas: (1) Defence Against Terrorism; (2) Countering other Threats to Security and (3) NATO, Partner and Mediterranean Dialogue Country Priorities. The types of meeting supported are generally "Advanced Study Institutes" and "Advanced Research Workshops". The NATO SPS Series collects together the results of these meetings. The meetings are co-organized by scientists from NATO countries and scientists from NATO's "Partner" or "Mediterranean Dialogue" countries. The observations and recommendations made at the meetings, as well as the contents of the volumes in the Series, reflect those of parti-cipants and contributors only; they should not necessarily be regarded as reflecting NATO views or policy.

Advanced Study Institutes (ASI) are high-level tutorial courses intended to convey the latest developments in a subject to an advanced-level audience

Advanced Research Workshops (ARW) are expert meetings where an intense but informal exchange of views at the frontiers of a subject aims at identifying directions for future action

Following a transformation of the programme in 2006 the Series has been re-named and re-organised. Recent volumes on topics not related to security, which result from meetings supported under the programme earlier, may be found in the NATO Science Series.

The Series is published by IOS Press, Amsterdam, and Springer, Dordrecht, in conjunction with the NATO Public Diplomacy Division.

Sub-Series

A.	Chemistry and Biology	Springer
B.	Physics and Biophysics	Springer
C.	Environmental Security	Springer
D.	Information and Communication Security	IOS Press
E.	Human and Societal Dynamics	IOS Press

http://www.nato.int/science
http://www.springer.com
http://www.iospress.nl

Series B: Physics and Biophysics

Nanostructured Materials for Advanced Technological Applications

Edited by

Johann Peter Reithmaier
Institute of Nanostructure Technologies and Analytics
University of Kassel
Germany

Plamen Petkov
Department of Physics
University of Chemical Technology and Metallurgy
Sofia, Bulgaria

Wilhelm Kulisch
Institute for Health and Consumer Protection
European Commission Joint Research Centre
Ispra, Italy

and

Cyril Popov
Institute of Nanostructure Technologies and Analytics
University of Kassel
Germany

 Springer

Published in cooperation with NATO Public Diplomacy Division

Proceedings of the NATO Advanced Study Institute on Nanostructured Materials
for Advanced Technological Applications
Sozopol, Bulgaria
1–13 June, 2008

Library of Congress Control Number: 2009920160

ISBN 978-1-4020-9915-1 (PB)
ISBN 978-1-4020-9914-4 (HB)
ISBN 978-1-4020-9916-8 (e-book)

Published by Springer,
P.O. Box 17, 3300 AA Dordrecht, The Netherlands.

www.springer.com

Printed on acid-free paper

CONTENTS

3. CHARACTERIZATION TECHNIQUES

4. PREPARATION TECHNIQUES

5. PREPARATION AND CHARACTERIZATION OF NANOSTRUCTURED MATERIALS

5.3. NANOCOMPOSITES

5.4. THIN FILMS

5.5. CHALCOGENIDES AND OTHER GLASS SYSTEMS

6. APPLICATIONS OF NANOSTRUCTURED MATERIALS

PREFACE

Nanoscience and Nanotechnology are experiencing a rapid development in many aspects, like real-space atomic-scale imaging, atomic and molecular manipulation, nano-fabrication, etc., which will have a profound impact not only in every field of research, but also on everyday life in the twenty-first century. The common efforts of researchers from different countries and fields of science can bring complementary expertise to solve the rising problems in order to take advantage of the nanoscale approaches in Materials Science. *Nanostructured materials*, i.e. materials made with atomic accuracy, show unique properties as a consequence of nanoscale size confinement, predominance of interfacial phenomena and quantum effects. Therefore, by reducing the dimensions of a structure to nanosize, many inconceivable properties will appear and may lead to different novel applications from nano-electronics and nanophotonics to nanobiological systems and nanomedicine. All this requires the contribution of multidisciplinary teams of physicists, chemists, materials scientists, engineers and biologists to work together on the synthesis and processing of nanomaterials and nanostructures, under-standing the properties related to the nanoscale, the design of nano-devices as well as of new tools for the characterization of nano-structured materials.

The first objective of the *NATO ASI on Nanostructured Materials for Advanced Technological Applications* was to assess the up-to-date achieve-ments and future perspectives of application of novel nanostructured materials, focusing on the relationships material structure ↔ functional properties ↔ possible applications. The second objective was the teaching and training of the participants in the scientific background of nanostruc-tured materials, the technologies for their preparation, investigation of their properties and possibilities for their applications. The third objective addressed the cross-border interaction and initiation of international and interdisciplinary collaborations between young scientists from NATO and Partner countries working in the field of nanoscience and nanotechnology, including a transfer of competencies and technology in order to meet one of the priorities of almost all Partner countries.

The ASI covered topics connected with the preparation of nanostructured materials, their characterization and (electro)chemical, mechanical, optical, optoelectronic and biotechnological applications. Thirty **lectures** were given by outstanding scientists from universities and research institutes who are experts in different fields of nanoscience and nanotechnology. In addi-tion, seventeen **thematic seminars** on specific topics were also included in the programme. Three **poster sessions**, namely "Nanostructured films – deposition, properties and applications", "Nanocomposite and hybrid materials – preparation, characterization and applications", and "Nanotechnology, nanosized materials and devices" were held during which the participants presented their current work and established closer contacts in less formal atmosphere.

Seventy seven participants coming from 18 NATO Countries (Bulgaria, Canada, Czech Republic, Denmark, Estonia, France, Germany, Greece, Hungary, Italy, Netherlands, Poland, Romania, Slovak Republic, Spain, Turkey, UK, USA), 9 Partner Countries (Armenia, Belarus, Kazakhstan, FYR Macedonia, Moldova, Russian Federation, Switzerland, Ukraine, Uzbekistan) and 1 Mediterranean Dialogue Country (Israel) insured that the overall objective of transfer of competence and technology in the field of nanostructured materials and their applications was indeed reached on a high level.

We would like to thank the NATO Science Committee for the financial support of the organisation of the ASI. The local organisation was actively supported by the Mayor of Sozopol, Mr. P. Reyzi and by Mrs. I. Vesselinova from the Foundation "26 Centuries Sozopol", whom we gratefully acknowledge.

Johann Peter Reithmaier   Plamen Petkov
Cyril Popov   Wilhelm Kulisch

Kassel – Sofia – Ispra
November 2008

In Memoriam Joe M. Marshall

With great sadness and dismay we learned that Joe Marshall died on 15.12.2008 in the age of 65. Joe was an excellent scientist and a good friend. One of his outstanding qualities was to work with students and young scientists and to motivate them. For almost 20 years he has supported the development of Bulgarian science. We therefore devote this volume to Joe.

Johann Peter Reithmaier   Plamen Petkov
Cyril Popov   Wilhelm Kulisch

Kassel – Sofia – Ispra
January 2009

1. GENERAL ASPECTS

NANOSTRUCTURED MATERIALS FOR ADVANCED TECHNOLOGICAL APPLICATIONS: A BRIEF INTRODUCTION

W. KULISCH*, R. FREUDENSTEIN, A. RUIZ,
A. VALSESIA, L. SIRGHI, J. PONTI, P. COLPO, F. ROSSI
*Nanotechnology and Molecular Imaging, Institute for Health
and Consumer Protection, European Commission Joint
Research Centre, Institute for Health and Consumer
Protection, Via Enrico Fermi, I-21020 Ispra (VA), Italy*

Abstract. In this contribution a short introduction to nanostructured materials for advanced technological applications is presented. A major aim is to demonstrate, on the one hand, the diversity of approaches, methods, techniques and solutions, which are used currently worldwide – but also by the authors of the contributions collected in this book – in the field of nanostructured materials, but also that, on the other hand, these diverse topics are based on the same principles, face similar problems, and bear similar prospects for future applications. For this reason, frequent reference is made to the contributions to this book. Some examples to illustrate current topics, advances and problems are taken from the recent work of the present home institute of the author, the NanoBioTech group of the IHCP at the JRC.

Keywords: nanostructured materials; nanotechnology; critical lengths; SPM techniques; AFM; colloidal lithography; microcontact printing; nanocomposites; nanoparticles; thin films; semiconductor devices; optoelectronic devices; memories; nanobiotechnology; cell–surface interactions; biosensors; nanotoxicology

1. Introduction

Nanotechnology, and alongside nanostructured materials, play an ever increasing role not only in science, research and development but meanwhile also in everyday's life, as more and more products based on nanostructured materials are introduced to the market. The reasons will become apparent not only in this article but also in each contribution to this book.

---

*wilhelm.kulisch@jrc.it

J.P. Reithmaier et al. (eds.), *Nanostructured Materials for Advanced Technological Applications*, 3
© Springer Science + Business Media B.V. 2009

Nanotechnology deals with materials with dimensions of nanometers only, i.e. nanostructured materials. It will be shown in the next section that the physical and chemical properties of materials undergo extreme changes if their sizes are of nanometer dimensions, which opens up a wide range of future, but partly already realized applications. There are at least two more forces driving towards nanotechnology. Most biomolecules and other bio-entities are of nanometer size; thus the nanoscale provides the best opportunity to study such bio-entities and their interactions with other materials. Another impetus is semiconductor industry which, by its ever-lasting demand for miniaturization, has been driven deeply into the nano-realm.

From the above it is clear that nanotechnology is not restricted to a single discipline; rather it requires an interdisciplinary approach, combining the knowledge, but also the methods of scientists from a variety of fields such as physics, chemistry, biology, medicine, engineering, and others. The contributors to this book, their different professions, and also their different topics serve to emphasize this interdisciplinary approach.[a]

The organization of this brief introduction follows that of the book; first, theoretical aspects of nano(structured) materials are discussed; then the problem arising concerning their characterization are addressed. Section 4 deals with the manifold of approaches to fabricate nanostructured materials. In Sect. 5 several types of nano(structured) materials are introduced, with an emphasis on those discussed in more detail in this book. Section 6 finally addresses the applications of nanostructured materials in such diverse field as electrochemistry, energy storage, semiconductor devices, memories, and biotechnology.

2. Theoretical Background

Nanotechnology deals with materials with dimensions in the nanometer range (<100 nm). They may consist of single, isolated nanoparticles, of assemblies of such nanoparticles, of thin films which are two-dimensional nanostructures, of composites in which the constituents are of nanosize dimensions (nanocomposites), of nanosized structures created by lithography and etching as for example semiconductor devices, or of structures created by self-organization of individual nanostructures. Irrespective of the nature of these nano-materials and their fabrication process, they all owe their interesting, sometime exciting properties to a very simple principle: *If the dimensions of materials/structures approach the nanoscale, tremendous*

---

[a] In this context the author would like to emphasize also the proceedings of the first NATO-ASI in Sozopol on "Functional Properties of Nanostructured Materials"[1].

changes of the physical, but also of the chemical properties take place. The latter is mainly due to the drastic increase of the surface to volume ratio which is shown schematically in Figure 1.

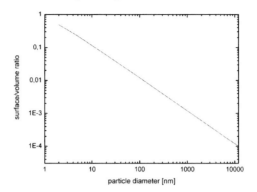

Figure 1. Surface/volume ratio for spherical particles as a function of the particle diameter. The surface was assumed to comprise 0.2 nm.

It can be seen that for a sphere radius of 1 nm, almost 50% of the material is situated at the surface (in nanotubes, all atoms are surface atoms!). This aspect will be revisited in Sect. 6.1.

In the realm of physics, there are likewise drastic changes to expect in the nanometer range. The reason can be expressed as follows: *The properties of a solid can change dramatically if its dimensions (or the dimensions of the constituent phases), the so-called size parameters, become smaller than some critical length associated with these properties.*[2-5] The following few examples, taken from electronics, optics, and mechanics may serve to illustrate this principle (for a more detailed discussion, the reader is referred to Refs. 2–6):

i) **Electric/optical effects**[b]: If the size of nanoparticles becomes smaller than the de Broglie wavelength and the mean free path of electrons (so-called quantum dots), amazing effects came to play a role. The band gap increases, and instead of a continuous density of states (DOS) as in the case of bulk materials (Figure 2a) there are discrete energy levels, similar to the case of atoms (Figure 2d). This has a considerable influence of the optical properties of nanoparticles, as the absorption shifts from the infrared to the visible range.

---

[b] Theoretical approches to electrical properties of nanostructured materials are presented in the two papers of Marshall and Main in Ch. 2.1. The papers of Sharlandjiev, and Turcan, Galemaov and Enaki in Ch. 2.2. deal with optical topics.

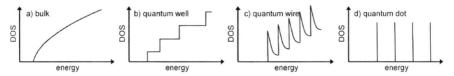

Figure 2. Density of states DOS (schematically) for (a) a bulk semiconductor and (b–d) quantum wells, quantum wires, and quantum dots, respectively.

Quantum confinement in quantum wells (one-dimensional), quantum wires (two-dimensional) and quantum dots (three-dimensional) and the resulting changes of the energy spectrum as shown in Figure 2 has also serious implications for the optoelectronic properties of nanostructured semiconductors, and is currently utilized in optoelectronic devices.[c]

ii) **Magnetic effects**: Ferromagnetic materials consist of domains with parallel magnetization (Weiss domains, Figure 3a). If a magnetic field H is applied, the magnetization of all domains takes the direction of the field and remains in this direction even if the outside field is removed. If the size of ferromagnetic nanoparticles becomes smaller than the critical domain size (10–20 nm), only one domain remains in the particle (Figure 3b). If again a magnetic field is applied, all particles will align according to this field (c), but if the field is removed, thermal motion will lead to a loss of orientation (d). This behaviour is similar to that of permanent magnetic dipoles in a paramagnet and is thus called super-paramagnetism. This effect sets, among others, an upper limit to the miniaturization of magnetic memories.[6] On the other hand, super-paramagnetic particles are envisioned to play an important role in nanobiotechnology and medicine.[7]

Another magnetic nanoeffect which is used presently in magnetic memories is the so-called giant magnetoresistive effect GMR. Depending on the details of the realization, the critical length is either the electron mean free path or the spin relaxation length. For further details, the reader is referred to Refs. 6 and 8.

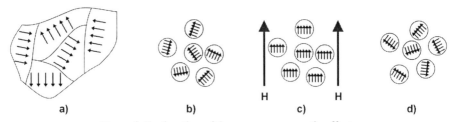

Figure 3. Explanation of the superparamagnetic effect.

---

[c] See in this context the paper of Reithmaier in Ch. 6.3.

iii) **Mechanical effects**: Deformation of ductile materials relies on the migration and multiplication of dislocations. The latter[d] can take place via the Frank–Read mechanism (Figure 4, left). If a dislocation is pinned between two obstacles of distance L (for example dispersed nanoparticles) a critical shear stress $\tau_c = GB/L$ is required to achieve a bowing radius of $R = L/2$ (G is the shear modulus and b the Burgers vector of the dislocation). After reaching this point, the half circle will grow larger automatically, forming a closed circle but with the original dislocation line still existing, which means that now two disclocations exist. If by structural constraints the distance L between the pinning points is limited (e.g. in nanocrystals or by a high density of dispersed nanopartices), very high stresses are required to activate a Frank–Read source, which causes a considerable hardening.

In brittle material failure takes place through the propagation of cracks. It can be shown (Griffith theory[9,10]) that a crack of length a propagates throughout the material if the critical stress $\sigma_c = \gamma E/\pi a$ (γ is the surface energy and E the Young's modulus) is reached. From this equation it is evident that the critical stress is the higher, the lower the length of pre-existing cracks. If the structure of a material constrains the maximum crack length to some nanometers, extremely high stresses would be required to cause catastrophic brittle failure.[11,12]

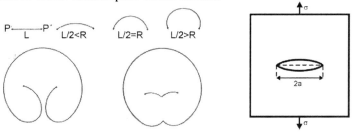

Figure 4. (Left) Frank–Read mechanisms of dislocation multiplication; (right) definition of Griffiths critical crack length.

3.   Characterization of Nanostructured Materials

3.1.   "CLASSICAL" TECHNIQUES

The characterization of nanostructured materials represents an enormous challenge as the dimensions and structures are beyond the resolution of

---

[d] For the migration of dislocations, also a mechanism, the so-called Hall-Petch[12] effect exists, which leads to considerable hardening at grain sizes in the nanometer range.

many standard techniques of materials sciences or thin film technology. For imaging techniques, but also other methods using photons, the Rayleigh criterion $R = 0.61\ \lambda/N$, where R is the resolution, λ the wavelength and N the numerical aperture, sets a strict limit to the resolution of e.g. optical microscopy. There are several possibilities to overcome this limit, e.g. the use of high energy electrons or ions as investigating tools. Techniques such as scanning electron microscopy (SEM) or transmission electron microscopy (TEM) are prominent examples of these approaches; with the latter atomic resolution is possible but sample preparation is time consuming and may introduce artefacts.

Two other examples of characterization techniques based on (high energy) ions and electrons, which yield not only images but other, e.g. chemical information, are Nano-SIMS (secondary ion mass spectrometry) with a spatial resolution of about 50 nm,[13] and XPS (X-ray photoelectron spectroscopy) and PEEM (photoemission electron microscopy) using synchrotron radiation,[14] which – in contrast to hard-to-focus conventional X-rays – can be focused to yield spatial resolutions of 100 nm or even lower.[e]

The ever increasing power of computers and their memories, which itself is a product of nanotechnology as will be discussed below in Sects. 6.2 and 6.3 has opened-up yet another approach as it allows fitting procedures and simulations of various sets of different types of measurements simultaneously, which would have been impossible 10 or 15 years ago. Such simulations may allow to gain information on the nanoscale from measurements performed on the macro- or microscale.[f,g]

At the NanoBioTech group, especially XPS and TOF-SIMS (Time-of-flight-SIMS) are use for the chemical characterization of surfaces. Neither of these methods is able to yield nanometer resolution laterally, but both are extreme surface sensitive. The information depth of XPS is a few monolayers; by varying the angle of incidence of the X-rays, even depth resolved information can be obtained. TOF-SIMS is restricted to the very surface; it is not a quantitative method but it can deliver information on the chemical entities present on a surface. Combining both methods gives a powerful tool to characterize the chemical bonding of the first few nanometers of a surface.

---

[e] Some further examples of using classical techniques in the nano-realm are described in the papers of Morgen and Veres in Ch. 3.

[f] In this context, the reader is referred to the papers on the Reverse Monte Carlo Simulations by Jovari and Kaban in Ch. 3.

[g] Another interesting technique, dynamic light scattering, is discussed by Yannopoulos and Bakaeva in Ch. 3.

3.2. SCANNING PROBE MICROSCOPIES

Finally, an almost new class of characterization techniques has to be emphasized, which allows the characterization of surfaces on the nanometer scale with respect to their topography, but also to a wide variety of other properties. These are the scanning probe microscopies (SPM), which have been rapidly developed after the invention of the scanning tunneling microscopy (STM) by Binnig and Rohrer in 1982,[15] and subsequently that of the atomic force microscope (AFM) by Binnig et al. in 1986.[16] All SPM techniques rely on the same principle: A probe or "tip" with an apex of nanometer dimensions is rastered by means of piezoelectric scanners with nanometer precision over a surface, while the interaction between tip and surface is measured. The difference between all the various techniques can be found in the type of interaction between tip and surface exploited which can rely e.g. on electrical, mechanical, optical, magnetic or thermal effects (for an overview the reader is referred to Refs. 2, 17–21). In the original SPM tunneling currents between a conductive surface and the atoms at the very end of a sharp metal tip are measured; as a consequence, the lateral resolution is extremely high; it is possible to observe the electron distribution around single surface atoms.

In the last years, especially AFM has developed as a routine technique available in many labs. In this method, the tip is situated at the end of a flexible cantilever (Figure 5). Any interaction between tip and surface leads to a bending or twisting of the cantilever which can be detected e.g. by a laser beam which is reflected by the cantilever onto a four-quadrant photo diode. The forces playing a role are, among others, van der Waals forces,

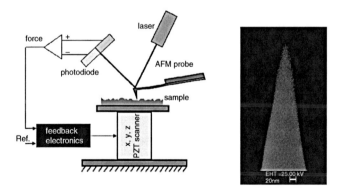

Figure 5. Left: Schematical drawing of an atomic force microscope; right: SEM image of a AFM tip (CGS 11 from NT-MDT, Russia).

electrostatic forces, capillary forces, etc. Depending on the nature of the tip (Figure 5), the lateral resolution can reach some nanometers, but the vertical one 0.1 nm.[h]

However, AFM can not only be used to obtain topography images. In fact, there is a plenty of further applications of this technique:

i) By measuring not only the bending but also the twisting of the cantilever, the friction between tip and surface can be determined.

ii) By measuring force/distance curves or nanoindentation curves, information e.g. on the hardness and the elastic properties of a material can be obtained. This can be done even with biological cells.[22,23] Likewise important is the possibility to investigate the interactions of biomolecules such as proteins[i] or even cells[24] attached to the tip with different (functionalized) surfaces.

An example is presented in Figure 6, showing topography and deflection images of a living cell. For details, the reader is referred to Ref. 23.

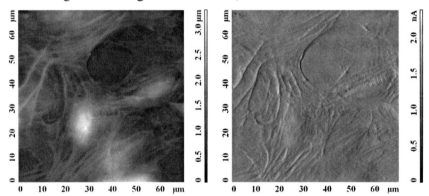

Figure 6. AFM topography (left) and deflection images of a living BALB cell in culture medium at 37°C. The white bump in the center is the nucleus.

iii) AFM or other SPM tips can be used to modify surfaces on the nanoscale. The most famous example is the IBM logo written of 35 xenon atoms on a nickel surface by means of an STM tip.[25] More important, it is among others possible to oxidize anodically in air along the pathway of the tip by applying a voltage between tip and sample,[26] or to use the tip to deposit locally material from metalorganic precursors.[27]

---

[h] Examples of such measurements obtained with UNCD surfaces are presented in the paper of Kulisch and Popov in Ch. 6.4.
[i] An example of this type of measurement can be found in the paper of Kulisch and Popov in Ch. 6.4.

Finally, it should be pointed out that SPM sensors such as the AFM tip shown in Figure 5 are in itself nanostructures, and that alongside with the rapid development of nanotechnology there has also been a rapid tremendous development of SPM sensors. This is but one example of synergistic effects between the development of fabrication and characterization techniques in nanotechnology.

4. Creation of Nanostructured Materials

There are basically two strategies to obtain nano(structured) materials: top-down and bottom-up.[2,28] The first starts from macroscopically structures which are subsequently structured by lithographical and etching techniques down to the nanometer scale. In contrast, the second makes advantage of some kind of self-organization to build up nanostructures from the very building blocks of nature, i.e. atoms and molecules.

The principles of these two approaches are compared schematically in Figure 7; Table 1 summarizes some techniques of both used to create nanostructured bulk materials or nanostructured thin films. The best example for the top-down techniques is the semiconductor industry (see Sect. 6.2) whereas Mother Nature herself can be regarded as providing brilliant examples of the bottom-up approach. Of course, meanwhile also hybrid techniques have been developed combining both approaches.[28]

It can be seen from Table 1 that there is an abundance of different techniques for both of the two approaches to produce nanostructured materials.[j]

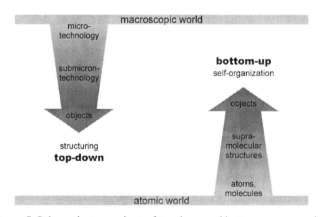

Figure 7. Schematic comparison of top-down and bottom-up approaches.

---

[j] Some special fabrication techniques relying on the use of ion or laser beams are introduced in the papers of Pivin and Stepanov in Ch. 4.

The NanoBioTech group at the JRC uses mainly three techniques to create micro- and nanostructures:

i)   Electron beam lithography
ii) Nanosphere lithography (NSL) or colloidal lithography
iii) Micro (nano) contact printing

TABLE 1. Examples of top-down and bottom-up techniques.

	Top-down	Bottom-up
Bulk materials	Mechanical attrition	Sol-gel
	Ball milling	Precipitation
	Ion implantation	Flame pyrolysis
		Electrodeposition
		Cluster assembly/consolidation
Thin films	Micro/nano structuring	Self-assembly
	Lithography	Self-limitation
	Etching	Self-alignment
		Selected area deposition
		Socused ion beam deposition

Whereas the first is rather common (replacement of photon by electrons to circumvent the Rayleigh criterion), the latter two are ingenious inventions of nanotechnology and should be introduced briefly in the following.

The colloidal lithography makes use of a mask formed by self-assembly of a monolayer of nanospheres.[32-34] One of many possible variants is shown schematically in Figure 8. It consists of five steps:

i)   A substrate is coated by a material A, the choice of which depends on the application in mind.
ii) A colloidal mask consisting of a monolayer of spherical nanoparticles is formed by self-assembly. Size, bulk composition and surface chemistry of the spheres depend again on the application. Typical materials are glass, silica or polymers such as polystyrene (PS); the size ranges from a few nanometers to some hundreds of nanometers. There are several strategies to achieve the assembly of the spheres, e.g. electrostatic adsorption, 2D-crystallization, Langmuir Blodgett techniques, etc.
iii) By a plasma etching step, the size of the spheres is reduced,[k] and the exposed parts of material A between them are removed.

---

[k] It should be noted that even if the size of the spheres remains unaltered, there is (two-dimensionally) uncovered space between them.

iv) A second material B is deposited on top of this structure; again the nature of the material and also the deposition method depend on the application.

v) Finally, the colloidal mask is removed, leaving a nanostructure of materials A and B on the substrate.

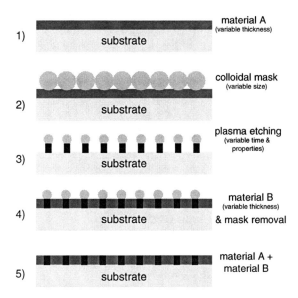

Figure 8. Schematic illustration of the fabrication of nanostructures by colloidal lithography.

Evidently, this is a hybrid of bottom-up and top-down approaches: Whereas deposition and structuring of materials A, B follow the top-down route, the formation of the colloidal mask takes place via a bottom-up step.

An example is presented in Figure 9 showing bio-adhesive nanodomes (material A, poly(acrylic acid)), separated by an antiadhesive (antifouling) polymer (material B, poly(ethylene glycol)). The masks consisted of polystyrene spheres of different sizes. For more details, the reader is referred to Ref. 35. A further example will be presented below in Sect. 6.4.

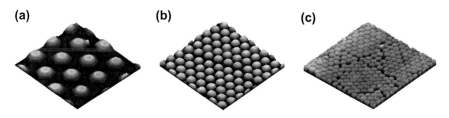

Figure 9. AFM images of nanodomes created by colloidal lithography. The sphere diameter was 1,000 (a), 500 (b) and 200 nm (c).

Microcontact printing (μCP) and meanwhile also nanocontact printing belong to a group of techniques which are summarized under the label "soft lithography". μCP has been developed by the group of Whitesides in the mid 1990s.[29] The basic principles are illustrated in Figure 10. A negative of the structure to be written is created on a durable material (usually silicon) to serve as a mould (master). Then a "stamp" is fabricated by casting a polymer into this mould. Usually, polydimethylsiloxane (PDMS) is used for this purpose; a viscous mixture of a pre-polymer and a curing agent is used for the casting process, which is then cured at slightly elevated temperatures. This PDMS stamp is soft and flexible (thus the name soft lithography) and can easily be remove from the master. In the following patterning process the stamp is coated with the material to be printed, e.g. with biomolecules as in the examples given below in Sect. 6.4.[30,31] The material can then be printed on the substrate of choice. This method has the advantage that the stamp can be used many times to create the same pattern, and the master to produce many stamps. The spatial resolution of this technique depends to a large extent on the resolution of the process by which the master is fabricated, but also on the properties of the PDMS used (presently some tens of nanometers, thus the name nanocontact printing).[29]

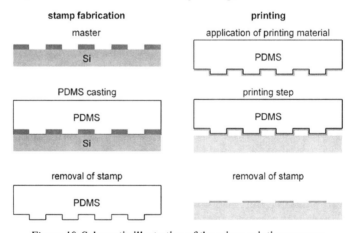

Figure 10. Schematic illustration of the micro-printing process.

One very interesting possibility to realize submicron resolution is to make use of the colloid lithography discussed above to produce the master (Figure 11a).[30] First, a monolayer of polystyrene beads is assembled on a silicon surface. Then the beads are reduced in size by plasma etching to define the size of the printing area. Onto this structure a SiO_x film is deposited. Finally, after mechanical removal of the PS beads by ultrasonic treatment, the negative of the printing stamp remains. Figure 11b shows AFM images of the master and the resulting PDMS stamp. Again, for examples the reader is referred to Sect. 6.4.

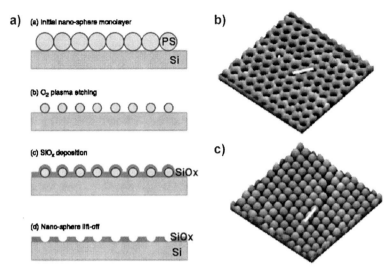

Figure 11. (a) Scheme of the process to fabricate PMDS stamps by colloidal lithography; (b) and (c) AFM images (5 μm × 5 μm × 80 nm) of the master and its replica in PMDS.

5. Types of Nanostructured Materials

There is a plenty of different types of nanostructured materials, for example

i) Nanoparticles[l]
ii) Special nanostructures such as nanotubes, fullerenes, nanowires, nano-rods, nanoonions, nanobulbs and related materials[m]
iii) Thin films with thicknesses below 100 nm, which itself can consist of nanomaterials
iv) Nanocomposites

Almost all of them are addressed in the contributions of this book. In the following, some of them are commented at least briefly.

5.1. NANOCOMPOSITES

Composite materials are engineered materials made from two or more constituent materials with significantly different physical or chemical properties, which remain separate and distinct on a macroscopic level within the

---

[l] Different types of nanoparticles and their applications are described in the papers collected in Ch. 5.2.

[m] Nanotubes and nanotube-containing nanocomposites are addressed in the paper of Maser in Ch. 5.1.

finished structure.[9] One of the most successful composites in history is concrete; other well-known examples are plywood and carbon fiber reinforced plastics. By combining two materials to a composite, physical or chemical properties are achieved which are not possible with a single, homogeneous material. It should not be concealed that also cost aspects have played a role in the development of nanocomposites, as one can combine an expensive material of distinct properties with a low-cost "filler" material.

Nanocomposites are nanotechnology's logical progression of composite materials. In nanocomposites, at least one of the constituents is of nanometer size (e.g. nanocrystallites, nanoparticles, nanotubes, etc.). Thus, in a certain sense nanocomposites also include porous materials, gels, co-polymers and colloids.[n] Thus they combine the properties of the host material with the exiting properties of nanomaterials discussed above in Sect. 4. By this approach, a new class of materials with extraordinary optical, magnetic, electrical and mechanical properties can be achieved.[o]

Another important aspect of nanocomposites is the fact that – compared to macroscopic composites – the surface (interface) to volume ratio is much higher, typically by at least one order of magnitude. As, for example, the mechanical properties of (nano)composites are to a large extent determined by the interface between the two components, this means that much stronger effects can be obtained by nanocomposites, or that much less amounts of the expensive reinforcement material (e.g. nanotubes, nanofibers) are necessary to obtain the same effect.

One special group of nanocomposites are nanostructures glasses (glasses are per definition amorphous, non-crystalline solids). In fact, glasses containing nanoparticles are probably the oldest man-made nanocomposites[38]: medieval stained church glasses contain gold or silver nanoparticles with diameters of 50–100 nm. At the present time, nanocomposite and other nanostructured glasses are – owing to their outstanding optical, electrical, mechanical and chemical properties – again intensively investigated.[p]

5.2. THIN FILMS

In general, thin films are used to render the (surface) properties of a bulk material in those cases where a change of the bulk properties is too expensive or simply impossible to achieve for physical or chemical reasons. In

---

[n] See in this context the papers of Boev and Berezovska in Ch. 5.3.

[o] Nanotube-containing nanocomposites are addressed in the paper of Maser in Ch. 5.1.

[p] Chalcogenide glasses are discussed, among others, in the contributions of Petkov and Petkova in Ch. 5.5, other glass systems by Tonchev and Ilieva also in Ch. 5.5.

addition, combinations of thin films may serve to achieve complex combinations of surface properties.

Thin films are two-dimensional nanostructures by definition. In the field of optics, thin films are defined as structures in which interferences effects don't play a role, i.e. they must be on the order of the wavelength of the light or even thinner. In the field of semiconductor electronics, active and passive layers of about 1.2 nm are currently investigated. Here, the films are moreover meanwhile structured laterally to dimensions of a few tens of nanometers (see Sect. 6.2 below). In recent times, besides homogeneous films also nanostructured films play an ever increasing role. Here, especially, three different types of films are to be mentioned[12]:

i)   Multilayer films consisting of stacks of individual films each of which possesses a thickness of a few nm only[39,40]
ii) Nanocomposite films consisting of nanocrystals embedded in an amorphous matrix[11,41]
iii) Nanocomposite films consisting of nanoparticles, nanotubes etc. embedded in an amorphous matrix

Nanostructured thin films make use of the same principles as the bulk nanocomposites discussed in the previous section; thus they combine the improved properties of nanomaterials (see Sect. 2) with the advantages of using thin films.[q]

Figure 12 shows as example a TEM image of an ultrananocrystalline diamond/amorphous carbon nanocomposite film consisting of diamond

Figure 12. TEM image of an ultrananocrystalline diamond/amorphous carbon nanocomposite film.

---

[q] Nanocrystalline hard films with a variety of properties are described in the paper of Ulrich in Ch. 5.4. The papers of Dimova-Malinosvka also in Ch. 5.4 address poly-Si and ZnO films.

nanocrystals of 4–5 nm diameter embedded in an amorphous carbon matrix.[42] Although the hardness of this material of ca. 40 GPa is below that of diamond single crystals or polycrystalline diamond films (80–100 GPa), the film possess a high toughness and exhibits a highly protecting effect on substrates such as silicon e.g. against scratching.[43]

6. Applications of Nanostructured Materials

Concerning the applications on nanostructured materials, the contributions in this book mainly concentrate on four topics:

i)   Chemical applications: hydrogen storage, batteries and fuel cells
ii) Semiconductor devices
iii) Memories
iv) Biotechnology

In the following, all of them are addressed at least briefly, with an emphasis on biotechnological applications, which is the main topic of the NanoBioTech group at the JRC.

6.1. ENERGY STORAGE, BATTERIES, FUEL CELLS

One of the outstanding properties of nanotubes and other kinds of nanoparticles is their extremely large surface area (see above), which can reach 2,800 m^2/g in the case of carbon nanotubes.[r] This means that ca. 2.5 g of nanotubes possess the same surface area as a soccer field (ca. 70×100 m^2). This makes nano-structured materials ideal for applications in chemical processes relying on surface processes, e.g. catalysis, but also for the storage of energy, especially in the form of hydrogen.

In fact, given the problems with fossil fuels (availability, air pollution etc.), a new technology emerges to provide power supplies e.g. for transportation, buildings, and portable electrical devices, which is called "hydrogen enocomy".[44] It consists of the production of hydrogen from renewable or fossil energy sources e.g. by electrolysis of water, use it as a transportable energy carrier, and then convert it at the point of use to electricity e.g. by means of fuel cells. In each of these steps nanostructured materials (especially carbon nanotubes), given their high surface to volume ratio, are envisioned to a play in future a most important role.[s] However, nanostructured

---

[r] This value was taken from the paper of Maser in Sec. 5.1.
[s] Hydrogen economy is the topic of the paper of Paunovic in Ch. 6.1.

materials will also have a great impact on other technologies for the production, conversion, storage and application of energy (e.g. batteries,[t] solar cells, etc.).

6.2. SEMICONDUCTOR DEVICES

Semiconductor electronics can with some right be regarded as nucleus and driving force of modern nanotechnology. Already in 1965, G.E. Moore from Intel predicted that the number of transistors on a integrated circuit will double each 2 years (Moore's law).[45,46] At that time, the structure widths were about 25 μm. From the upper diagram in Figure 13 it can be seen that Moore's law held more or less up to now. Submicron structures

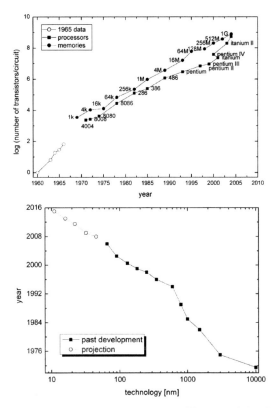

Figure 13. Above: Moore's law and the development of integrated circuit complexity in the past 4 decades.[2] Below: Development of the generations of semiconductor devices.

---

[t] Batteries addressed in the papers of Stankulov and Ketterer in Ch. 6.1

(i.e. nanostructures) were realized already in 1989 (Intel 486 CPU: 800 nm); semiconductor industry is now producing circuits of the 65 nm generation[u] while preparing for the next generations of 45 and 32 nm (see the lower diagram in Figure 13).

This development is the more astonishing as three basic main pillars have remained unchanged over more than 40 years of development although each of them has been proposed to be not suited for this level of miniaturization:

i)   The semiconductor material of choice is still Si despite the partly inferior properties as compared e.g. to III/V semiconductors such as GaAs and its family, which is almost exclusively used for optoelectronic devices.[v]

ii) Photolithography is still the major tool for structuring. As a consequence of the Rayleigh criterion, presently UV radiation of 193 nm is used. Concurring technologies such as ion beam or electron beam lithography – although regarded as promising – could not replace photon-based lithography up to now.

iii) The circuits are still based on the complementary metal oxide semiconductor (CMOS) logic.

Nevertheless, there are limits to this development (which was mainly driven by commercial reasons, i.e. cost aspects[45,46]). They are – one the one hand – related to lithography, but one the other hand closely to the very principles of nanotechnology discussed above in Sect. 2: the change of material properties and/or the applicability of physical laws if the dimensions became smaller than some critical lengths. In the case of semiconductor electronics, there are quite a number of such limits to be taken into account, which does not only regard single transistors but also the circuits. As a detailed discussion is beyond the scope of this article, the reader is referred to Refs. 47, 48.

Thus it seems that further miniaturization of semiconductor devices must completely change the approaches used in all previous generations with great skill, but also with rapidly increasing costs. With other words, we are on the verge of a change from microelectronics (although driven well

---

[u] The term refers to the average half pitch of a memory cell. Printed line-widths can be as low as 25 nm.

[v] Reasons for the success of silicon are, on the one side, the mastery to grow large ingots of single crystalline silicon (currently with a diameter of 300 mm) and, on the other side, the excellent properties of its native oxide (SiO_2) used as passivation but also for active layers, which can be produced by thermal oxidation, and which hitherto has found no match by native oxides of other semiconductor materials.

below the micron limit) to real nanoelectronics. But on the other hand, nanotechnology is ready to provide new solutions to realize semiconductor devices on the nanoscale by utilizing effects only operating on the nanoscale: nanotube transistors, single electron transistors, transistors based on small organic molecules or larger biomolecules, nanoparticle quantum dots, and photonic devices are among the approaches studied presently for their use in future nanoelectronics. It is, however, beyond the scope of this contribution to discuss these developments in details. For further information, the reader is referred to the "International Technology Roadmap for Semiconductors".[49]

As already pointed out, the development sketched above was solely based on silicon as the semiconducting material. There is, however, one extremely important type of devices which can not be realized with silicon as it is an indirect semiconductor. These are the optoelectronical devices such as light emitting diodes, lasers, detectors, etc. Here, direct binary, ternary and quaternary semiconductors from the GaAs or InP families are the materials of choice. But again, nanotechnology and thus nanostructured materials have led to a tremendous development also in this field, making use of quantum dots, quantum wells, etc. in which the spatial confinement of the electrons utilized.[w]

Finally, a very important synergetic effect should be emphasized here. The increasing computing power and the increasing computer memories allow the use of ever improving fabrication and characterization techniques. In other words, the development of the next generation of semiconductor devices is based on the application of the present or the last one.

6.3. MEMORIES

When one of the present authors (W.K.) started his "computer career" in 1983, the only data storage device was the famous 5 1/4 in. floppy disk with a capacity of 110 kB, while the internal memory of his Commodore Pet computer was 32 kB. Today, he uses CDs with a capacity of 700 MB, DVDs with 4 or more GB, and memory sticks with a capacity of 16 GB. The capacity of hard disks for his computer is meanwhile approaching TB.

In the case of memories, there are always two aspects which have to be kept in mind: The data have to be stored in a medium by some kind of effect, but that also have to be written and read within a reasonable time, which at least in the case of reading is defined by real time reproduction of music, films etc. As in the case of semiconductor devices, higher storage

---

[w] See the paper of Reithmaier in Ch. 6.2.

densities have been achieved by reducing the sizes of a bit, and similar to semiconductor devices, the technologies are currently approaching limits by reaching the nanometer region with the consequences emphasized throughout this contribution. In the case of technique using ferromagnets as storage medium (such as hard disks), for example, a lower limit of the size of a bit is given by the superparamagnetic effect discussed in Sect. 2.

For compact discs and their successors which rely on optical readout, the wavelength of the lasers used has been reduced from infrared (CD) over red (DVD) to blue (Blue-ray discs). The pit size for the latter is 160×320 nm^2. However, future improvement will very probably require completely new approaches. One possibility is the use of optical near-field techniques for which the Rayleigh criterion is not applicable, or the use of three dimensional approaches, e.g. holographic techniques. In the former case, the reading head must be situated a few nanometers only from the storage medium.

A large research effort is presently devoted to the so-called phase-change materials, which exhibit drastic differences of their electrical (resistivity) and optical properties (refractive index) properties, depending on their structure (crystalline or amorphous). As a change of the structure can be easily achieved in very short times by heating with a laser beam either to the crystallization or to the melting temperature, data can be written optically, while the readout can take place either electrically or optically.[x,y]

6.4. NANOBIOTECHNOLOGY

6.4.1. Overview

Nanobiotechnology has emerged as one of the most important applications of nanotechnology. There are several reasons driving this development:

i)   Many biomolecules and bio-entities are of nanometer size (Figure 14). Even cells, which are in the micrometer range, consist of nanosized component. Thus, the nanometer range is best suited to study the interactions between bio-entities and non-biological materials.

ii) Adhesion, growth, proliferation and viability of cells strongly depend not only on the chemical nature of the substrate but also on its nano-structure.[50,51,z]

---

[x] Phase change materials are discussed in the paper of Lencer and Wuttig in Ch. 6.2.

[y] Future applications of chalcogenide glasses for memory applications are discussed in the paper of Petkov in Ch. 5.5 and that of Pradel in Ch. 6.2.

[z] The effect of cell nanotopography on cell proliferation is addressed in the contribution of Şaşmazel in Ch. 6.4.

iii) Biosensors will play an ever increasing role in modern medicine. For the development of such sensors, an understanding of the interactions between biomolecules and surfaces on the nanoscale is imperative. On the other hand, devices such as labs-on-a-chip[52] require the sensors themselves to be of micro- or even nanometer size.

iv) Nanoparticles and other nanostructured materials are envisioned to play an important role in drug delivery, i.e. the transport of drugs and other medical substances to the very place of use in the body, and also in medical imaging.[53]

v) Nature herself very often uses nanostructured approaches to realize astonishing material properties. The famous lotus effect is but one example. The nacre of abalone shells is formed by a laminated structure consisting of layers of inorganic material ($CaCO_3$) of about 0.5 μm thickness which are separated by very thin organic layers of less than 10 nm. Such a structure is simultaneously hard, strong, and tough.[54] Of course, nature uses exclusively "bottom-up" strategies to obtain these and many other effects. One important aim of current research is to copy such approaches of nature; such materials are call biomimetic.

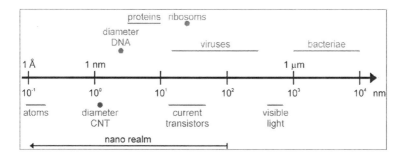

Figure 14. Dimensions of biological and non-biological entities. Note the logarithmic scale.

The remainder of this section will be devoted to examples serving to illustrate these points; they are taken from the present work of the NanoBio-Tech group at the IHCP.

6.4.2. Surface Functionalization

The control of the interaction between material surfaces and biomolecules is essential for the design of high performance biosensors and medical devices. For example immunosensors require not only a high density of biomolecules but also to keep them in an active state and in the correct orientation in order to guarantee specificy, sensitivity, and selectivity. On the other hand, for cell-based devices the surfaces must host cells, guarantee their integrity, and promote their growth and proliferation.

One of the most important aims is therefore to provide surfaces with well-defined physical, chemical, biological and structural properties which allow the immobilization of biomolecules or cells. On the other hand, the surfaces to be used are to a large extent defined by the task of the device under consideration. For biosensors, silicon and various polymers, but also diamond[aa] and optical waveguide materials such as Ta_2O_5 are of importance. For other medical devices, the range of materials is even wider. Usually, the first step of immobilization consists of surface functionalization, i.e. providing the surface with well-defined chemical functionalities which allow subsequent immobilization of biomolecules by well-established chemical processes. In many cases it is necessary to carry out this in a patterned way to achieve several functionalities on the same substrate; moreover, for many applications these patterns should be of nanometer dimensions only.

One way to achieve at least a first step towards this aim are plasma treatments. Silicon surfaces can be rendered OH-terminated by applying oxygen plasmas; likewise, diamond surfaces can be rendered H-, OH- or CF_x-terminated by simple plasma processes.[55] A more general solution, which is almost independent of the substrate used is to deposit a thin (10–20 nm) polymer film which provides the required functionality. By choice of the precursor and the plasma parameter, properties such as functionality, wettability, fouling characteristics, and others can be varied in wide ranges.

As an example, Figure 15 shows the contact angles against water and the adsorption of bovine serum albumin (BAS, a heavily fouling protein)

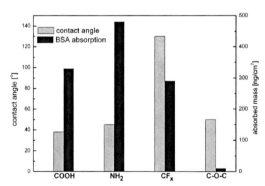

Figure 15. Contact angles and adsorption of BSA (pH 7.5, 45 mg/l) of various plasma polymers. Precursors: COOH: acrylic acid; NH_2: allylamine; CF_x: octafluorocyyclobutane; C–O–C: diethylene glycol dimethyl ether.

[aa] The functionalization of ultrananocrystalline diamond films is discussed in the paper of Kulisch and Popov and that of Vasilchina in Sec. 6.4.

for thin plasma polymer films bearing different functional groups. It can be seen that both properties can be varied in wide ranges. The contact angle is highest for the teflon-like film with CF groups, while the PEO-like (poly(ethylene oxide)) film with C–O–C groups is strongly antifouling.

Another approach to achieve different surface functionalizations – again if required in a patterned way – is to make use of self-assembled monolayers (SAMs) of organic molecules bearing a special functional group. For example, organothiols (molecules with a –SH group at one end) spontaneously organize a SAM on a gold surface. As thiols with a wide variety of functionalities exist (e.g. –CH$_3$, –COOH, –NH$_2$, etc.), by this method easily surface properties such as wettability, fouling character and others can be adjusted. As the thiols only react with gold coated surfaces, patterning can easily be achieved by a patterned gold layer, e.g. created by colloidal litho-graphy. SAMs can also be obtained e.g. by organosilanes on oxidized Si or diamond surfaces.

An example[34] involving colloidal lithography is shown in Figure 16. First a SAM of mercaptohexadecanoic acid (HS(CH$_2$)$_{15}$COOH, MHD) is formed on a gold surface. On top of this a monolayer of polystyrene beads with diameters of 500 nm is formed. In a subsequent etching step the beads are reduced in size and the exposed MHD is removed. Then a SAM of hexadecanethiol (HS(CH$_2$)$_{15}$CH$_3$, HDT) is formed on the freshly exposed gold surface between the MHD island. Finally, the PS beads are removed ultrasonically, leaving a surface patterned with two different functionalities (MHD: COOH; HDT: CH3) and also with different heights ($\Delta l \approx 0.4$ nm).

Yet another approach of nanopatterning bioadhesive spots in an anti-fouling environment by plasma processes and electron beam modification is described in Ref. 56.

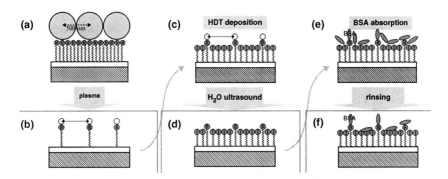

Figure 16. Experimental procedure to produce MHD/HDT nanopatterns.

6.4.3. Interactions of Biomolecules and Cells with Surfaces

It has already been pointed out that adhesion, growth, proliferation and viability of cells depend not only on the physical, chemical and biological properties of a surface but also on its nanostructure.[50,51] A great deal of work of the NanoBioTech group is therefore devoted to the immobilization of biomolecules and cells on nanopatterned surfaces created by the techniques described above and also in Sect. 4. These investigations are of importance for basic studies as well as for applications. For basic investigations the use of nanopatterned substrates allows:

i) In the case of biomolecules such as peptides, enzymes or proteins the study of the thermodynamics and kinetics of adsorption processes[34] and

ii) In the case of cells to investigate the influence of the nanostructure on cell adhesion, migration and spreading mechanisms, as well as to study the metabolisms of cells and their interaction between one another and with surfaces[31]

On the other hand, such nanopatterned surfaces are of course of great importance for applications such as biosensors, chemical sensors and tissue engineering. Another interesting effect is that nanostructured surfaces with a contrasted chemical response (e.g. adhesive/antifouling) leads to an increase of the recognition sensitivity e.g. between antigen/antibody pairs as compared to unstructured surfaces.[34]

In the following, but a few examples are presented to demonstrate these approaches. The surface patterned with two different thiols discussed above has been exposed to BSA (Figure 16e). By means of AFM measurements and other techniques it was proven that the BSA molecules almost exclusively attach on the COOH-terminated MHD (Figure 16f).

A similar result was obtained with the bioadhesive nanodomes surrounded by an antifouling environment created by colloidal lithography as discussed above in Sect. 4 (Figure 8). As can be seen in Figure 17, BSA molecules will attach exclusively on the bioadhesive nanodomes.

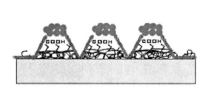 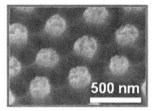

Figure 17. Immobilization of BSA proteins on PAA nanodomes, which were created by nanosphere lithography and which are separated by an anti-fouling PEG. Left: Schematical drawing; right: SEM image.

Whereas the examples discussed hitherto rely on colloidal lithography the following stem from micro/nano contact printing.

Figure 18 shows AFM images of poly-L-lysine (PLL, a small polypeptide of the essential amino acid L-Lysine) stamped on an antifouling PEO-like polymer (see above). The height of the PLL areas is about 1.1 nm.[57]

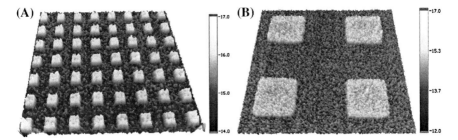

Figure 18. Thickness maps of PLL stamped on plasma-deposited PEO-like at pH 7 calculated from ellipsometric measurements. The thickness is given in nm with respect to the Si surface. Size of the maps: (a) 1,500 x 1,690 μm and (b) 380 × 390 μm.

Figure 19 shows in the upper row fluorescence images of FITC-labeled PLL patterns, in the lower TOF-SIMS images of the structures. The PPL areas are clearly visible in the CN-images, while the PEO regions could be visualized by the C_3–H_3–O pictures.[57]

Figure 19. Fluorescence microscopy images of the PLL patterns. Long inter-patch distances (a) are suited to study stem cell maintenance and long migration while close islands (b) and islands connected by lines (c) are used for monitoring short migrations and outgrowth projections. ToF-SIMS mapping showed good chemical contrast between PLL positive areas (d, f) recognized by C–N, and PEO-like positive background (e, g) ascribed to C_3–H_3–O.

For the stamping of PLL, also the nanosized stamps produced by colloidal lithography (Figure 11) can be used as is evident from Figure 20.[30]

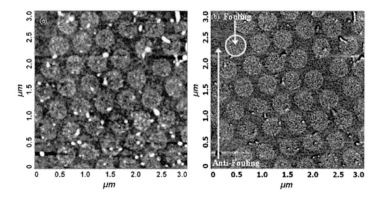

Figure 20. Tapping AFM topography (a) and phase lag (b) images of poly-L-lysine nano-patterns on PEG created by nanocontact printing using 50 gr. PDMS stamps.

Finally, human umbilical cord neural stem cells (HUCB-NSC) have been incubated on larger-space PLL patterns.[57,58] The results are shown in Figure 21. It can be seen that the cells attach to the pattern. After 2 days, they are still confined inside the PLL areas. But after 7, and even more after 9 days, they migrate out of the islands and even travel between them. Moreover, they form at their periphery long extensions. It is obvious that such studies are of great advantage in the understanding of cell–cell and cell–substrate interactions.

6.4.4. Biosensors

A biosensor is an analytical device which incorporates a biological sensing element which is directly coupled to a physicochemical transducer. Its signals – either discrete or continuous – are proportional to the concentration of the target analyte. The interaction between recognition element and analyte takes place on the molecular scale; it can be detected as a local change in mass or of the electrical or optical properties. Biomolecules such as enzymes and antibodies or even cells with highly specific recognition abilities are used as recognition elements. Among others, piezoelectric materials (e.g. in quartz crystal microbalances), electrodes or optical waveguides[bb] are used as transducer, yielding a measurable electrical or optical signal as a consequence of the biomolecule–analyte reaction.

[bb] The paper of Kulisch in Sec. 6.4 deals with Ta_2O_5 films as waveguides for biosensors.

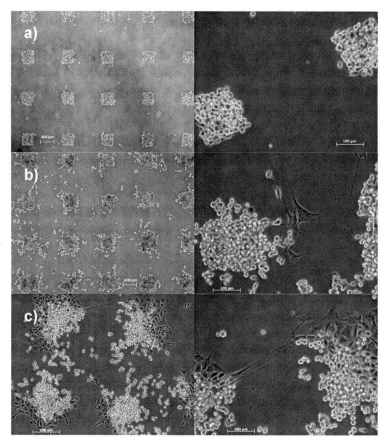

Figure 21. HUCB neural stem cells incubated on PLL patterns for 2 days (a), 7 days (b) and 9 days (c).

6.4.5. Nanotoxicology

The unique and diverse physicochemical properties of nanoscale materials suggest that also their toxicological properties may differ from those of the corresponding bulk materials. The potential occupational and public exposure – through inhalation, oral ingestion, dermal absorption or by injection – of manufactured nanoparticles (mNPs) with particles size ≤100 nm probably will increase in the near future, due to the ability of nanomaterials to improve the quality and performance of many consumer products as well as the development of therapeutic strategies and tests. However, there is still a lack of information about the impact of mNPs on environment and on human health, as well as of reliable data on risk assessment.[59,60]

To overcome these problems a research plan on the toxicology of nanoparticles has developed at the IHCP, based on an integrated approach

combining cell-based *in vitro* assays with specific radiochemical and physicochemical facilities.[61]

The research consists in the selection of mNPs, based on industrial and
biomedical interest, their synthesis, and the assessment of their physico-
chemical characterization and toxicological profile using *in vitro* systems,
which are relevant for human exposure and represent the different exposure
routes (derma, blood, lung, intestine, liver and kidney).

In particular, a first screening on basal cytotoxicity is carried out with
human cheratinocytes (HaCaT), human alveolar basal epithelial cells
(A549), human intestinal epithelial barrier cells (Caco-2 cell lines), dog
kidney cell lines (MDCK), hepatocellular carcinoma (HepG2), and peripheral blood mononuclear cell (PBMC) by standard *in vitro* methods, an
approach chosen also to reduce the number of experiments with living
animals.[62]

These *in vitro* methods are adapted and optimized for nanotoxicology; if
necessary new *in vitro* tests are developed. In this context, standardized *in
vitro* tests such as vital staining methods (Neutral Red uptake and MTT
assay) and colonigenic assays (colony forming efficiency CFE) are compared. The results showed that CFE is the most promising *in vitro* method.
It consists of seeding a fixed number of cells, expose them to mNPs for 72 h
and then maintain them in culture for 7 days (Figure 22).[63] No dye is used
for the analysis; the colonies formed stem from the surviving cells. The
cytotoxicity is expressed as percentage of a control sample. This test permits
to observe the basal cytotoxicity without considering specific toxicological
effects, and also allows the cells to eventually repair themselves after the
exposure, or die to by apoptosis or necrosis.

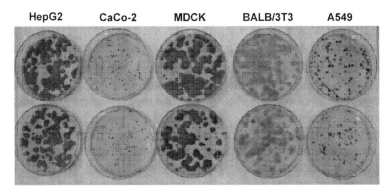

Figure 22. CFE test carried out with different cell lines. The petri dishes in the upper row
contain colonies of untreated cells (control), those in the lower cells exposed for 72 h to 100
µg/ml MWCNTs (no cytotoxicity is observed excepted for the A549 cells). The colonies
show different morphologies related to the cell lines. In general larger colonies are formed
by HepG2, MDCK and Balb/3T3, smaller by CaCo-2 and A549.

Since the behaviour and effects of nanoparticles in biological systems is not yet well understood, no effect can be neglected. In fact it was observed by SEM and optical microscopy that mNPs interact with cells, are internalised and seem to have a high biopersistence (Figure 23).

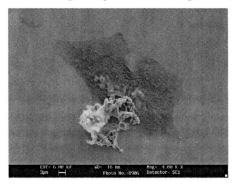

Figure 23. SEM image of A549 cells exposed for 72 h to 100 μg/ml MWCNTs. The cells were fixed using formaldehyde and dried under a laminar flow hood. In the picture two cells are interacting with a MWCNT aggregate which was not detached during the sample preparation.

The uptake can be quantified using nanomaterials radiolabelled by neutron, deuteron or proton irradiation (using the cyclotron facility at the JRC). In the case of Au and Co mNPs, it could be demonstrated that the irradiation procedures do not change the physicochemical structure of the mNPs.[64]

As it was observed that standard methods are not useful to study the intracellular distribution of mNPs, new well-standardized protocols to separate the different intracellular organelles fractions (nuclei, mitochondria, lysosome, endoplasmic reticulum) and to measure the presence of mNPs inside them were developed. In addition the intracellular fate is observed by TEM.

Inflammation, genotoxicity, carcinogenic potential and toxicity to specific organs must be considered and studied, even if no cytotoxicity is observed. Immunotoxicity and genotoxicity are tested in human peripheral blood mononuclear cells (PBMC). The carcinogenic potential is assessed by immortalised mouse fibroblasts (Balb/3T3 cell line, Figure 24) by the cell transformation assay (*in vitro* model to study carcinogenic potential of chemicals).

For *in vitro* organ toxicity studies, specific end-points such as epithelial integrity, measured by trans epithelial electrical resistance (TEER), and epithelial adsorption measured by radiotracers are studied for Caco-2 and MDCK cells.

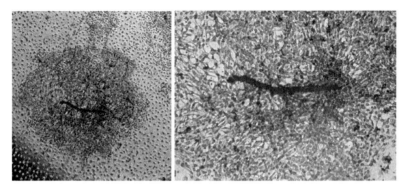

Figure 24. Morphologically transformed colony of Balb/3T3 cells after 72 h exposure to COOH-functionalized MWCNTs (10 µg/ml). The transformed cells are organized in a dense multilayer, randomly orientated, are invasive into the monolayer, and the colony has the morphology of a focus edge.[65] In the centre of the transformed colony a CNT filament is observable. Around the transformed colony, the cells have the morphology of a typical control sample (monolayer).

ACKNOWLEDGEMENTS

W.K. likes to thank his colleagues G. Ceccone, R. Colognato, F. dos Santos, and D. Gilliland from the NanoBioTech group at the IHCP of the Joint Research Centre for providing their results used as examples in this paper and discussing them with him.

References

1. R. Kassing, P. Petkov, W. Kulisch, and C. Popov (Eds.), *Functional Properties of Nanostructured Materials* (Springer, Berlin, 2006).
2. R. Kassing and W. Kulisch, in R. Kassing, P. Petkov, W. Kulisch, and C. Popov (Eds.), *Functional Properties of Nanostructured Materials* (Springer, Berlin, 2006), p. 3.
3. P. Moriarty, Rep. Prog. Phys. 64, 297 (2001).
4. J. Jortner and C.N.R. Rao, Pure Appl. Chem. 74, 1491 (2002).
5. R.W. Siegel, Nanophase materials, in G.L. Trigg (Ed.), *Encyclopedia of Applied Physics* (VCH Publishers, Weinheim, 1994) vol. 11, p. 173.
6. R.D. Shull, Int. J. Iron Steel Res. 14, 69 (2004).
7. W. Zhou, P. Gao, L. Shao, D. Caruntu, M. Yu, J. Chen, and C.J. O'Connor, Nanomed. Nanotech. Biol. Med. 1, 233 (2005).
8. F.J. Himpsel, T.A. Jung, and P.F. Seidler, *IBM J. Res. Dev.* 42, 33 (1998).
9. J.C. Anderson, K.D. Leaver, R.D. Rawlings, and J.M Alexander, *Material Science* (Chapman & Hall, London, 1990).
10. G. Gottstein, *Physikalische Grundlagen der Materialkunde* (Springer, Berlin, 2001).
11. S. Veprek, J. Vac. Sci. Technol. A 17, 2401 (1999).
12. W. Kulisch, in R. Kassing, P. Petkov, W. Kulisch, and C. Popov (Eds.), *Functional Properties of Nanostructured Materials* (Springer, Berlin, 2006), p. 113.

13. S. Lozano-Perez, M.R. Kilburn, T. Yamada, T. Terachi, C.A. English, and C.R.M. Grovenor, J. Nucl. Mater. 374, 61 (2008).
14. O. Renault, N. Barret, A. Bailly, L.F. Zagonel, D. Mariolle, J.C. Cezar, N.B. Brookes, K. Winkler, B. Krömker, and D. Funnemann, Surf. Sci. 601, 4727 (2007).
15. G. Binnig and H. Rohrer, Helv. Phys. Acta 55, 726 (1982).
16. G. Binnig, C.F. Quate, and C. Gerber, Phys. Rev. Lett. 56, 930 (1986).
17. N.J. DiNardo, *Nanoscale Characterization of Surfaces and Interfaces* (Verlag Chemie, Weinheim, 1994).
18. R. Kassing and E. Oesterschulze, in B. Bushan (Ed.), *Micro/Nanotribology and Its Application* (Kluwer NATO ASI Series, Dordrecht, 1997).
19. H.J. Güntherodt and R. Wiesendanger (Eds.), *Scanning Tunneling Microscopies I* (Springer, Berlin, 1995); H.J. Güntherodt and R. Wiesendanger (Eds.), *Scanning Tunneling Microscopies II* (Springer, Berlin, 1995).
20. G. Friedbacher and H. Fuchs, Pure Appl. Chem. 71, 1337 (1999).
21. E. Meyer, S.P. Jarvis, and N.D. Spencer, MRS Bulletin 29, 443 (July 2004).
22. L. Sirghi and F. Rossi, Appl. Phys. Lett. 89, 243118 (2006).
23. L. Sirghi, J. Ponti, F. Broggi, and F. Rossi, Eur. Biophys. J. 37, 935 (2008).
24. K.F. Chong, K.P. Loh, S.R.K. Vedula, C.T. Lim, H. Sternschulte, D. Steinmüller, F.S. Sheu, and Y.L. Zhong, Langmuir 23, 5615 (2007).
25. D.M. Eigler and E.K. Schweizer, Nature 344, 524 (1990).
26. Y. Kaibara, K. Sugata, M. Tachiki, H. Umezawa, and H. Kawarada, Diamond Relat. Mater. 12, 560 (2003).
27. D.S. Saulys, A. Ermakov, E.L. Garfunkel, and P.A. Dowben, J. Appl. Phys. 76, 7639 (1994).
28. B.K. Teo and X.H. Sun, J. Clust. Sci. 17, 529 (2006).
29. Y. Xia and G.M. Whitesides, Angew. Chem. Int. Edit. 37, 550 (1998).
30. A. Ruiz, A. Valsesia, F. Bretagnol, P. Colpc, and F. Rossi, Nanotechnology 18, 505306 (2007).
31. A. Ruiz, L. Cereotti, L. Buzanka, M. Hasiwa, F. Bretagnol, G. Ceccone, D. Gilliland, H. Rauscher, S. Coecke, P. Colpo, and F. Rossi, Microelectron. Eng. 84, 1733 (2007).
32. J.C. Hulteen and R.P. Van Duyne, J. Vac. Sci. Technol. A 13, 1553 (1995).
33. E. Oesterschulze, G. Georgiev, M. Müller-Wiegand, A. Georgieva, and K. Ludolph, J. Vac. Sci. Technol. B 21, 2496 (2003).
34. A. Valsesia, P. Colpo, T. Meziani, P. Lisboa, M. Lejeune, and F. Rossi, Langmuir 22, 1763 (2006).
35. A. Valsesia, P. Colpo, T. Meziani, F. Bretagnol, M. Lejeune, F. Rossi, A. Bouma, and M. Garcia-Parajo, Adv. Funct. Mater. 16, 1242 (2006).
36. H.W. Kroto, J.R. Heath, S.C. O'Brien, R.F. Curl, and R.E. Smalley, Nature 318, 162 (1985).
37. S. Iijima, Nature 354, 56 (1991).
38. H.-J. Freund, Surf. Sci. 500, 271 (2002).
39. X. Chu and S.A. Barnett, J. Appl. Phys. 77, 4403 (1995).
40. H. Holleck and V. Schier, Surf. Coat. Technol. 76–77, 328 (1995).
41. A.A. Voevodin and J.S. Zabinski, J. Mater. Sci. 33, 319 (1998).
42. C. Popov, W. Kulisch, S. Boycheva, K. Yamamoto, G. Ceccone, and Y. Koga, Diamond Relat. Mater. 13, 2071 (2004).
43. W. Kulisch, C. Popov, S. Boycheva, L. Buforn, G. Favaro, and N. Conte, Diamond Relat. Mater. 13, 1997 (2004).
44. S.A. Sherif, F. Barbir, and T.N. Veziroglu, Sol. Energy 78, 647 (2005).
45. G.E. Moore, Electronics 38(8), 114–117 (April 19, 1965).

46. G.E. Moore, Proc. SPIE 2437, 2 (1995).
47. J.D. Meindl, Q. Chen, and J.A. Davis, Science 293, 2044 (2001).
48. G.F. Cerofolini, G. Arena, M. Camalleri, C. Galati, S. Reina, L. Renna, D. Mascolo, and V. Nosik, Microelectron. Eng. 81, 405 (2005).
49. http://www.itrs.net/
50. P.K. Chu, J.Y. Chen, L.P. Wang, and N. Huang, Mat. Sci. Eng. R 36, 143 (2002).
51. A.I. Teixeira, G.A. Abrams, P.J. Bertics, C.J. Murphy, and P.F. Nealey, J. Cell Sci. 116, 1881 (2003).
52. J. Tanaka, K. Sato, T. Shimizu, M. Yamamoto, T. Okano, and T. Kitamori, Biosensor. Bioelectron. 23, 449 (2007).
53. O.M. Koo, I. Rubinstein, and H. Onyuksel, Nanomed. Nanotechnol. Biol. Med. 1, 193 (2005).
54. I.A. Aksay, M. Trau, S. Manne, I. Homna, N. Yao, L. Zhou, P. Fenter, P.M. Eisenberger, and S.M. Gruner, Science 273, 892 (1996).
55. C. Popov, W. Kulisch, S. Bliznakov, G. Ceccone, D. Gilliland, L. Sirghi, and F. Rossi, Diamond Relat. Mater. 17, 1229 (2008).
56. F. Bretagnol, L. Sirghi, S. Mornet, T. Sasaki, D. Gilliland, P. Colpo, and F. Rossi, Nanotechnology 19, 125306 (2008).
57. A. Ruiz, L. Buzanka, D. Gilliland, H. Rauscher, L. Sirghi, T. Sobanski, M. Zychowicz, L. Ceriotti, R. Bretagnol, S. Coecke, and P. Colpo, Biomaterials 29, 4766 (2008).
58. A. Ruiz, L. Buzanka, L. Ceriotti, R. Bretagnol, S. Coecke, and P. Colpo, J. Biomater. Sci. Polym. Edit. 19, 1649 (2008).
59. E. Bergamaschi, O. Bussolati, A. Magrini, M. Bottini, L. Migliore, S. Bellucci, I. Iavicoli, and A. Bergamaschi, Int. J. Immunopathol. Pharmacol. 19, 3 (2006).
60. V. Colvin, Nature Biotechnol. 21, 1166 (2003).
61. J. Ponti, Paper presented at the ICONOTX 2008, February 5–7, Lucknow, India. Nanotoxicology 2, S 35 (2008).
62. T. Hartung, S. Bremer, S. Casati, S. Coecke, R. Corvi, S. Fortaner, L. Gribaldo, M. Halder, S. Hoffmann, A.J. Roi, P. Prieto, E. Sabbioni, L. Scott, A. Worth, and V. Zuang, ATLA 32, 467 (2004).
63. J. Ponti, L. Ceriotti, B. Munaro, M. Farina, A. Munari, M. Whelan, P. Colpo, E. Sabbioni, and F. Rossi, ATLA 34, 515 (2006).
64. J. Ponti, R. Colognato, F. Franchini, S. Gioria, F. Simonelli, K. Abbas, C. Uboldi, C.J. Kirkpatrick, U. Holzwarth, and F. Rossi, Paper presented at the NANOTOX 2008 Conference, Zurich, September 7–10, 2008.
65. IARC/NCI/EPA Working Group, Cancer Res. 45, 2395 (1985).

2. THEORETICAL APPROACHES

2.1. ELECTRICS AND ELECTRONICS

LOW TEMPERATURE VARIABLE-RANGE HOPPING IN DISORDERED SOLIDS: THE FAILURE OF THE MOTT "$T^{-1/4}$" MODEL AND THE DEMONSTRATION OF A NEW PROCEDURE

J.M. MARSHALL[1], C. MAIN[*2]
[1]*Emeritus Professor, School of Engineering,
University of Wales Swansea, U.K.*
[2]*University of Dundee, Division of Electronic Engineering
and Physics, Dundee DD1 4HN, U.K.*

Abstract. Recent developments in nanostructured and related materials have rekindled interest in the interpretation of their electronic properties. In particular, this includes the interpretation of low-temperature electrical conductivity data, in the "variable range" hopping (quantum mechanical tunnelling) regime. In this paper, we show how and why the "Mott $T^{-1/4}$" interpretive technique fails. We then advance a simple and effective alternative procedure. This is validated using computer simulation data and then successfully applied to previously published experimental data.

Keywords: disordered semiconductors; density of states; variable range hopping; Mott's $T^{-1/4}$ law; computer simulation; analytical techniques

1. Introduction

The understanding of the properties of non-crystalline solids is greatly hindered by the difficulty of treating the effects of disorder in a rigorous manner. There is currently no equivalent of the Bloch theorem and associated concepts that allowed elegant analyses of the properties of single crystal materials.

As far as the present authors are aware, *all* techniques so far advanced for interpreting experimental data for disordered semiconductors involve important assumptions and simplifications. In these circumstances, it

---

[*]c.main@dundee.ac.uk

becomes extremely valuable to assess the validity and limitations of such approaches. Here, computer-based numerical modelling has proved particularly effective. In addition to allowing checks to be made concerning the underlying assumptions, it can provide valuable new insights into the details of the transport processes etc. In this paper, we employ such a technique to examine hopping (quantum mechanical tunnelling) transport in the low-temperature "variable range" transport regime. This allows us to illustrate the major inconsistencies that arise from the application of Mott's "$T^{-1/4}$" model,[1] and also to demonstrate that these are eliminated via the use of a simple new analytical procedure.

The effectiveness of the new procedure is then further demonstrated via its application to previously published experimental data.

2. The Monte Carlo Computer Simulation Procedure

At temperature T, the rate of carrier jumps from a site at energy E to one at E', over a distance r, is taken to be described via the Miller-Abrahams[2] expression:

$$v_j = v_0 \exp(-2r/r_0) \left\{ \begin{array}{ll} \exp(-(E'-E)/kT), & (E' > E) \\ 1, & (E' \le E) \end{array} \right. \tag{1}$$

Here, v_0 is the "attempt to hop" frequency (taken as 10^{12} Hz throughout this paper), r_0 is the localization range for the sites, and k is the Boltzmann constant. At finite applied fields, $E'-E$ includes any additional field-induced potential difference between the sites. At low fields, as in our present simulations, the resulting carrier mobility becomes field independent.

An array of 960,000 hopping sites, randomly distributed in a volume of $96 \times 100 \times 100$ units was used in the simulations. The present studies employed both an energy-independent and an exponentially energy-dependent density of states (DOS). The former featured a site density of $N(E) = N(E_F) = 10^{20}$ cm^{-3}eV^{-1}, with the sites having randomly assigned energies in the range $E = E_F \pm 0.25$ eV where E_F is the Fermi energy. This range was chosen to be well in excess of kT, at all temperatures investigated, while avoiding excessive dilution of the array by including sites irrelevant to the low-temperature transport process. The latter had the form $N(E) = N(E_F) \exp((E-E_F)/kT_0)$, with energies measured upwards relative to E_F and with $N(E_F) = 10^{20}$ cm^{-3}eV^{-1} and $T_0 = 500$ K. The site energy range was $0.2 \geq (E-E_F) \geq -0.1$ eV. For each site, its Fermi–Dirac occupation probability, $f(E)$, at the temperature under consideration was employed, in conjunction with a pseudo-random number, to select whether or not it was occupied. Sites defined as occupied were taken to be unavailable for initial excess

carrier generation or for subsequent hopping transitions. For each unoccupied site, the 16 nearest *unoccupied* neighbours (i.e. those with the highest
values of v_j, using $r_0 = 4 \times 10^{-8}$ cm and an applied electric field of $F = 10^4$
V/cm) were identified. We separately confirmed that increasing F by an
order of magnitude did not alter the resulting calculated mobilities. Periodic
boundary conditions were used in the selection of the neighbors, in all three
hopping dimensions.

Individual carriers were then initially generated on randomly-selected
unoccupied sites, and were allowed sufficient simulation events to equili-
brate with and to drift within the array. Their resulting net displacements in
the field direction were then combined to determine the average drift mobility
μ. The individual steps involved in the Monte Carlo procedure for such
simulations are described in more detail in Ref. 3 for example. Finally, the
resulting mobility value was used in conjunction with the charge carrier
concentration, $n = \int N(E) f(E) \, dE$ (easily calculated for the known $N(E)$ distributions), to obtain the electrical conductivity in the form $\sigma = ne\mu$. Before
presenting the results in Sect. 5, we will first indicate the source of the
major inconsistencies that arise when the Mott $T^{-1/4}$ model is applied, and
then describe our new analytical procedure for interpreting the data.

3.   The Mott "$T^{-1/4}$" Procedure

We will demonstrate below that the Mott $T^{-1/4}$ model yields major inconsistencies when applied to both our simulation results and previously published data. The root cause of these anomalies is that the original version of
the Mott model[4] used a parameter, r_{max}, to represent two *very different* quantities, the dominant hopping distance and the radius of a sphere *within*
which such hopping occurs. Subsequently,[2] a nominal attempt was made
to correct this by re-defining r_{max} as $\int_0^R r^3 dr \, / \int_0^R r^2 dr = 75\%$ of the sphere

radius R. This, however, totally neglected the effect of tunneling over
varying distances *within* the sphere. The correct expression should be
$r_{max} = \int_0^R r^3 \exp(-2r/r_0) dr \, / \int_0^R r^2 dr$. This only approaches the 75% value for

$R/r_0 \ll 1$, which is clearly incompatible with "variable range hopping". For
$R/r_0 > 3$, r_{max} falls to less than 1% of the sphere radius!

The revised Mott model,[1] including the expected temperature depend-
ence of its pre-exponential term, predicts a gradient, for a $\ln(\sigma T^{1/2})$ vs. $T^{-1/4}$
plot, of $B = 1.66/(r_0^3 \, k \, N(E_F))^{1/4}$, giving the possibility for calculating $N(E_F)$.
We will present the resulting values, plus the further implications, in Sects.
5 and 6, together with corresponding information derived using published
experimental data. We will also show that the new procedure described in
Sect. 4 resolves the major anomalies inherent within the Mott model.

4. Analytical Procedure for Interpreting Electrical Conductivity Data

From experimental studies, the available parameters are the magnitude σ and activation energy E_σ, of the conductivity at any given temperature T. However, the latter involves both the activation energy of the mobility E_μ, and the temperature dependence of the carrier concentration $n(T)$. Therefore, the effect of the latter component must first be addressed. Moreover, almost all equilibrated charge carriers can be expected to occupy states closer to the Fermi level than any measured value of E_σ. Thus, only limited information regarding the concentration and energy distribution of these deeper states can be inferred *directly* from the conductivity data, and some form of extrapolation is required. However, all that we initially require here is the *temperature dependence* of the carrier concentration, rather than its precise magnitude. Now consider two specific cases:

(i) A constant DOS ($N(E_\mu)$) between the energy E_μ and a similar energy below the Fermi level: The integrated concentration of carriers (assumed to be solely electrons below) over these states, via Fermi–Dirac occupation statistics, is then $n(T) = N(E_\mu)\, kT \ln(2)$ and so varies linearly with temperature. Hence, a modified Arrhenius plot of $\ln(\sigma/T)$ vs. $1/T$ can be employed to obtain E_μ.

(ii) A DOS that decays exponentially from E_μ to E_F and below, with a characteristic temperature T_0: $n(T)$ is then proportional to $((1/kT)-(1/kT_0))^{-1}$. Since we may expect (and will demonstrate for all data below) T_0 to be significantly greater than T in the low-temperature hopping regime, the procedure in (a) can again be employed.

Other forms of $N(E)$ are obviously possible, although perhaps less physically probable. However, having confirmed by inspection that other types of sufficiently weak temperature dependence yield extremely small changes in the calculated parameters, we consider it justified and sufficient to adopt the procedure in (a), hereafter.

Having determined E_μ as above, we now envisage that the rate-limiting transport process for a carrier normally occupying a thermally equilibrated deeper site close to E_F must involve traversing a typical distance r_c and also gaining the necessary energy to access a site close to E_μ.

To identify an appropriate value, r_c can be adjusted until the calculated conductivity is equal to the experimental one, within a specified level of accuracy. The computation steps for each temperature then comprise:

(i)   Calculation of E_μ, as above.

(ii) Choice of an initial seeding value of r_c. Here, one can use $r_c = r_0$, since the final iterated values of r_c are obviously expected to be greater than or equal to this.

(iii)   Calculation of the resulting drift mobility, for carriers moving via intermediate states close to E_μ, as $\mu = e\, v_0 r_c^2 / 6kT \exp(-2r_c/r_0) \exp(-E_\mu/kT)$.

(iv)   Calculation of the volume of a sphere of radius r_c, as $V = 4\pi r_c^3/3$.

(v)   Assumption that this sphere must contain *just one* site at an energy
close to E_μ (otherwise, a smaller or larger value of r_c would automatically arise). Thus, calculation of the DOS at this energy, as
$N(E_\mu) \sim 1/(VE_\mu)$.

(vi)   Estimation of the carrier density as $n \sim N(E_\mu)kT$ (*see comment below).

(vii)   Calculation of the resulting electrical conductivity as $\sigma = ne\mu$.

(viii) Iterative adjustment of r_c, until the calculated conductivity equals the
measured one, within the specified accuracy limit (one part in 10^3
hereafter).

*Note that step (vi) implicitly assumes that the DOS calculated in step
(v) remains constant from the energy E_μ down to and below the Fermi level.
Further refinement could be made here (e.g. an initial estimate of the energy
variation of the DOS, extrapolated to E_F, could be used in step (vi), in additional iterations). However, we simply adopt this assumption below.

Note also that the above steps could be collected into the solution of the
equation $\sigma = a\, r_c^{-1} \exp(-b\, r_c)$. However, we have found that this can give
rise to instabilities or a lack of convergence. Thus, we below confine our-
selves to the use of the above step-wise iterative procedure. This can readily
be executed in (e.g.) a very simple computer program, to give extremely
rapid convergence in terms of r_c. *An example of the core elements of such
code is available from the corresponding author, upon request.*

5.   The Computer Simulation Data and Their Analyses

Figure 1a shows the temperature dependence of the computer simulated and
fitted conductivities, for the energy-independent DOS, while Figure 1b
shows the calculated DOS and r_c parameters. The former includes a fit to a
$\ln(\sigma) = A - BT^{-1/n}$ relationship, with the resulting value of n being 8.13 –
well distant from the value of 4 arising from the Mott model. Note also that
the fitting accuracy is primarily determined by the values of E_μ calculated
using adjacent conductivity data points. Thus, although smoothing or curve
fitting could be desirable in practice, we have not employed it here.

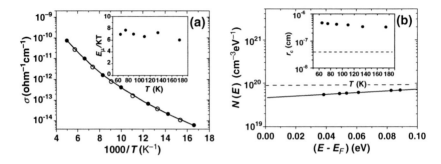

Figure 1. Temperature dependence of the d.c. electrical conductivity for the energy-independent simulation (a) (filled circles – data points; open circles – fitted σ values), and the DOS calculated using the new procedure (b). The solid line in (a) is a fit as described in the text. That in (b) is an exponential relationship, with a characteristic temperature of $T_0 = 2,500$ K, while the dashed line indicates the true DOS. The inset in (a) shows the temperature dependence of E_μ relative to kT. That in (b) is the temperature dependence of r_c, with the dashed line indicating the value of r_0.

From Figure 1b, the DOS values calculated via the new procedure are within a factor of 2 of the correct ones. They do, however, deviate slightly from the true energy distribution, in yielding a very slowly rising DOS. This is in no respect a consequence of the inclusion of the full Fermi–Dirac statistics in the present study. Virtually identical results were obtained in Ref. 5, in which a zero-temperature approximation was employed to specify the number of unoccupied states close to E_F. This is understandable, since the inset in Figure 1a shows the energies, to which the DOS values correspond, to be at least $6kT$ above E_F in all cases. These minor discrepancies must thus involve residual limitations in the present version of the analytical procedure itself. However, we consider that an accuracy of a factor of two or better is highly acceptable, in comparison to many previous attempts to analyze such data (see Sect. 6 for some examples).

Figure 2a displays the simulated conductivity data for the exponential DOS, Figure 2b the calculated DOS values. These are slightly less that the true ones, but again only by a small factor. They yield a value of $T_0 = 416 \pm 5$ K, compared to the actual value of 500 K (i.e. to a slightly steeper gradient of the DOS vs. energy). Both observations are consistent with those for Figure 1a, b. Indeed, assuming an identical influence of the analytical procedure upon the energy dependence of the calculated DOS, the corrected value of the characteristic temperature would be $T_0' = 1/(416^{-1} - 2,500^{-1}) = 499 \pm 5$ K. Again, the activation energies are much larger than kT, so that Fermi–Dirac occupation statistics do not reduce the density of available hopping sites close to these energies.

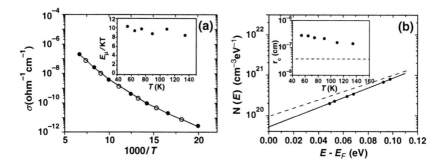

Figure 2. Temperature dependence of the d.c. electrical conductivity for the energy-dependent simulation (a) (filled circles – data points; open circles – fitted σ values), and the DOS calculated using the new procedure (b). The solid line in (a) is a guide to the eye. That in (b) is a fit of the data to an exponential DOS, with a characteristic temperature of $T_0 = 416$ K, while the dashed line indicates the true DOS. The inset in (a) shows the temperature dependence of E_μ, relative to kT. That in (b) shows the temperature dependence of r_c, with the dashed line indicating the value of r_0.

It is important to note here that:

(i) The simulation units were normalized to a mean inter-site distance of unity. Thus, if we had selected a different value of $N(E_F)$, then (with the other simulation parameters remaining the same) r_0 would have changed correspondingly. This would leave the subsequent comparison between the calculated and true values of $N(E_F)$ totally unaffected.

(ii) In the above cases, the r_c values remained appreciably higher than r_0 at all investigated temperatures. Should this criterion cease to be satisfied, the premises of our analytical procedure would obviously become invalid. This will happen when carrier transport is departing from the low-T "variable range" hopping regime, and becoming dominated by another process such as trap-limited transport. In a more detailed study,[6] we show that this also causes an abrupt rise in the calculated $N(E)$, to physically unrealistic values.

The values of $N(E_F)$ calculated using the Mott model are 10^{19} and 4.6×10^{18} cm^{-3}eV^{-1} for the energy-independent and exponential DOS, respectively. These may not appear *too* incompatible with the true values of 10^{20} cm^{-3}eV^{-1}. However, we will show below that much more major discrepancies are revealed when the implications of the model are examined in fully appropriate detail.

6. Published Experimental Data and Their Analyses

Figure 3a shows data obtained by Paul and Mitra[7] for specimens of r.f. sputtered amorphous silicon, subjected to various post-deposition annealing treatments. These data are of particular value here, since such early specimens contained much higher densities of defect states close to E_F than have been achieved using subsequent fabrication techniques (and thus featured a much higher contribution due to hopping within such states). Figure 3b shows the densities of states calculated using our new procedure. Here and below, the assumed underlying parameters were $r_0 = 10^{-7}$ cm and $v_0 = 10^{12}$ Hz.

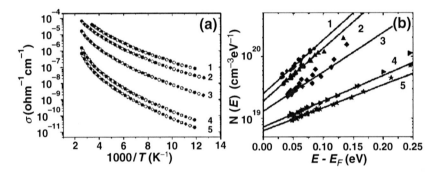

Figure 3. Experimental data (a) (filled circles – data points; open circles – fitted σ values), and the DOS calculated using the new procedure (b), for specimens of r.f. sputtered amorphous silicon,[7] as deposited at 300 K and after annealing for 24 h at 383, 523, 653 and 723 K (1–5 respectively). The solid lines in (b) use the fitted values of $N(E_F)$ and T_0.

Figure 3b displays fits of the lowest energy (i.e. lowest temperature) data to assumed exponential forms of the DOS. The resulting values of $N(E_F)$ are 2.5×10^{19}, 1.9×10^{19}, 1.3×10^{19}, 7.3×10^{18} and 6.4×10^{18} cm^{-3}eV^{-1} for specimens 1–5 respectively. The corresponding values of T_0 are 580, 590, 795, 1,220 and 1,370 K. Correction of these, as surmised from the simulation analyses, yields T_0' values of 760, 770, 1,160, 2,380 and 3,030 K. The values of $N(E_F)$ are all physically reasonable, as is the effect of annealing at progressively higher temperatures in reducing both $N(E_F)$ and the energy dependence of the DOS.

We now turn to the *detailed* consequences of applying the Mott $T^{-1/4}$ model. As mentioned above, the model, including the expected temperature dependence of the pre-exponential term, predicts a gradient, for a $\ln(\sigma T^{1/2})$ vs. $T^{-1/4}$ plot, of $B = 1.66/(r_0^3 \, k \, N(E_F))^{1/4}$. Discounting the inappropriate 3/4 correction, it additionally predicts a dominant hopping distance of $r_{max} = 3^{1/4}/(2\pi N(E_F)kT/r_0)^{1/4}$, and a hopping activation energy of $W = 3r_{max}^3/(4\pi N(E_F))$. The usual expression for the carrier mobility then becomes $\mu = (ev r_{max}^2/6kT) \exp(-2r_{max}/r_0) \exp(-W/kT)$. The carrier density should be $n \sim N(E_F) kT$, yielding an electrical conductivity of $\sigma \sim ne\tilde{\mu}$.

Now consider the data for Specimen 1 in Figure 3. The force-fitted value of B is 69, yielding $N(E_F) \sim 3.9$ x 10^{18} cm^{-3}eV^{-1}. At 100 K, this yields $r_{max} \sim 1.1 \times 10^{-6}$ cm and $W \sim 0.042$ eV. The pre-exponential term in the expression for the mobility is ~ 13 cm^{2}V^{-1}s^{-1}. However, the two exponential terms reduce this to 3.2×10^{-11} cm^{2}V^{-1}s^{-1}, yielding a conductivity of 1.7×10^{-13} Ω^{-1}cm^{-1}. Since the experimental value at this temperature is 2.0×10^{-8} Ω^{-1}cm^{-1}, the discrepancy is both clear and totally unacceptable. Table 1 displays these results, plus corresponding ones for other data sets. The major inconsistencies between the predicted and actual σ values are obvious. More detailed analyses of these and other experimental data, including much more recent ones for microcrystalline silicon,[8] can be found in Ref. 6. Similar or greater inconsistencies emerge in all cases.

TABLE 1. Values of Mott's "$T^{-1/4}$" parameter, B, plus the resulting values of $N(E_F)$ (cm^{-3}eV^{-1}) and σ_{calc} (Ω^{-1}cm^{-1}), at the temperatures indicated, for various data as presented above. σ_{true} is the true electrical conductivity at the temperature in question.

	B	$N(E_F)$	T	σ_{cal}	σ_{true}
E-independent DOS simulation	107	1.0×10^{19}	80	4.5×10^{-21}	6.5×10^{-14}
Exponential DOS simulation	132	4.6×10^{18}	80	1.1×10^{-25}	3.9×10^{-10}
Ref. 7, Specimen 1	69	3.9×10^{18}	100	1.7×10^{-13}	2.0×10^{-8}
Ref. 7, Specimen 3	76	2.7×10^{18}	100	8.8×10^{-15}	8.5×10^{-10}
Ref. 7, Specimen 5	91	1.3×10^{18}	100	1.8×10^{-17}	7.3×10^{-12}

7. Further Studies

Our new analytical procedure reproduces the density of localized states employed in simulation studies within a factor of 2 or better. It does, however, yield a slightly increased energy dependence of the calculated DOS for the two simulation studies. It would be rewarding if this could be eliminated by further refinement of the procedure, and this possibility will be explored.

Finally, additional experimental measurements on un-hydrogenated (i.e. highly disordered) amorphous semiconductors, over the widest possible temperature range, would be of great value in the further evaluation of the procedure. These will also be performed and reported upon subsequently.

8. Conclusions

A new procedure for interpreting the electrical conductivity associated with hopping at low temperatures in disordered semiconductors has been advanced.

Its validity and effectiveness has been demonstrated using data generated via Monte Carlo simulations of hopping within a large array of sites. It has also been applied to published experimental data, for which it yielded entirely credible and self-consistent values of the density of states at the Fermi level, and of the energy distribution of states above this. The results were also fully consistent with the expected effects of specimen annealing.

In contrast, the Mott "$T^{-1/4}$" model yielded major inconsistencies. Even when these were not directly evident from the estimated $N(E_F)$ values, they became immediately apparent when the resulting associated parameters were used to compare the predicted and experimental values of the electrical conductivity.

ACKNOWLEDGEMENTS

The authors gratefully acknowledge the stimulation provided by the late Professor Vladimir Arkhipov, via his incisive papers concerning hopping and many other topics, and for encouraging our initial approach to the present study. Vladimir will be sadly missed within our scientific community! We also thank Dr. Kostadinka Gesheva for her insistence that the deficiencies implicit in the Mott "$T^{-1/4}$" model should be addressed, as attempted herein.

References

1. N.F. Mott and E.A. Davis, *Electronic Processes in Non-Crystalline Materials*, 2nd Edition (Clarendon Press, Oxford, 1979).
2. A. Miller and E. Abrahams, Phys. Rev. 120, 745 (1960).
3. J.M. Marshall and V.I Arkhipov, J. Optoelectron. Adv. Mater. 7, 43 (2005).
4. N.F. Mott, Phil. Mag. 19, 333 (1969).
5. J.M. Marshall, J. Optoelectron. Adv. Mater. 9, 84 (2007).
6. J.M. Marshall and C. Main, J. Phys.: Condens. Mat. 20, 285210 (2008).
7. D.K. Paul and S.S. Mitra, Phys. Rev. Lett. 31, 1000 (1973).
8. G. Ambrosone, U. Coscia, A. Cassinese, M. Barra, S. Restello, V. Rigato and S. Ferrero, Thin Solid Films 515, 7629 (2007).

ON THE VALIDITIES AND LIMITATIONS OF TRANSIENT PHOTO-DECAY TECHNIQUES FOR IDENTIFYING THE ENERGY DISTRIBUTION AND RELATED PROPERTIES OF LOCALIZED STATES IN DISORDERED SEMICONDUCTORS

J.M. MARSHALL[1], C. MAIN[*2]
[1]*Emeritus Professor, School of Engineering, University of Wales Swansea, U.K.*
[2]*University of Dundee, Division of Electronic Engineering and Physics, Dundee DD1 4HN, U.K.*

Abstract. A computer simulation technique is employed to calculate the transient photo-decay characteristics for a disordered semiconductor featuring an exponential tail of localized states plus a narrow Gaussian feature of adjustable height. The resulting data are subjected to analysis via the "pre-transit" $(1/I(t).t)$, "post-transit" $(I(t)t)$ and "Fourier transform" procedures. It is shown that all three options can detect the *presence* of the Gaussian component. However, even when the peak height of this becomes comparable to or less than that of the local exponential background, the pre-transit procedure consistently miscalculates its energy. In contrast, the post-transit and Fourier transform procedures correctly identify this energy, and also provide improved resolution of the energy distribution and other properties of the localized states.

Keywords: disordered semiconductors; density of states; transient photoconductivity; computer simulation; analytical techniques

1. Introduction

It has long been appreciated that in the case of trap-limited band transport in disordered semiconductors, transient photo-decay measurements have the *potential* to provide information on the energy distribution, and possibly

---

[*]c.main@dundee.ac.uk

J.P. Reithmaier et al. (eds.), *Nanostructured Materials for Advanced Technological Applications*, 49
© Springer Science + Business Media B.V. 2009

other characteristics, of the localized states involved. In particular, an exponential density-of-states (DOS) distribution yields (after the initial carrier trapping regime) a featureless power-law decay in the current $I(t)$ with elapsed time after carrier generation t, while the presence of additional structure in the DOS can give rise to significant deviations from this.[1,2]

In the "pre-transit" regime of the photo-decay response (i.e. in the absence of carrier losses by arrival at an extraction electrode or by recombination), a simple procedure was proposed (Ref. 3, following the model advanced in Ref. 4) for interpreting *suitable* experimental data. In this "pre-transit" technique (following completion of the initial free carrier trapping regime), the DOS $g(E_{th})$ $(\text{cm}^{-3}\text{eV}^{-1})$ at the energy $E_{th} = kT \ln(\nu t)$ was predicted to be

$$g(E_{th}) = (I(t=0) \, g(E_c) \, \nu^{-1})/(I(t)t). \qquad (1)$$

Here, T is the temperature, ν the "attempt to escape" frequency for trapped carriers, $I(t = 0)$ the transient photocurrent at zero time (i.e. the product of the number of excess carriers per unit specimen thickness, the electronic charge, the free carrier mobility and the applied electric field), and $g(E_c)$ the density of states at the mobility edge separating extended from localized states. The rationale for the choice of E_{th} arose from the concept of a "thermalization energy", such that the release time constant for states at this depth, $\nu^{-1}\exp(E_{th}/kT)$, is equal to the elapsed time t. It was envisaged that shallower levels will have had sufficient time to achieve quasi-thermal equilibrium with the extended transport states, while deeper traps will not. For the case of a slowly decaying exponential DOS, this was predicted to yield a reasonably sharp peak in the trapped carrier density close to E_{th}[4] leading straightforwardly[3] to the above analytical procedure.

Although this procedure worked well in the case of a *purely exponential* DOS, it soon became clear that problems arose in the case of more complex energy distributions. Indeed, for a very highly structured DOS, it could even yield the complete inverse of the true energy distribution (e.g. Ref. 2).

The resulting impression was that the pre-transit procedure *should* be acceptable, provided any structure in the DOS was "not very pronounced", but would become invalid for more highly structured distributions. However, to the best of our knowledge, the definition of "not very pronounced", and indeed whether the pre-transit analysis is valid *even then*, has not been properly explored. It was one of the main objectives of the present study to make such an evaluation.

We will also examine the viability of the "post-transit" procedure (Ref. 3 and references therein), which predicts the relationship

$$g(E_{th}) = C \, I(t) \, t. \qquad (2)$$

Here, C is a constant subsequently[5] quantified as $2\ g(E_c)/(Q_0\ t_0\ \nu)$, where Q_0 is the total charge and t_0 the free carrier transit time. However, since at least one of these parameters (t_0) may well not easily be estimated from experimental data, it is more common in practice to perform normalized calculations of the DOS, using such data.

A subsequent alternative[6,7] to the above analytical techniques, although admittedly less straightforward to apply, involved a Fourier transformation of the current vs. time measurements into the frequency (ω) domain, followed by an analysis of the resulting data to yield the DOS. Here, it is important to note that although the resulting energy scale, defined as $E(\omega) = kT \ln(\nu/\omega)$, may *appear* equivalent to that for the thermalization energy E_{th} the analysis carries no implication of, or indeed requirement for, a peak in the occupied state density close to this energy. Rather, the procedure employs the *full* available time range of the experimental data in performing the transform for any particular value of ω and hence $E(\omega)$.

2. The Computer Simulation Procedure and Results for the "Pre-Transit" Case

The model DOS employed in this study, as shown in Figure 1a, featured an exponential bandtail of the form $g_e(E) = g(E_c) \exp(-E/kT_o)$, with $g(E_c) = 4 \times 10^{21}$ cm^{-3}eV^{-1} and with $T_o = 600$ K as the characteristic temperature for the exponential component. To this was added a relatively narrow Gaussian component of the form $g(E_{pk}) \exp(-0.5((E-E_{pk})/E_w)^2))$, where $E_{pk} = 0.35$ eV was the peak energy of the feature and $E_w = 0.025$ eV its width parameter. The peak height $g(E_{pk})$ was varied via a multiplying factor M with respect to

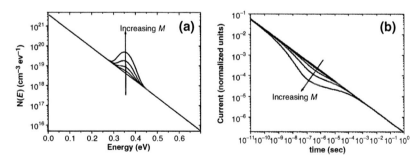

Figure 1. (a) DOS employed in the simulation study, and (b) the resulting computed transient photocurrents, normalized to the current at zero time and with no carrier losses due to arrival at an extraction electrode or to recombination. The values of M are 0, 0.3, 1, 3 and 10, increasing as indicated, and the temperature used for the simulations was 300 K.

the value of the background exponential component at a depth of 0.35 eV, i.e. $g(E_{pk}) = M g(E_c) \exp(-0.35/kT_o) = 4.5 \times 10^{18} M$ cm^{-3}eV^{-1} for the above parameters.

Figure 1b shows the resulting transient photocurrents, as computed at a measurement temperature of 300 K, using the technique outlined in Ref. 8, in the "pre-transit" regime (i.e. with no carrier losses due to arrival at an extraction electrode or to recombination).

Figure 2 displays the DOS values, as calculated using the pre-transit procedure, for the cases M = 0.3, 1, 3 and 10. It also shows iterative fits, with the parameters $g(E_c)$, T_o, $g(E_{pk})$, E_{pk} and E_w all being totally free variables. The inset shows the contributions of the various components to the computed DOS for the case M = 1. While the fitting procedure gives excellent agreement for M = 3 or less, it can be seen that the fit is not quite so precise for the case M = 10. Note that with the normalized currents in Figure 1, the conversion expression for the pre-transit procedure is reduced to $g(E) = (1/I(t)t) (g(E_c)/v)$, with $g(E_c) = 4 \times 10^{21}$ cm^{-3}eV^{-1} and $v = 10^{12}$ Hz, as employed in all of the present simulations.

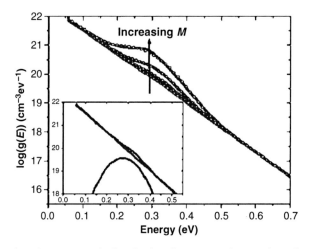

Figure 2. Density of states, as calculated using the pre-transit procedure, for the cases M = 0.3–10. The solid line shows the data, and the open circles are fits to an exponential bandtail plus a Gaussian feature. The inset graph indicates the individual fitted components for the case M = 1.

Of course, although the technique of fitting data to a known functional DOS works effectively here, it is unlikely to be applicable when interpreting actual experimental data, where an assumption of the form of the DOS would be required. Its value in the present study is to assist in the evaluation of the viabilities of the pre-transit, and (below) post-transit and FT techniques, under such "idealized" circumstances.

Note also that since one or more of the parameters $I(t = 0)$, $g(E_c)$ and v in Eq. (1) would almost certainly not be known in practice, such a quantitative conversion would not be possible in the case of experimental data. However, the first two parameters only influence the absolute magnitude of the calculated $g(E)$, while inaccuracy in the assumed value of v will also displace the energy scale uniformly, but not influence the actual forms of the curves or the resulting values of E_{pk} and E_w.

Table 1 summarizes the data arising from such fits. The calculated fitting accuracies in the values of E_{pk} are better than ± 1 meV for $M \leq 3$, even in the case of the barely-detectable deviation of the current from the pure power-law form for $M = 0.3$ (see Figure 1b). The poorer fit for $M = 10$ results in a somewhat smaller calculated value of E_{pk}, as shown in the table.

TABLE 1. Fitting parameters for the DOS data in Figure 2.

M	$g(E_c)$ (cm^{-3}eV^{-1})	T_o (K)	$g(E_{pk})$ (cm^{-3}eV^{-1})	E_{pk} (eV)	E_w (eV)
0.3	2.3×10^{22}	590	9.6×10^{18}	0.283	0.049
1	2.5×10^{22}	598	3.7×10^{19}	0.274	0.052
3	2.5×10^{22}	596	1.3×10^{20}	0.263	0.055
10	2.7×10^{22}	593	6.0×10^{20}	0.246	0.061

Critically, it is clear that all values of E_{pk} are significantly smaller than the true value of 0.35 eV. Moreover, they all lie in the general vicinity of the local *minimum* of the DOS in Figure 1a for $M = 10$ (i.e. ~0.28 eV). Here, we again note the finding in Ref. 2 that sufficient structure in a DOS can completely invert its form, as calculated using the pre-transit analysis. *It now seems that this situation applies to much smaller values of M than was previously anticipated!*

Although there is a *weak* trend towards the correct value of 0.35 eV as M falls, it is clear that this might (*if at all!*) only be approached when the Gaussian component is so small as to be undetectable! Indeed, contrary to prior notional expectations, the fitted value of E_{pk} remains close to 0.28 eV, *even for the case M = 0.3!*

The fitted widths of the Gaussian component E_w are approximately twice the correct value of 0.025 eV. Such "kT broadening" is a well known consequence of the fact that this analytical technique assumes that all carriers trapped in a state at energy E, with a release time constant $\tau = v^{-1}\exp(E/kT)$, are released after *exactly* this dwell time, rather than over an appropriate probability distribution of times around it.

The values of T_o for the exponential component of the DOS are close to the true value of 600 K. However, the values of $g(E_c)$ are about a factor of 6 larger than the true value of 4×10^{21} cm^{-3}eV^{-1}. This illustrates a further limitation of the procedure. The states at E_{th} can never be in *complete* quasi-thermal equilibrium with the extended ones, while free carriers in the latter are still being lost to deeper-lying traps. This reduces the current relative to that in Eq. (1), and thus raises (*inter alia*) the calculated values of $g(E_c)$.

In respect of the pre-transit procedure, we may thus conclude that:

(i) Although this can provide an indication of the *presence* of the Gaussian feature, it gives a significant error in the peak energy of such a feature.

(ii) Unexpectedly, at least in terms of previous notional impressions, this situation persists even when the additional height of the feature *falls below that of the background DOS.*

3. Computer Simulation Results for the "Post-Transit" Case

Figure 3a shows data corresponding to those in Figure 1b, but with a finite and short free carrier transit time. The simulation parameters were $g(E_c) = 4 \times 10^{21}$ cm^{-3}eV^{-1}, $Q_0 = 1.6 \times 10^{-9}$ C, $t_0 = 2 \times 10^{-11}$ s and $\nu = 10^{12}$ Hz, giving the conversion parameter $C = 2\ g(E_c)/(Q_0\ t_0\ \nu)$ in (2) as 2.5×10^{29} cm^{-3} eV^{-1}C^{-1}.

Figure 3b presents the resulting calculated DOS data, and Table 2 shows the associated fitting parameters. For $M \leq 3$, the values of $g(E_c)$ and T_0 are very close to the correct figures of 4×10^{21} cm^{-3}eV^{-1} and 600 K, respectively. The agreement is less precise for the case $M = 10$, where the large Gaussian component significantly influences the current at shorter times,

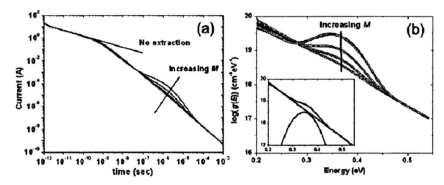

Figure 3. (a) Computer simulated current at $T = 300$ K, for values of M of 0.3, 1, 3 and 10, increasing in the direction indicated. In this case, the simulation parameters (see text) were set to yield a short transit time. The graph also shows the current that would be obtained in the case of an infinitely thick specimen (i.e. with no carrier extraction). (b) Density of states, as calculated using the post-transit procedure, for the cases $M = 0.3$–10. The solid line shows the data, and the open circles are fits to an exponential bandtail plus a Gaussian feature. The inset graph indicates the individual fitted components for the case $M = 1$.

and thus the corresponding DOS. All of the E_{pk} values are now within 1% of the correct one of 0.35 eV. The E_w values are now approximately 40% larger than the true value of 0.025 eV. This indicates a reduced degree of thermally-induced broadening, to $\sim kT/2$. This broadening also influences the fitted heights of the Gaussian component. The $g(E_{pk})$ value should be 4.5×10^{18} cm^{-3}eV^{-1} for the case $M = 1$, and proportionately lower or higher for the other cases. This effect is approximately what would be expected if the total area under the Gaussian component is preserved in the presence of the broadening.

TABLE 2. Fitting parameters for the DOS data in Figure 3b.

M	$g(E_c)$ (cm^{-3}eV^{-1})	T_o (K)	$g(E_{pk})$ (cm^{-3}eV^{-1})	E_{pk} (eV)	E_w (eV)
0.3	3.6×10^{21}	603	9.8×10^{17}	0.342	0.036
1	3.3×10^{21}	607	3.1×10^{18}	0.344	0.035
3	3.8×10^{21}	598	8.5×10^{18}	0.349	0.033
10	1.9×10^{22}	642	2.7×10^{19}	0.348	0.034

Overall, and in dramatic contrast to the case of the pre-transit analysis, we conclude that the post-transit procedure is an effective one for revealing the DOS, irrespective (subject to quite limited thermally-induced broadening) of the degree of structure present within it.

4. Fourier Transform Analysis of the Simulation Data

Figure 4a, b show the results of an analysis of the pre- and post-transit data using the Fourier transform method, and Table 3 presents the fitted parameters. It can be seen that the iterative fits are again very good, and the

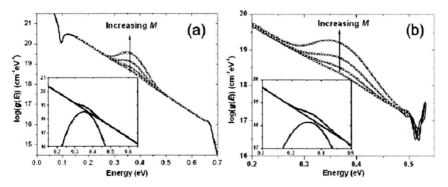

Figure 4. Densities of states, as calculated using the Fourier transform analysis of (a) the pre-transit data and (b) the post-transit data, for the cases $M = 0.3$–10. The solid lines show the data, and the open circles are fits to an exponential bandtail plus a Gaussian feature. The inset graphs indicate the individual fitted components for the case $M = 1$.

calculated peak energies of the Gaussian component are extremely close to the correct value of 0.35 eV. The degree of thermal broadening is comparable to that obtained using the post-transit analysis, and is obviously a significant improvement upon that for the pre-transit case.

TABLE 3. Fitting parameters for the DOS data in Figure 4a, b.

M	$g(E_c)$ (cm^{-3}eV^{-1})	T_0 (K)	$g(E_{pk})$ (cm^{-3}eV^{-1})	E_{pk} (eV)	E_w (eV)
		Figure 4a – pre-transit data			
0.3	4.2×10^{21}	604	1.0×10^{18}	0.345	0.037
1	4.2×10^{21}	604	3.2×10^{18}	0.348	0.036
3	4.2×10^{21}	604	9.4×10^{18}	0.349	0.035
10	4.3×10^{21}	603	3.1×10^{19}	0.350	0.035
		Figure 4b – post-transit data			
0.3	2.2×10^{21}	605	3.7×10^{17}	0.359	0.030
1	2.3×10^{21}	598	1.4×10^{18}	0.354	0.036
3	2.6×10^{21}	597	4.3×10^{18}	0.353	0.036
10	2.8×10^{21}	597	1.5×10^{19}	0.351	0.036

The distortions of the calculated DOS at the extremes of the energy ranges are due to the truncation of information about the current at very short and very long times. We have recently found that improved apodization techniques can significantly reduce this effect. It will also be noted that the values of $g(E_c)$ and $g(E_{pk})$ for the post-transit data are about 50% of those expected. This is due to subtle differences between the effects of carrier losses by recombination (as assumed in the Fourier transform procedure) and by arrival at an extraction electrode (as in the present simulations).

5. The Potential Value of Measurements at Different Temperatures

We now turn to the effect of the measurement temperature upon the data extracted using the three techniques. The thermalization energy concept gives a time corresponding to the peak energy of $t_{th} = v^{-1}\exp(E_{th}/kT)$, which can be re-written as $\ln(t_{th}) = \ln(v^{-1}) + E_{th}(1/kT)$. Thus, in our case, a plot of $\ln(t_{pk})$ vs. $(1/kT)$ should have a gradient of E_{pk}, and an intercept of $\ln(v^{-1})$. The potential value of such a plot is to eliminate the need to *assume* a value of v in determining the DOS, thereby not only giving an improved value for the peak energy of any feature under examination, but also offering an indication of the *nature* of the localized states, since their capture cross section σ and thus their state (charged or neutral when empty) can be inferred from v, via detailed balance considerations, i.e.,

$$\sigma = v \, / \, (g(E_c) \, kT \, v_{th}), \tag{3}$$

where v_{th} is the thermal velocity of the charge carriers (weakly temperature dependent but expected to be $\sim 10^7$ cm s^{-1} at normal temperatures).

We have thus examined the temperature dependence of the time corresponding to the value of E_{pk} for the case $M = 1$, for the case of no carrier losses by arrival at an extraction electrode or recombination. The results are shown in Figure 5. For the pre-transit analysis, the intercept is -27.15, yielding $v = 6 \times 10^{11}$ Hz, which is acceptably close to the true value of 1×10^{12} Hz. However, the gradient yields $E_{pk} = 0.265 \pm 0.002$ eV, which is similar to the values in Table 1. This is obviously again incompatible with the true value of 0.35 eV and is also close to the energy of the local minimum (at ~ 0.28 eV) in the DOS for larger values of M. For the FT analysis, the corresponding values are $E_{pk} = 0.35$ eV and $v = 1.03 \times 10^{12}$ Hz. The results of a similar analysis of simulation data with a very short transit time, as interpreted using both the post-transit and the FT procedure, yield similar close agreement.

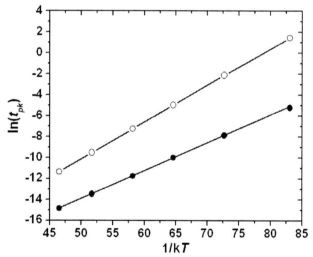

Figure 5. Inverse temperature dependence of the time t_{pk} corresponding to E_{pk}, as determined using the pre-transit (filled circles) and FT techniques (open circles), for simulation data generated with no carrier losses by completion of transit or recombination.

We have also examined a limited number other cases (specifically T_0 values of 400 and 800 K with other parameters as above), and have established that the inaccuracies arising from the pre-transit analytical procedure are similar to those presented above, in terms of the resulting values of E_{pk} etc. Thus, we can state that even though this problem may not be *totally* universal, the potential error in the energy placement of any significant feature in the DOS is sufficient in itself to invalidate any *confident* use of the procedure.

6. Conclusions

We have demonstrated that even for the case of a *very* small deviation from an exponential DOS, as induced by the addition of a narrow Gaussian component, application of the pre-transit $(1/I(t)t)$ analytical technique yields a significant error in respect of the peak energy of this feature.

If our results can be generalized to other forms of DOS, and we see no reason why this should not be the case, then this technique (despite its attractive simplicity) appears inapplicable in respect of any attempted quantitative study of localized state distributions in disordered semiconductor materials.

In contrast, we have also demonstrated that both the post-transit and Fourier transform techniques can yield accurate representations of the DOS, and therefore remain valid.

References

1.   J.M. Marshall and R.A. Street, Solid State Commun. 50, 91 (1984).
2.   J.M. Marshall and C. Main, Phil. Mag. B 47, 471 (1983).
3.   J.M. Marshall, Rep. Prog. Phys., 46, 1235 (1983).
4.   T. Tiedje and A. Rose, Solid State Commun. 37, 49 (1980).
5.   G.F. Seynhaeve, R.P. Barclay, G.J. Adriaenssens and J.M. Marshall, Phys. Rev. B 39, 10196 (1989).
6.   C. Main, R. Brüggemann, D.P. Webb and S. Reynolds, Solid State Commun. 93, 401 (1992).
7.   C. Main, Mat. Res. Soc. Symp. Proc. 467, 167 (1997).
8.   C. Main, J. Berkin and A. Merazga, in M. Borissov, N. Kirov, J.M. Marshall and A. Vavrek (Eds.), *New Physical Problems in Electronic Materials* (World Scientific, Singapore, 1991), p. 55.

2.2. OPTICS

COMPOSITE FILMS WITH NANOSIZED MULTILAYER PARTICLES: NUMERIC MODELLING OF THEIR OPTICAL RESPONSE

P. SHARLANDJIEV[*]
Central Laboratory of Optical Storage and Processing of Information, Bulgarian Academy of Sciences, Acad. G. Bonchev St., bl. 101, Sofia PS-1113, P.O. Box 95, Bulgaria

Abstract. Core-shell nanosized particles are intensively studied because of their potential for many novel applications. Their optical properties show the presence of resonances at certain wavelengths. Herein, we present a numeric analysis of multilayer spherical particles, randomly dispersed in a dielectric host matrix. Each particle is assumed to consist of four concentric spheres. The overall optical response of the composite films is investigated.

Keywords: core-shell nanoparticles; nanocomposites; optical resonances

1. Introduction

Core-shell nanosized particles are of particular interest for new nanotechnological applications because of their layered structure, which gives a new degree of freedom for the engineering design. Recently Olenburg et al.[1] reported a fabrication method of layered nanospheres. Each particle consists of a silica core (50–200 nm) and a thin (6–20 nm) gold shell. Tunable optical resonances are obtained in the visible and NIR by controlling both the size of the core and the shell thickness. This is the main advantage of core-shell particles. Besides, their optical response is more sensitive to the environment refractive index than homogeneous particles of the same diameter and material. On the theoretical side, after the pioneering work of Lorenz and Mie on light scattering by spheres, solutions for scattering by particles with other simple forms were found.[2,3] In the mid of the twentieth century,

[*]pete@optics.bas.bg

J.P. Reithmaier et al. (eds.), *Nanostructured Materials for Advanced Technological Applications*, 61
© Springer Science + Business Media B.V. 2009

several exact but iterative and inexplicit expressions were proposed for core-shell particles.[4] The development of numeric techniques and the use of computers solved the problem of the description of light scattering by multi-layer spheres.

Glass materials with embedded metal particles have been studied for centuries. In terms of nanoscience, these materials are considered as composite layers (CL) with nanosized particles. Although an exact description of the optical response of a single particle (or ensembles of particles) against a dielectric background is available, as well as that of as thin (thick) homogeneous layer, the case of CLs is usually treated with the help of effective medium approximations (EMA). A proper effective permittivity is assigned to a heterogeneous film in such a way that the CL can be treated as homogeneous. Usually, this is done by some 'mixing rules' (Maxwell-Garnett, Brugemann, etc.) without analysis of their domain of validity.[5]

Herein, we present a numeric analysis of spherical multilayer particles, randomly dispersed in a dielectric layer. Conditions for the appearance of morphology dependent resonances are found. The overall optical response of the CLs is investigated. We discuss also the case that the effective index of the film can be tailored to very low values.

2. Particle Model and Computational Procedures

We consider a multilayer particle, which consists of four concentric spheres. The core (sphere 1, radius 40 nm) and the 2nd shell (sphere 3, radius 54 nm) are supposed to be dielectric (real refractive index of 1.45). The 1st shell is metallic (sphere 2, radius 44 nm, gold), while the 3rd shell (sphere 4, radius 58 nm) is silver. The complex permittivities of Au and Ag are derived from experimental data of the bulk materials.[6] The thicknesses of the metallic shells are kept below the skin depth, so that the radiation "sees" the whole of the particle structure. The optical behaviour of the single particle is analysed by the exact vector Maxwell equations by the method of separation of variables. The scattered fields are estimated in terms of infinite series of the electric and magnetic fields. The expansion coefficients, also called multipole electric $\{a_n\}$ and magnetic $\{b_n\}$ modes, are determined by the boundary conditions of the equations. The modes depend on the sphere radii and the relative matrix/particle permittivity. Expressions for the $\{a_n, b_n\}$ modes can be found in specialized literature.[3,4] They are evaluated by iterations over the successive sphere boundaries with the help of special Hankel and Bessel functions and their derivatives. Resonances (sharp minima/maxima) in the internal and scattered fields are related to the poles of the expansion coefficients $\{a_n, b_n\}$. Explicit equations for the electric and magnetic modes of a heterogeneous sphere are given in

Ref. 7. We used them to develop computer codes for the estimation of $\{a_n, b_n\}$ and the optical response of the particles.

In fact, by computer calculations we passed from exact to rigorous solutions. This means that postponing of the truncation of the infinite series can generate wrong results due to the limited precision of the machine. The solution of this problem is to monitor the convergence of the series with $\{a_n, b_n\}$ coefficients. In our simulations, 11 terms are highly sufficient for the mode evaluations at the resonance wavelengths. Keeping more than 30 terms generates wrong results.

In Figure 1, the real parts of the electric dipole and quadrupole (multiplied by 10) modes are shown. The wavelength W dependences show morphology resonances related to the poles in the $\{a_n\}$ modes, which depend on the 1st kind Bessel and 2nd kind Hankel functions of large complex arguments.

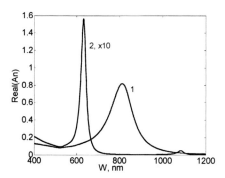

Figure 1. Real part of the electric (1) dipole and (2) quadropole mode (multiplied by 10).

The effect of a_2 cannot be neglected. The magnetic modes show no resonances but b_1 and b_2 are significantly different from zero. Next, we assume that identical particles are randomly dispersed in a thin dielectric film. The filling factor F, which is defined as the number of particles per unit volume, is kept low (0.05–0.2). This means that there is no multiple scattering between individual spheres. The last step is to evaluate an effective refractive index of the heterogeneous film. All EMA approaches are based on electric dipole approximations, where only the term a_1 is retained in the series expansion of the amplitude of the electric field. As a result, scattering in the forward and backward directions are equal and the effective CL permittivity does not depend on the particles size. None of these assumptions is valid in our case, where the contribution of the higher electric and the magnetic modes cannot be ignored. We follow a suggestion of C. Bohren (based on an idea of William Doyle[5]) for the effective relative permittivity ε and the effective relative permeability μ of CLs:

$(\varepsilon\ \mu)^{1/2} = 1 + a*S(0^o);$ $(\varepsilon)^{1/2} - (\mu)^{1/2} = a*S(180^o),$ with $a = i*2*\pi*F/k^3/v,$

where S is the scattering amplitude in the forward/backward direction, F the filling factor, k the wave vector, and v the particle volume. In such a way the effective refractive CL index $N_{eff} = (\varepsilon\ \mu)^{1/2}$ is defined and evaluated. We consider a thin film matrix with a refractive index $N_m = 1.45$, cladded between two half spaces with refractive indices of 1 and 1.55.

3.   Results and Discussion

Most of the simulations were performed on multilayer spheres which can be formally described by a self-obvious notation: {R1: 40 nm, 1.45; R2: 44 nm, Au; R3: 54 nm, 1.45; R4: 58 nm, Ag}. Let us consider that the particles are embedded in 150 nm thin film and treat the real refractive index of the film as parameter ($N_m = 1.33, 1.45, 2$). The extinction efficiency E of the particle is presented in Figure 2.

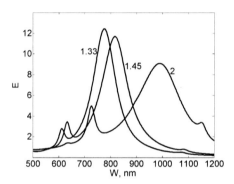

Figure 2. Particle extinction efficiency. The refractive indices of the matrix are shown next to the curves.

There is a strong red shift of the main resonance maximum from 780 to 1,000 nm with N_m increasing from 1.33 to 2. Secondary maxima due to the electric quadrupole mode are present as well. With the increase of N_m, the strength of the resonance decreases and the half-band half-width (HBHW) increases. If we treat the radius of the second sphere R2 as a parameter with values of 4, 12 and 20 nm, we observe a net blue shift of the resonance of the CL (described again by F = 0.2 and $N_m = 1.45$). The transmittance minimum of the layer (T < 8%) shifts from 820 to 650 nm with the increase of R2. The evaluated CL reflectance R possesses antireflection characteristics from 400 to 650 nm (R = 0 at 560 nm for a 4 nm shell). Similar results are obtained when treating the last shell as parameter (R4 = 58, 62, 66 nm). The thickness of the dielectric spacer (R3) also affects the spectral position of the transmittance minima. For R3 = 54, 56, 64 nm, a red shift is observed from 820 nm for the thinnest shell to 890 nm for a shell of 20 nm.

In Figure 3 we present results for the evaluation of CLs with particles of the main configuration, where the parameter is the filling factor (F = 0.05, 0.2, 0.3). The real part of the effective refractive index of the layer decreases from its initial value of 1.45, passes through a minimum at 760 nm, then there is a sharp increase. The imaginary part of N_{eff} stays positive (maximum at 810 nm) and increases from 0.5 to 3 for F = 0.3. In this case, the HBHW of the transmittance is ~100 nm and the minimum T ~ 0%.

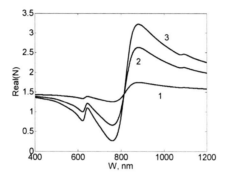

Figure 3. Real part of the effective refractive index of CLs. (1) F = 0.05, (2) F = 0.2, (3) F = 0.3.

It can be seen that the real part of the effective refractive index can reach significantly low values at the spectral position of the morphology resonances. At the same time, these resonances are accompanied by considerable absorption. By changing the filling factor of CLs with multilayer spheres, different effective refractive indices can be designed. The question is: How low values of N_{eff} can be reached? Even more, it can be speculated that "amorphous photonic" core-shell nanostructures are strong candidates for the synthesis of so-called "left-handed metamaterials". Metamaterials, known as "left-handed" or negative refractive index (NRI) materials, are defined by their simultaneously negative permittivity and permeability. Metamaterials do not exist in nature but have been successfully constructed as photonic subwavelength crystals and demonstrated to operate in the terahertz part of the spectrum. While a negative permittivity can be easily supplied by variety of metals, the problem is to obtain a negative permeability. In series of simulations we have "obtained" negative refractive indices by slight modifications of main configuration of the CLs. We diminished the particle core to 20 nm and increased the filling factor to 0.3. Two spectral bands with a NRI open up at 590–600 nm and a larger one at 1,050–1,120 nm. There, the transmittance is very low, but (T + R) > 0; the HBHW at 1,100 nm is ~50 nm. These results are rigorous with the only assumption that the evaluation of the effective refractive index with the help of the amplitude scattering matrix of the particle is still valid. The situation is even more controversial for intrinsically non-magnetic structures, as the ones discussed here. On the other hand, the problem of the optical response of CLs can be solved without recourse to effective refractive index approaches, i.e. by direct calculations with the radiation transfer equation.[2,3] However, in this case an important feature of CLs is lost, because this equation is based on field intensity propagation, where in the forward and backward directions the light scattering from all particles is coherent.

4. Conclusions

We have developed rigorous machine codes for the evaluation of electric and magnetic modes of multilayer spheres, which are based on the Volkov algorithm.[7] Numeric simulations for particles consisting of four concentric spheres, revealed that the pole locations of the modes are closely related to physical phenomena within the particles or in their immediate vicinity. In the range of our numeric simulations, magnetic modes cannot be ignored. At resonant illumination of a nanosized composite structure of the kind discussed above, the peak location is a sensitive function of the morphology of the particle (spheres size and relative complex refractive indices), the host layer refractive index, and the film filling factor. Important trends in the strength and width of the resonances are identified numerically. The optical responses of composite layer obtained by effective medium concepts are inappropriate for larger particles, such as the multilayer spheres in our case. Even the evaluation of the effective refractive index of CLS with the help of the amplitude scattering matrix of the particle must be used with extreme caution, because it can lead to unphysical results. In principle, this situation can be checked by finite-difference time-domain methods (FDTD) or other powerful computational techniques.

ACKNOWLEDGEMENT

This work was partially supported by the National Science Foundation at the Ministry of Education of Bulgaria by grant D01-377/2006.

References

1. S. Olenburg, J. Jackson, S. Westcott and N. Hales, Appl. Phys. Lett. 75, 2897 (1999).
2. U. Kreibig and M. Voller, *Optical Properties of Metal Clusters* (Springer, Berlin, 1995).

3.   C. Bohren and D. Huffman, *Absorption and Scattering of Light by Small Particles* (Wiley, New York, 1998).
4.   R. Bhandari, Proc. SPIE 540, 500 (1985).
5.   C. Bohren, J. Atmos. Sci. 36, 880 (2004).
6.   P. Johnson and R. Christy, Phys. Rev. B 6, 4370 (1972).
7.   N. Volkov and Yu. Kovach, Atmos. Ocean. Phys. 26, 381 (1990).

TWO-PHOTON LASING IN MICRO-CAVITIES

N. ENAKI, M. ŢURCAN[*]
Center of Optoelectronics, Institute of Applied Physics, Academy of Sciences of Moldova Academiei St. 5, Chisinau MD-2028, Republic of Moldova

Abstract. In this report we propose a model interaction Hamiltonian, which takes into account four-photon processes in the interaction of dressed fields with atoms in a micro-cavity. Using the method of elimination of virtual states, we obtain the effective interaction Hamiltonian which describes the simultaneous generation of photon pairs. In the good-cavity limits the master equation for the laser field is obtain. Cooperative phenomena between the multi-mode aspects of two-photon generation are taken into account.

Keywords: quantum optics; photon generation; coherent emission

1. Introduction

The coherent two-photon generation of light is in the center of attention of many experimental and theoretical studies in the last time. Taking into account the experimental realization of two-photon lasers and masers reported in Refs. 1 and 2, in this paper a new model of two-photon generation is proposed, which takes into account one- and two-photon losses from the cavity as well as cooperative phenomena between the degenerate and non-degenerate aspect of two-photon lasing.

2. Model Hamiltonian and Master Equation

We propose a model interaction Hamiltonian which takes into account four-photon processes in the interaction of external dressed fields with atoms in a micro-cavity. The effective Hamiltonian,

_____
[*]tmaryna@yahoo.com

J.P. Reithmaier et al. (eds.), *Nanostructured Materials for Advanced Technological Applications,*   69

$$H_I^{eff} = \sum_v \{ G(z)b(t)b(t)e^{2ikpx} |e(t)\rangle\langle g(t)|$$

$$+ G^*(z)b^+(t)b^+(t)e^{-2ikpx} |g(t)\rangle\langle e(t)| \}, \quad (1)$$

describing the simultaneous generation of photon pairs, was obtained using the method of elimination of virtual states. Here $G(z)$ is interaction constant which depends on the dressed field

$$G(z) = \left[\left(\frac{1}{\omega_{vg} - \omega_P} + \frac{1}{\omega_{ve} - \omega_D} \right) * \frac{(\vec{d}_{ve}, \vec{v})(\vec{d}_{vg}, \vec{g})^2 (\vec{d}_{vg}, \vec{v})}{\hbar^3 (\omega_{vg} - \omega_D)(\omega_{vg} - \omega_P)} \right.$$

$$+ Ibed(e - g)] * \vec{E}_D^+(t,z)\vec{E}_D^+(t,z);$$

$b(t)$ and $b^+(t)$ are the annihilation and creation photon operators in the cavity; $\vec{E}_D^+$ and $\vec{E}_D$ the complex components of the full wave vector of the dressed field; $|e(t)\rangle\langle g(t)|$ and $|g(t)\rangle\langle e(t)|$ are the two-photon excitation and de-excitation operators of the atomic subsystem, which describe the transitions of an atom from the ground to excited states, and vice versa. In Figure 1, the excited state $|e\rangle$ is considered as a virtual state $|v\rangle$.

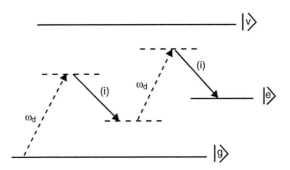

Figure 1. Scattering diagram showing the two-photon Raman process.

In the good-cavity limits the master equation for the laser field can be obtained. Indeed using the method of elimination of atomic variables proposed in Refs. 3 and 4 we can obtain the following master equation for the density matrix of bi-photons in the cavity

$$\frac{\partial W(t)}{\partial t} = k[I^- W(t), I^+] + i\chi[I^+ (1 + pI^- I^+)^{-1} W(t), I^-]$$

$$+ u_1[I^+ (1 + pI^- I^+)^{-1} W(t), I^-] \quad (2)$$

$$+ u_2[I^+ [I^-, I^+ (1 + pI^- I^+)^{-1} W(t)(1 + pI^- I^+)^{-1}], I^-] + H.c.$$

Here k describes the losses of bi-photons (pairs of entangled photons) from the micro-resonator; $\chi = 2|G|^2 N\sigma_o\,(\omega - 2\Omega)/[(\omega - 2\Omega)^2 + \gamma^2]$ is the renormalization energy of biphotons; $u_1 = 2|G|^2 N\sigma_o\gamma/[(\omega - 2\Omega)^2 + \gamma^2]$ represents the generation rate of photon pairs; $u_2 = T/N\sigma_o\,(u_1^2 + \gamma^2)$ is the nonlinear damping constant and finally $p = 2Tu_1/N\sigma_o$.

We introduced here the bi-photon operators belonging to the su(1,1) commutation algebra

$$[I^+, I^-] = -I_z, \quad [I_z, I^\pm] = \pm I^\pm,$$

where for the degenerate case the new operators are $I^+ = b^{+2}/2$, $I^- = b^2/2$, and $I_z = (b^+ b + bb^+)/4$; for non-degenerate two-photon processes, in which two different photons are generated, we can introduced the new boson representation: $I^+ = b^+ a^+$, $I^- = ba$ and $I_z = (b^+ b + aa^+)/2$. From the master equation (2) it is not difficult to observe that in the process of bi-photon exchanges the Kasimir vector is conserved: $I^2 = j(j - 1)$, where $j = 1/4$ and $j = 1/2$ for the degenerate and non-degenerate cases, respectively. The number j represents the possibilities in which the bi-photons can be generated in the modes of the cavity and depends on the mode structure of the resonator.

3. Master and Fokker–Planck Equations for Two-Photon Lasing in Micro-cavities with Losses in Pairs

Let us introduced the complex P representation for the su(1,1) algebra in a similar way as it was introduced by the authors in Ref. 5 for Bose quantum oscillators

$$W = \int_D \Lambda(\alpha, \beta) P(\alpha, \beta) d\mu(\alpha, \beta); \quad \Lambda(\alpha, \beta) = \frac{|\alpha\rangle\langle\beta^*|}{\langle\beta^* | \alpha\rangle}. \tag{3}$$

Here, $d\mu(\alpha, \beta)$ is the integration measure and D is the domain of integration. The projection operator $\Lambda(\alpha, \beta)$ is analytic in (α, β). In this section we are interested in the complex P-representation in which $d\mu(\alpha, \beta) = d\alpha d\beta$. Here α and β are regarded as complex variables, which are integrated on individual contours C and C'. The coherent state for su(1,1) can be defined in the following form

$$|\alpha\rangle = (1 - |\alpha|^2)^j \exp(\alpha I^+)|0\rangle, \quad \langle\beta^*| = (1 - |\beta|^2)^j \langle 0| \exp(\beta I^-),$$

the scalar product of which is

$$\langle\beta^* | \alpha\rangle = \frac{(1 - |\alpha|)^{2j}(1 - |\beta|)^{2j}}{(1 - \alpha\beta)^{2j}}.$$

A similar representation was introduced in Ref. 6 for the parametrical amplification of photon pairs. Using this method, the normalization of the complex P-representation can be found in the similarities with the diagonal P-representation. In this case many proprieties of P- and Q-functions are conserved for su(1,1) algebra. Using the following actions of the operators I^+, I^-, I^z on the coherent state of the su(1,1) algebra

$$I^+|\alpha\rangle = (1-|\alpha|^2)^j \frac{\partial}{\partial \alpha} \exp(\alpha I^+)|0\rangle,$$

$$I^-|\alpha\rangle = (1-|\alpha|^2)^j (\alpha^2 \frac{\partial}{\partial \alpha} + 2j\alpha)\exp(\alpha I^+)|0\rangle,$$

$$I^z|\alpha\rangle = (1-|\alpha|^2)^j (\alpha \frac{\partial}{\partial \alpha} + j)\exp(\alpha I^+)|0\rangle,$$

one can obtain the Fokker–Planck equation. Below we discuss two situations: (a) below the threshold and (b) above the threshold.

BELOW THE THRESHOLD

Let us first discuss the emission below the threshold, $k > u_1$. In this situation one can neglect the terms proportional to the square value of the amplification coefficient u_1 considering $u_1 \gg u_2$. Using the $P(\alpha,\beta)$-representation (2), the master equation takes the following form in this approximation

$$\int_{c'} d\alpha \int_c d\beta \frac{\partial}{\partial t} P(\alpha, \beta, t) \exp(\alpha I^+)(1 - \alpha\beta)^{2j} |0\rangle\langle 0| \exp(\beta I^-)$$

$$= \int_c d\alpha \int_c d\beta P(\alpha, \beta, t)(1 - \alpha\beta)^{2j} k \left(\alpha^2 \frac{\partial}{\partial \alpha} + 2j\alpha\right)\left[\left(\beta^2 \frac{\partial}{\partial \beta} + 2j\beta\right) - \frac{\partial}{\partial \alpha}\right]$$

$$+ (i\chi + u_1) \frac{\partial}{\partial \alpha}\left[\frac{\partial}{\partial \beta} - \left(\alpha^2 \frac{\partial}{\partial \alpha} + 2j\alpha\right)\right] \exp(\alpha I^+)|0\rangle\langle 0|\exp(\beta I^-) + H.c.$$

$$(4)$$

Following the method of partial integration of the right hand side of this equation one can found another expression for Eq. (4). Taking into account this notations, the following Fokker–Planck equation it is obtained

$$\frac{\partial}{\partial t} P(\alpha, \beta) = 2[jk - (j-1)u_1][\frac{\partial}{\partial \alpha} \alpha P(\alpha, \beta) + \frac{\partial}{\partial \beta} \beta P(\alpha, \beta)]$$

$$+ 2\frac{\partial^2}{\partial \alpha \partial \beta}[k\alpha^2\beta^2 + u_1]P(\alpha, \beta) \qquad (5)$$

$$- (k + u_1) \times [\frac{\partial^2}{\partial \beta^2}\beta^2 P(\alpha, \beta) + \frac{\partial^2}{\partial \alpha^2}\alpha^2 P(\alpha, \beta)].$$

The Fokker–Planck equation (5) is obtained in the first order of approximation of the interaction constant $|G|^2$. The stationary solution $\partial p / \partial t = 0$ of this equation can be found easily:

$$P(\alpha, \beta) = D \frac{\left| k\alpha\beta - u_1 \right|^{2(j-1)}}{\left| 1 - \alpha\beta \right|^{2j}}.$$

Here D is the integration constant. This stationary solution describes the amplification of a two-photon lasing effect in the cavity stimulated by the pumping field. Passing again to the density matrix, one can found the following expresion

$$W = \int_D \Lambda(\alpha, \beta) P(\alpha, \beta) d\mu(\alpha, \beta); \quad \Lambda(\alpha, \beta) = \frac{|\alpha\rangle\langle\beta^*|}{\langle\beta^* | \alpha\rangle}. \tag{6}$$

ABOVE THE THRESHOLD

Let us discuss the other situation where $u_1 \sim u_2$. In this case we obtain the following expression, similar to the Fokker–Planck equation

$$\frac{\partial}{\partial t} P(\alpha, \beta) = 2[jk - (j-1)u_1][\frac{\partial}{\partial \alpha}\alpha P(\alpha, \beta) + \frac{\partial}{\partial \beta}\beta P(\alpha, \beta)]$$

$$+ 2\frac{\partial^2}{\partial\alpha\partial\beta}[k\alpha^2\beta^2 + u_1]P(\alpha, \beta) \tag{7}$$

$$- (k + u_1) \times [\frac{\partial^2}{\partial\beta^2}\beta^2 P(\alpha, \beta) + \frac{\partial^2}{\partial\alpha^2}\alpha^2 P(\alpha, \beta)].$$

Using the notation $P(\alpha,\beta,t)(1-\alpha\beta)^{2j} = F(x)$, where $(x = \alpha\beta)$, in the stationary case the following equation for the function $F(x)$ is obtained.

$$\frac{\partial^3 F(x)}{\partial x^3}x^2 - \frac{\partial^2 F(x)}{\partial x^2}Ax - \frac{\partial F(x)}{\partial x}(P_2 - P_1 x) - F(x)B = 0. \tag{8}$$

Here the following notation is used:

$$A = 3(2j-1); \quad P_2 = 2(j-1) + \frac{u_1}{u_2}; \quad P_1 = \frac{1}{u_2}; \quad B = \frac{2k(j-1)}{u_2}$$

In the general case it is impossible to solve Eq. (8). Considering that $P_1 = B$, one can reduce this equation to a second order differential equation. Indeed in this case (8) becomes

$$\frac{d}{dx}[F''(x)x^2] - F''(x)[(A+2)x] - F'(x)(P_2 - P_1 x) - F(x)B = 0 \tag{9}$$

After the integration of this equation one can obtain the second order differential Eq. (10):

$$x^2 F'' - [(A+2)x]F' + (P_1 x - A - 2 - P_2)F = 0, \tag{10}$$

where a = -A + 2; b = P_1; c = a - P_2.

In accordance with the analytical representation of the solution of this equation in Ref. 7 we found the solution, which is represented through the special functions:

$$F = y = x^{\frac{1-a}{2}} Z_v(2\sqrt{P_1} x^{\frac{1}{2}}). \tag{11}$$

Here $v = \sqrt{(1-a)^2 - 4c}$; Z_v is the arbitrary cylindrical function, which can be represented through the Bessel function[8] in the following way

$$Z(v) = C_1 J_v(x) + C_2 J_{-v}(x),$$

Here C_1 and C_2 are constants; J_v and J_{-v} are Bessel functions.

This solution was very well studied in the literature.[9–13] In particular it describes the amplification of two-photon lasing effect[9,10] in the cavity stimulated by a pumping field. Passing again to the density matrix (6) one can found all quantum correlations between the bi-photons.

4. Conclusions

In this paper, using the methods of elimination of the operators for virtual state, the interaction Hamiltonian of atoms with a pumping field and a generation field in the cavity was obtained. Using the P-representation for su(1,1) biboson fields, the Fokker–Planck equation for the two-photon laser field was obtained. The solution for this equation gives us the possibility to solve the problem of quantum behavior of bi-photon fields in the lasing process.

ACKNOWLEDGEMENTS

Support for this work has been provided by the Young Scientist Grant No. 08.819.05.03F of the Academy of Sciences of Moldova.

References

1. O. Pfister, W.J. Brown, M.D. Stenner, and D.J. Gauthier, Phys. Rev. Lett. 86, 4512 (2001).
2. M. Brune, J.M. Raimond, and S. Haroche, Phys. Rev. A 35, 154 (1987).

3.   N. Enaki and V. Eremeev, Opt. Commun. 247, 381 (2005).
4.   N. Enaki, JETP, 98, 783 (1990).
5.   P.D. Drummond and C.W. Gardiner, J. Phys. A: Math. Gen. 13, 2353 (1980).
6.   N. Enaki and V.I. Ciornea, Physica A 340, 436 (2004).
7.   E. Kamke, *Differentialgleichungen* (Akademische Verlagsgesellschaft, Leipzig, 1959).
8.   N.N. Lebedev, *Special Functions and Their Applications* (Dover, New York, 1972).
9.   D.J. Gauthier, Q. Wu, S.E. Morin, and T.W. Mossberg, Phys. Rev. Lett. 68, 464 (1992).
10. H.M. Concanon, W.J. Brown, J.R. Gardiner, and D.J. Gauthier, Phys. Rev. A 56 (1997).
11. M. Poelker and P. Kumar, Opt. Lett. 17, 399 (1992).
12. A. Hemmerich, C. Zimmerman, and T.W. Hänsch, Phys. Rev. A 72, 625 (1992).
13. F.S. Cataliotti, R. Scheunemann, T.W. Hänsch, and M. Weitz, Phys. Rev. Lett. 87, 113601 (2002).

INTERFERENCE OF COOPERATIVE RESONANT FLUORESCENCE FROM TWO DISTANT SYSTEMS OF RADIATORS

E. GALEAMOV[*], N. ENAKI
*Center of Optoelectronics, Institute of Applied Physics,
Academy of Sciences of Moldova, Academiei St. 5,
Chisinau MD-2028, Republic of Moldova*

Abstract. The possibility to realize collective resonance fluorescence of extended systems of radiators in strong travelling and standing laser waves is demonstrated. The influence of a regular arrangement of the atoms in interaction with the laser pulses on the fluorescence intensity is analyzed. Taking into account that such interference can be realized in two distant atomic (or solid-state) ensembles, we analyzed collective interference phenomena between the fluorescent fields in the far-field detection region. These effects depend on the exchange integrals between cold atoms through a vacuum fluorescence field. The fluorescence spectrum of such atoms was found. The influence of the sample geometry on the coherent phenomena between two distant systems of radiators is also presented in this paper.

Keywords: quantum optics; distant radiators; resonant fluorescence; travel and standing laser waves

1. Introduction

There are reports on similar coherence interference experiments of spontaneous emission of light from two distant solid-state ensembles of atoms that are coherently excited by a short laser pulse.[1-4] For example, in Ref. 1 the entanglement between two distant atomic ensembles located in distinct apparatuses on different tables was observed. Quantum interference in the detection of a photon emitted by one of the samples projects the otherwise

---

[*]egaleamov@gmail.com

independent ensembles into an entangled state with one joint excitation stored remotely in 105 atoms at each site. Another experiment[2] dealt with ensembles of erbium ions doped into two LiNbO$_3$ crystals with channel wave guides, which are placed in the two arms of a Mach–Zehnder interferometer. The light emitted spontaneously after the excitation pulse shows first-order interference. By a strong collective enhancement of the emission, the atoms behave as ideal two-level quantum systems; no "which-path" information is left in the atomic ensembles after emission of a photon. This results in a high fringe visibility of 95%, which implies that the observed spontaneous emission is highly coherent.

The aim of this paper is to study the collective phenomena in the presence of a dressed laser field in resonance with one active line of cold atoms. As the atoms are localized in space, an interesting effect can appear in the standing wave formed in the process of reflection of the laser light by the mirrors of native material. The collective resonance fluorescence of extended systems of atoms in the standing wave is discussed. The influence of a regular arrangement of the atoms in interaction with the standing wave of the cavity on the fluorescence intensity is analyzed. The collective interference phenomena in the far-field detection region are studied also. These effects depend on the exchange integrals between two-level radiators through a vacuum fluorescence field. The interference of the fluorescence field as a function of the position of two cold atoms in the standing wave resonator was obtained. A system of equations that describes this cooperative effect was obtained.

2. The Master Equation for Atomic Ensembles Dressed by Coherent Field

Performing the diagonalization of the free atomic parts of the Hamiltonian as in the Ref. 5, one can obtain the dressed effective Hamiltonian of a system of radiators and radiation field in standing and travelling waves

$$\widetilde{H} = \widetilde{H}_0 + \widetilde{H}_I, \quad \text{where} \tag{1}$$

$$\widetilde{H}_0 = \sum_k \hbar(\omega_k - \omega)a_k^+ a_k + \hbar \sum_{j=1}^{N} \Omega_j D_{zj}(t), \tag{2}$$

is the free dressed part of the system. The interaction Hamiltonian part $\widetilde{H}_I$ in the rotation wave approximation relative to the Rabi frequency can be expressed in the following way:

$$\tilde{H}_I = \frac{\omega_0 d}{2c} \sum_k \sum_{j=1}^{N} g_k \left\{ \left(-\Delta_{1j} D_j^+(t) + \Delta_{2j} D_j^-(t) + 2\Delta_{3j} D_{zj}(t) \right) a_k \right.$$

$$\left. \exp[i(\mathbf{k} - (1-\lambda)\mathbf{k}_0, \mathbf{r}_j)] + h.c. \right\} \tag{3}$$

The new dressed atomic operators D_j^+, D_j^-, D_{zj} are expressed through the old atomic operators according to the relations:

$$D_j^+ = -\frac{i\Delta_{2j}}{2} R_j^- e^{-i(1-\lambda)(\mathbf{k}_0, \mathbf{r}_j)} - \frac{i\Delta_{1j}}{2} R_j^+ e^{i(1-\lambda)(\mathbf{k}_0, \mathbf{r}_j)} + \Delta_{3j} R_{zj},$$

$$D_j^- = [D_j^+]^+, \quad D_{zj} = \frac{\delta}{\Omega_j} R_{zj} - \frac{i\Delta_{3j}}{2} \left\{ R_j^- e^{-i(1-\lambda)(\mathbf{k}_0, \mathbf{r}_j)} - h.c. \right\} \tag{4}$$

Here ω_0 is the frequency of the transition, R_j^+ (R_j^-) are the operators of creation (annihilation) of the excited state of the j-th atom, a_k^+ (a_k) are the radiation field creation (annihilation) operators. We take into account two situations: the travelling wave case $\lambda = 0$, for which the value of the Rabi frequency $\Omega_j = (\delta^2 + \Omega_0^2)^{1/2}$ doesn't depend on the position of the atoms, but the operators contain the phaze position factors $\exp(\pm i k_0 r_j)$, and the standing wave $\lambda = 1$, for which the counter propagate phases of the waves in the resonator is contained in the Rabi frequency $\Omega_j = (\delta^2 + 4\Omega_0^2 \sin^2(k_0 r_{0j}))^{1/2}$, $\Delta_{1j} = 1 + (\delta/\Omega_j)$, $\Delta_{2j} = 1 - (\delta/\Omega_j)$, $\Delta_{3j} = \tilde{\Omega}_j / \Omega_j$, D_j^+, D_j^-, and D_{zj} are new quasi-spin operators connected with the transition processes between quasi-levels of split states of the j-th atom in the strong laser field. The Hamiltonians $\tilde{H}_{i1}$ and $\tilde{H}_{i2}$ describe the processes of collective generation of light with frequencies $\omega + \Omega_j$ and $\omega - \Omega_j$ in the strong external laser field. The light with frequencies $\omega + \Omega_j$, $\omega - \Omega_j$ is generated in the process of transition from the quasi-level $\hbar\Omega/2$ to the quasi-level $-\hbar\Omega/2$, while the light with the frequency ω is scattered by the atom without transitions between new quasi-levels.

Using the method of elimination of field operators,[5] we can obtain the following system of equations for dressed atomic correlators

$$X(t) = \langle D_{zA}(t) \rangle, \quad Y(t) = \langle D_A^+(t) D_B^-(t) \rangle \text{ and } Z(t) = \langle D_{zA}(t) D_{zB}(t) \rangle:$$

$$\frac{dX(t)}{dt} = -\frac{\Delta_{1A}^2}{4} J_{AA}^a [X(t) + 1/2] + \frac{\Delta_{2A}^2}{4} J_{AA}^s [1/2 - X(t)]$$

$$- \left(\Delta_{1A} \Delta_{1B} I_{AB}^a - \Delta_{2A} \Delta_{2B} I_{AB}^a \right) Y(t)$$

$$\frac{dY(t)}{dt} = -\frac{1}{4} \left(\Delta_{2A}^2 J_{AA}^s + \Delta_{1B}^2 I_{AA}^a + 2 \left(\Delta_{3A}^2 + \Delta_{3B}^2 \right) J_{AA}^e \right) Y(t) + \left(\Delta_{1A} \Delta_{1B} J_{AB}^a \right.$$

$$+ \Delta_{2A} \Delta_{2B} I_{AB}^s \right) Z(t) - \frac{1}{2} \left(\Delta_{2A} \Delta_{2B} I_{AB}^s - \Delta_{1A} \Delta_{1B} J_{AB}^a \right) X(t) - 2\Delta_{3A} \Delta_{3B} J_{AB}^e Y(t)$$

$$\frac{dZ(t)}{dt} = -\frac{1}{4}\left(\Delta_{1A}^2 J_{AA}^a + \Delta_{1B}^2 I_{AA}^a + \Delta_{2A}^2 J_{AA}^s + \Delta_{2B}^2 I_{AA}^s\right)Z(t) + \left(\Delta_{1A}^2 J_{AA}^a + \Delta_{1B}^2 I_{AA}^a\right) \tag{5}$$

$$-1/2[\Delta_{2A}^2 J_{AA}^s + \Delta_{2B}^2 I_{AA}^s]\big)X(t) - \left(\Delta_{1A}\Delta_{1B}\left(J_{AB}^a + I_{AB}^a\right) + 2\Delta_{2A}\Delta_{2B}\left(J_{AB}^s + I_{AB}^s\right)\right)Y(t).$$

Here it is considered that the atoms A and B are situated in equivalent positions in the standing wave (in an oscillating loop or in a node of the wave); the notations

$$J_{AA}^\alpha = \frac{4}{3}\frac{\omega_0^2 \omega_\alpha d^2}{\hbar c^3} \tag{6}$$

$$J_{AB}^\alpha = \frac{\omega_0^2 \omega_\alpha d^2}{\hbar c^3}\left\{1 + \cos^2 \xi \frac{\partial^2}{\partial u^2} - \frac{\sin^2 \xi}{2}\left(1 + \frac{\partial^2}{\partial u^2}\right)\right\}\frac{\sin u}{u}\bigg|_{u=\omega_\alpha D/c} \tag{7}$$

are used, where $\omega_\alpha = \{\omega, \omega - \Omega_{A,B}, \omega + \Omega_{A,B}\}$. For $\alpha, \beta = s$ we have the fluorescent emission at the Stokes frequency $\omega_S = \omega - \Omega_{A,B}$, for $\alpha, \beta = a$, the frequency is $\omega_S = \omega + \Omega_{A,B}$, and for $\alpha, \beta = e$ we have the spontaneous scattering effect without change of the frequency. The coefficient J_{AA}^α represents the decay rate at the frequency ω_α, the coefficient J_{AB}^α the cooperative decay rate between the atoms; it describes the mutual influence of the spontaneous emission of light by the atoms in the process of resonance fluorescence. The solution of the system of Eq. (5) describes the collective resonant spontaneous emission of two cold atoms situated at the distance D in a dressed standing wave.

3.   Discussions and Results

Let us study the case when the distance $D > \lambda$. The atoms A and B are situated in different loops (or nodes) of the standing wave. Introducing the autocorrelation function $F(D) = \tau_0 J_{AB}^0$,

$$F(D) = \frac{3}{4}\left\{1 + \cos^2 \xi \frac{\partial^2}{\partial u^2} - \frac{\sin^2 \xi}{2}\left(1 + \frac{\partial^2}{\partial u^2}\right)\right\}\frac{\sin u}{u}\bigg|_{u=\omega_\alpha r_{AB}/c}, \tag{8}$$

we can examine two possibilities of the atomic orientation of the dipole moments. (a) A parallel orientation of the prepared dipole moments for both radiators relative to the direction of the atoms $\mathbf{r}_{AB}$. This corresponds to the angle $\xi = n\pi$, where n is integer. In this case the autocorrelation function $F(D)$ decreases with the distance

$$F_\parallel(D) = \frac{3}{2}\frac{1}{(k_0 D)^2}\left\{\frac{\sin k_0 D}{k_0 D} - \cos k_0 D\right\}. \tag{9}$$

(b) By preparing with laser π-pulses the orientation of the atomic dipole moments perpendicular to r_{AB} ($\xi = \pi/2$,), one can obtain another value for the autocorrelation function:

$$F_\perp(D) = \frac{3}{4}\frac{1}{(k_0D)^2}\left\{k_0D\sin k_0D - \frac{\sin k_0D}{k_0D} + \cos k_0D\right\}. \qquad (11)$$

It can be seen that the autocorrelation function of two radiators placed at a distance $D > \lambda$ is larger in the perpendicular case (11). Indeed, neglecting the last two terms in Exp. (11), the autocorrelation function becomes $F_\perp(D) = \frac{3}{4}\sin(k_0D)/k_0D$, which corresponds to the situation $F_\perp(D) \gg F_\parallel(D)$. For the perpendicular case, a new system of equation is obtained from Exp. (5):

$$\frac{dX(\tau)}{d\tau} = -\frac{1}{2}X(\tau), \quad \frac{dY(\tau)}{d\tau} = -(2F_\perp(D) - 3/2)Y(\tau) + 2F_\perp(D)Z(\tau),$$

$$\frac{dZ(\tau)}{d\tau} = F_\perp(D)Y(\tau) - Z(\tau). \qquad (12)$$

The solution of this system is

$$X(\tau) = -\frac{\delta}{2\Omega_j}\exp[-\tau/2]|_{\delta=0} = 0,$$

$$Y(\tau) = \frac{1}{8}\left(\frac{F_\perp - 1/4}{q} + 1\right)\exp\left[(F_\perp - 5/4 + q)\tau\right] + \frac{1}{8}\left(1 - \frac{F_\perp - 1/4}{q}\right)\exp\left[(F_\perp - 5/4 - q)\tau\right],$$

$$Z(\tau) = \frac{1}{8}(F_\perp/q)\exp[(F_\perp - 5/4 + q)\tau] - \frac{1}{8}(F_\perp/q)\exp[(F_\perp - 5/4 - q)\tau], \qquad (13)$$

where $q = (3F_\perp^2 - F_\perp/2 + 1/16)^{1/2}$. To determine the propagation direction of the radiation emitted by the atoms A and B placed at the distance D, we calculate the photon rate $dN_k(t)/dt$, which is proportional to the intensity I of the light. Using the solution of the Heisenberg equation for electromagnetic field operators,[6] we obtain the following equation for the photon rate in the standing wave case $\lambda = 1$, and for the far-field detection region (it is assumed that $k \perp r_{0A} - r_{0B}$):

$$\frac{dN_k(t)}{dt} = \frac{2\pi\omega_0^2 d^2}{\hbar^2 c^2}g_k^2(\Delta_{1\alpha,\beta}^2\delta[\omega_k - (\omega + \Omega_{\alpha,\beta})] - \Delta_{2\alpha,\beta}^2\delta[\omega_k - (\omega - \Omega_{\alpha,\beta})])X(t)$$

$$+ (\Delta_{1\alpha,\beta}^2\delta[\omega_k - (\omega + \Omega_{\alpha,\beta})] + \Delta_{2\alpha,\beta}^2\delta[\omega_k - (\omega - \Omega_{\alpha,\beta})])/2 + \Delta_{3\alpha,\beta}^2\delta[\omega_k - \omega]$$

$$+ \frac{1}{2}(\Delta_{1\alpha,\beta}^2\delta[\omega_k - (\omega + \Omega_{\alpha,\beta})] + \Delta_{2\alpha,\beta}^2\delta[\omega_k - (\omega - \Omega_{\alpha,\beta})])Y(t) + \Delta_{3\alpha,\beta}^2\delta[\omega_k - \omega]Z(t).$$

$$(14)$$

Integrating over the module of wave vector **k**, one can obtain the following expression for the photon emission rate in the direction perpendicular to the vector $\mathbf{r}_{AB}$ and the solid angle $\Delta\Omega$

$$\dot{\mathbf{N}}\Delta\Omega = \int_{\Delta\Omega} d\Omega_0 \int \mathbf{k}^2 d\mathbf{k} \frac{dN_k(t)}{dt} \frac{V}{(2\pi)^3}. \tag{15}$$

According to the solution (13) the following expression is obtained:

$$\dot{\mathbf{N}}\Delta\Omega = \frac{1}{\tau_0} \frac{3}{2\pi} \frac{\delta}{\Omega} X(t) + \frac{1}{4}\left(1 + \delta^2/\Omega^2\right)\left(Y(t) + 1\right) + \frac{\tilde{\Omega}^2}{\Omega^2}\left(Z(t) + 1/2\right). \tag{16}$$

4. Conclusions

In this paper the cooperative spontaneous emission of two cold atoms situated at a distance D in a standing wave of a laser field was studied. The closed system of equations that describes the collective resonant spontaneous emission of two radiators in the dressed standing wave was obtained. The mutual influences of the atoms were studied from a symmetrical point of view. From the proposed approach follows that, in collective resonant spontaneous emission of two cold atoms, the sign of the collective decay rate (7) strongly depends on the distance between the atoms. For small distances $D << \lambda$, $J^\alpha_{AB} = J^\alpha_{AA}$, and for large $D > \lambda$ tends to $J^\alpha_{AA} \sin(k_\alpha D)/k_\alpha D$.

Another interesting effect is the dependence of the Rabi frequency on the atomic position in the standing wave. If radiator "A" is situated in a oscillating loop, the Rabi frequency $\tilde{\Omega}_j$ takes the maximal value; in the case when radiator "A" is situated in a node of the wave, the Rabi frequency takes a minimal value. This is connected with the behavior of the function $4(\mathbf{d},\mathbf{E}_0)\cos(r_j,k_0)/\hbar$; if $r_j,k_0 = n\pi$ the value of $|\tilde{\Omega}_j|$ is maximal; if $r_j,k_0 = (n+1/2)\pi$ the Rabi frequency $|\tilde{\Omega}_j|$ takes a zero value.

ACKNOWLEDGEMENTS

This work has been supported by the Young Scientist Grants No. 07.408. 31INDF and 08.819.05.03F of the Academy of Sciences of Moldova.

References

1. C.W. Chou, H. de Riedmatten, D. Felinto, S.V. Polyakov, S.J. van Enk, and H.J. Kimble, arXiv:quant-ph/0510055v1.
2. M. Afzelius, M.U. Staudt, H. Riedmatten, C. Simon, S.R. Hastings-Simon, R. Ricken, H. Suche, W. Sohler, and N. Gisin, N. J. Phys. 9, 413 (2007).

3. J. Beugnon, M.P.A. Jones, J. Dingjan, B. Darquie, G. Messin, A. Browaeys, and P. Grangier, Nature (London) 440, 779 (2006).
4. P. Maunz, D.L. Moehring, M.J. Madsen, R.N. Kohn Jr., K.C. Younge, and C. Monroe, eprint arXiv:quant-ph/0608047.
5. N. Enaki, Sov. J. Opt. i Spectr. 66, 1076 (1989).
6. N. Enaki and E. Galeamov, Int. J. Theor. Phys. 47, 911 (2008).

2.3. FURTHER TOPICS

MODELLING THE COHESIVE ENERGY OF CHALCOGENIDE NANOPARTICLES

S.S. DALGIC[*]

*Department of Physics, Faculty of Arts and Sciences,
Trakya University, Güllapoglu Campus 22030 Edirne, Turkey*

Abstract. A new technique for the optimization of the potential parameters of the modified analytic embedded atom method (MAEAM) which is used for the simulation and theoretical calculation of metal nanoparticles has been developed in this work. A size dependent cohesive energy model is proposed in order to describe the melting evolution and diffusion behaviour of chalcogenide nanoparticles. The melting process of nanoparticles, which occurs in three stages, has been analysized using molecular dynamics simulation and liquid state theories. The proposed scheme is applied to two different EAM function sets validated for outer and inner shells, respectively. The MAEAM potential parameters have been determined by fitting the cohesive energy and other physical properties at each level. This model has been applied to Se nanoparticles as a test case. It is compared with other theoretical models as well as the available experimental data.

Keywords: nanoparticles; cohesive energy; melting process; interatomic potential

1. Introduction

Recently, the modelling of the cohesive energy plays an important role in understanding the melting process of small particles, since it is not only of scientific interest, but also bears some technological implications. As the dimensions of materials reduce to the nanoscale, they exhibit a series of unique thermodynamic properties. In order to explain these special thermodynamic properties, different models for the cohesive energy and the melting temperature have been developed,[1–9] such as the liquid drop model,[1] the latent heat model,[2] the BOLS model,[3] the surface area difference model,[4]

---
[*] serapd@trakya.edu.tr

J.P. Reithmaier et al. (eds.), *Nanostructured Materials for Advanced Technological Applications*, 87
© Springer Science + Business Media B.V. 2009

the bond energy model,[5] the size dependent cohesive energy and melting temperature model proposed by Qi and co-workers,[6] etc. Zhao and co-workers[9] have also proposed a model for three distinctive melting mechanisms in isolated nanoparticles.

In this paper, based on previous studies[6,8,9] the cohesive energy model is proposed to describe the melting process of chalcogenide nanoparticles. It has been observed experimentally that the melting phenomena of small particles are determined by the liquid surface skin resulting from surface premelting. One is that an ensemble of small particles in the melting region is a mixture of solid-like and liquid-like forms. There is a dynamic coexistence of two states. This has been observed in simulation experiments and some theories for metallic systems[9] and is called intermediate melting. On the other hand, Raman scattering, spectroscopy and calorimetry experiments on several families of chalcogenide glasses have shown that there are three distinct phases of self-organized network glasses. The intermediate phase has been discovered in chalcogenide glasses by Boolchand and co-workers.[10] Therefore, it is of great importance to understand the melting behavior of chalcogenide nanoparticles with three distinct types of states. Due to the technological importance, selenium nanoparticles have been chosen as a test case. Neutron scattering studies on selenium nanoparticles have been reported by Johnson and co-workers.[11] Up to now, there is no theoretical study on melting process of chalcogenide nanoparticles. In the following, the melting evolution and diffusion behavior of selenium nanoparticles, divided in three spherical shell regions, near the melting point have been studied with the proposed model by molecular dynamics calculations using the modified embedded atom method (MAEAM), which was extended to liquid state calculations of chalcogenides.[12] It has been found that the diffusion behavior of nanoparticles below the bulk melting point is dominated by atoms placed in the intermediate shell region.

2. Theory and Method

2.1. MODELLING COHESIVE ENERGY AND MELTING TEMPERATURE

A model has been developed to account for the size dependent cohesive energy and melting temperature. This model can deal with thermodynamic properties of spherical nanocrystals (nanoparticles) with free surfaces and with non-free surfaces (embedded in a matrix). For this reason, three spherical shell regions with the same thickness a_0 are considered, labelled as A, B, and C, starting from the outmost shell to the core region.

It was assumed that the total number of atoms of the nanocrystal is the sum of interface atoms and interior atoms where the former denote the first layer of the nanocrystals. It is labeled as A layer. The B shell denotes the region where the atoms are of liquid-like and solid-like form. The C layer denotes the core region where the atoms are solid-like. For free-standing nanoparticles, the interface atoms become the surface atoms.

The cohesive energy E_{coh} of each shell of the nanoparticles can be expressed as the sum of the bond energy of the total atoms given by[6]

$$E_{coh} = E_0\left\{1 - \frac{1}{2}s(1+k)\left[1 - \frac{1}{2}pq(1+c)\right]\right\} \quad (1)$$

$$T_m = T_{mb}\left\{1 - \frac{1}{2}s(1+k)\left[1 - \frac{1}{2}pq(1+c)\right]\right\} \quad (2)$$

where E_0 is cohesive energy per atom at T = 0 K, T_{mb} the melting temperature of the bulk material, k a parameter to estimate the surface relaxation, c the cohesive energy of per atom of the nanocrystal given as $c = E_M / E_0$ where E_M is the cohesive matrix energy. In this work we have defined E_M and s as

$$E_M = \frac{D^3}{d^3}(1-s)E_0, \quad s = 4\alpha\frac{d}{D} \quad (3)$$

where d and D are the diameters of atoms and nanoparticles, respectively. α is a shape factor defined as 1 for spherical nanoparticles. The modeling parameters k, p and q in the above equations are determined for each layer.

2.2. INTERATOMIC INTERACTIONS

A molecular dynamic simulation method with a modified embedded atom potential was used to investigate the melting process. The total energy of the system in the MAEAM model[12] can be written as,

$$E_{tot} = \sum E_i \quad (4)$$

$$E_i = \sum F(\rho_i) + \frac{1}{2}\sum \phi(r_{ij}) + \sum M(P_i) \quad (5)$$

where E_i and E_{tot} are the single atom and total energy of the model system respectively. $F(\rho_i)$ is the energy to embed an atom with an electron density ρ_i, $\phi(r_{ij})$ is the pair potential between atoms i and j with the separation distance r_{ij}, $M(P_i)$ is a modification term which describes the energy change caused by the non-spherical distribution of electrons and the deviation from the linear superposition of the atomic electron density. The pair potential in Eq. (5) used in the MD simulations can be given as

$$\phi(r) = \sum_{n=-1}^{6} k_n \left(\frac{r}{r_{1e}} \right)^n . \tag{6}$$

The MAEAM potential functions have been parameterized using the procedure given in previous studies.[12] In order to compare the applicability of the potential functions, the Doyama-Kogure EAM (DK-EAM) model[13] has been studied first with a Foiles-type effective pair potential approximation given as

$$\phi_{eff}(r) = \phi(r) + 2F'(\rho)f(r) , \tag{7}$$

where $\phi(r)$, f(r), and $F'(\rho)$, are the potential functions of the DK-EAM model.[13]

2.3. SIMULATION PROCEDURE

The initial configurations of spherical nanoparticles with 512 atoms are extracted from a random distribution. During heating in order to get an energy-optimized structure at a given temperature, molecular dynamic calculations are carried out for a constant volume and a constant temperature, controlled by a Nose Thermostat (NVT ensemble). For the integration of the classic equations of motion, the Leapfrog Verlet algorithm is used. The system is stable in 0.1 ns. At several temperatures during the heating the radial distribution functions g(r) for each spherical shell have been calculated.

With the effective pair potential known, integral equations are able to provide us the liquid structure of metals and alloys. In our structural calculations, one of the integral equation theories, which has been shown to be a very reliable theory of liquids, namely the variational modified hypernetted chain (VMHNC) has been carried out.[12]

3. Results and Discussion

To explain the melting process of Se nanoparticles, the model parameters of the cohesive energy have been chosen first. The shape of Se nanoparticles observed experimentally[11] is spherical and their diameter 4 nm. The input parameters used in our calculations are given in Table 1.

TABLE 1. Input parameters for Se nanoparticles

	E_0 (eV)	T_{mb} (K)	D (nm)	d (nm)	a_0 (nm)
Se	2.46	493	4.0	0.4276	0.436

In Table 1, E_0 is the cohesive energy of the nanoclusters at T = 0 K, T_{mb} the melting temperature of the corresponding bulk material, D the diameter of the spherical nanoparticles, d, the diameter of the atoms and a_0 the lattice constant. The model parameters used in order to calculate the cohesive energy and melting temperature, and the values obtained are given in Table 2.

TABLE 2. Model parameters used for the calculation of the cohesive energy and the melting temperature. The values obtained are given also.

	A layer	B layer	C layer
Shape	Spherical	Spherical	Spherical
Free standing	Yes	Yes	Yes
Surface relaxation	Yes	No	Yes
α	1	1	1
k	0.5	0	0.5
p	0	0	0.5
q	0	0	1.0
E_{coh} [eV]	1.67	1.93	2.065
T_m [K]	335	387	414

The dependence of the cohesive energy and melting temperature on the particle size is shown in Figure 1. It can be seen that the both decrease linearly with the reciprocal of the nanoparticles size. In our calculations the number of atoms in the system is 818, while the MD calculations are carried out with 512 atoms.

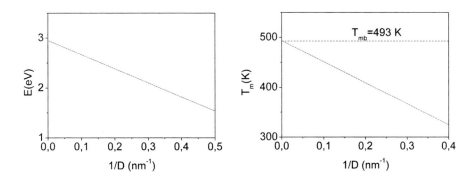

Figure 1. Size dependence of cohesive energy (left) and melting temperature (right) of Se nanoparticles.

For each layer, the parameters of the potential functions were fitted with the values calculated by Eq. (1) by combining the two equations for the cut-off procedure and the equation of the equilibrium condition, the equation for the cohesive energy and the equation for minimizing the configurational free energy (details are given in our previous work[12]). The obtained model parameters of the MAEAM and EAM for Se are listed in Tables 3 and 4, respectively. The calculated effective potentials are shown in Figure 2. The thus constructed MAEAM and EAM interatomic potentials for liquid Se are used as input data in our VMHNC and MD calculations. The static properties of liquid Se nanoparticles have been calculated for the following thermodynamic states: $T = 335$ K, $\rho = 0.0319$ atoms/Å^3, $T = 387$ K, $\rho = 0.0314$ atoms/Å^3; and $T = 430$ K; $\rho = 0.0311$ atoms/Å^3, respectively. Figure 3a, b show the calculated $g(r)$ for the A and B shell regions obtained by VMHNC and MD calculations.

TABLE 3. The MAEAM potential parameters for Se.

Parameter	A-layer	B-layer	C-layer
n	0.600000	0.600000	0.600000
F_0(eV)	1.114000	1.289330	1.376666
α (eV)	0.290029	0.335762	0.358092
K_{-1}(eV)	33.42901	38.88269	41.40754
k_0(eV)	−147.0388	−171.0270	−182.1326
k_1(eV)	273.8899	318.5728	339.2594
k_2(eV)	−281.7882	−327.7597	−349.0428
k_3(eV)	173.5135	201.8208	214.9261
k_4(eV)	−64.03857	−74.48950	−79.3227
k_5(eV)	13.12017	15.26063	16.25158
k_6(eV)	−1.150550	−1.338260	−1.425160

TABLE 4. The DK-EAM potential parameters for Se.

Parameter	A-layer	B-layer	C-layer
A_1(ev/Å^2)	6205.9643	6279.0878	6320.4185
$A_2(\text{Å}^{-2})$	0.7858100	0.7758100	0.7699100
$C_1(\text{Å}^{-1})$	8.4680200	8.6816250	8.6547580
$C_2(\text{Å}^{-1})$	1.8251200	1.8466250	1.8587800
D(eV)	0.3060400	0.3970400	0.420400
r_{c1} (Å)	5.9800000	5.9800000	5.9800000
r_{c2} (Å)	7.4290000	7.4290000	7.4290000

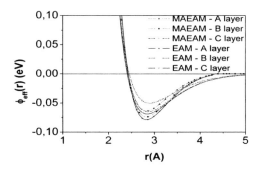

Figure 2. Effective pair potentials for Se nanoparticles.

It appears in Figure 3a, b that the calculated VMHNC radial distribution functions are in a good agreement with the MD results. The calculated $g(r)$ in each shell shows the correct behaviour with the temperature. EAM results shown in Figure 3 are obtained by DK-EAM model.

In order to determine the temperature dependence of the diffusion co-efficient, the calculated coefficients of Se nanoparticles are plotted versus the inverse temperature in Figure 4a–c for the shells A, B and C.

Figure 4 shows the Arrhenius plot for the diffusion of Se atoms placed in each layer with the same thickness. It reveals that the melting first occurred at the surface overlayer; after the outer shell of the particles became a liquid layer, melting of the whole particles started from this liquid layer to the core region very quickly. This behavior can clearly be seen in the B layer due to the intermediate melting shell region. This can be identified by the calculations of the pair distribution functions $g(r)$ given in Figure 3.

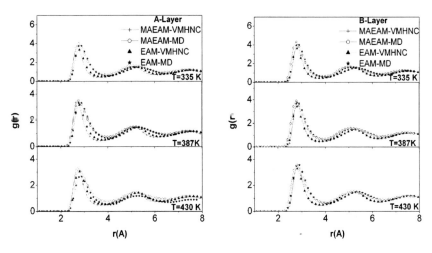

Figure 3. The calculated $g(r)$ functions of Se nanoparticles for the A and B layers.

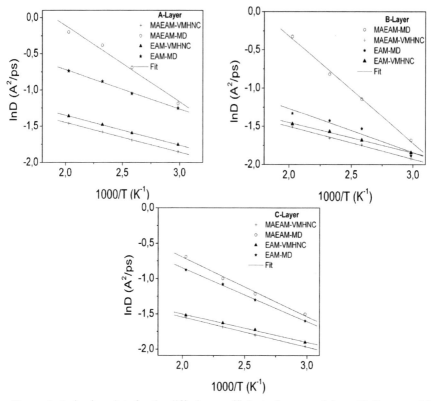

Figure 4. Arrhenius plots for the diffusion coefficient of nanoparticles with the same thickness of A, B and C layers.

4. Conclusions

The melting behaviour of Se nanoclusters of 512 atoms is characterized by a pre-melting stage dominated by the diffusion of Se atoms from the surface toward the centre of the nanoclusters. A model has been developed to account for the size dependence of cohesive energy and melting processes of Se nanoparticles. This model can deal with the thermodynamic properties of nanoparticles (spherical and non-spherical) with free or non-free surface s(embedded in a matrix). A decrease of the cohesive energy of the nanoparticles has been predicted, and the conditions of intermediate melting were obtained. It was found that the present theoretical results are consistent with the available theoretical models. It has been noted that the intermediate melting occurs within a finite critical particle-size region. A further investigation, in particular theoretical approaches, of the mechanism of the intermediate melting is of urgent need. This work will progress on this line.

References

1. K. K. Nanda, S. N. Sahu and S. N. Behera, Phys. Rev. A66, 013208 (2002).
2. Q. Jiang, J. C. Li, B. Q. Chi, Chem. Phys. Lett. 366, 551 (2002).
3. Chang Q. Sun, Y. Wang, B. K. Tay, S. Li, H. Huang and Y. B. Zhang, J. Phys. Chem. B106, 10701 (2002).
4. W. Qi, M. P. Wang, M. Zhou and W. Y. Hu, J. Phys. D: Appl. Phys. 38, 1429 (2005).
5. W. H. Qi, M. P. Wang, G. Y. Xu, Chem. Phys. Lett. 372, 632 (2003).
6. W. Qi, M. P. Wang, M. Zhou, X. Q. Shen, X. F. Zhang, J. Phys. Chem. Solids 67, 851 (2006).
7. M. Attarian Shandiz, A. Safaei, S. Sanjabi and Z. H. Barber, J. Solid-State Commun. 145, 432 (2008).
8. W. Hu, S. Xiao, J. Yang and Z. Zhang, Eur. Phys. J. B45, 547 (2005).
9. S. J. Zhao, S. Q. Wang, D. Y. Cheng and H. Q. Ye, J. Phys. Chem. B105, 12857 (2001).
10. P. Boolchand, D. G. Georgiev, B. Goodman, J. Optoelectron. Adv. Mater. 3, 703 (2001).
11. J. A. Johnson, M. L. Saboungi, P. Thiyagarajan, R. Csencsits and D. Meisel, J. Phys. Chem. B103, 59 (1999).
12. S. S. Dalgic and S. Sengul, J. Optoelectron. Adv. Mater. 9, 3028 (2007).
13. M. Doyama and Y. Kagure, Comput. Mater. Sci. 14, 80 (1999).

QUANTUM-DIMENSIONAL EFFECTS IN THE POSITRON

ANNIHILATION SPECTRA OF La–Ba–Cu–O SUPERCONDUCTORS

A.A. PAIZIEV*

Positron Physics Laboratory, Arifov Institute of Electronics
Uzbek Academy of Science, 100125 Tashkent, Uzbekistan

Abstract. This paper is devoted to explain the anisotropy phenomena of angular distribution annihilation photons (ADAP) in mono-crystalline La–Ba–Cu–O high temperature superconductors (HTSC) with alternating conductive (Cu–O(1)) and nonconductive (La–O(2)) layers. It is shown that the basic contribution to the anisotropy of ADAP spectra is connected with positron annihilation in the conductive Cu–O(1) layers. To describe the behavior of conduction electrons in the Cu–O(1) layer, the 2D model of a free Fermi gas was used. In the transverse direction, electron states are described in model one-dimensional potential well. It is shown that in the transverse direction the ADAP spectra possess several "plateaus" corresponding to positions of allowed electron states, while in longitudinal direction the ADAP spectra posses a parabolic form.

Keywords: electron–positron annihilation; anisotropy; high-temperature superconductors; quantum dimensional effects

1. Introduction

For the first time the effect of anisotropy in ADAP spectra of strongly non-uniform layered electronic systems was considered for quantized metal films (QMF) in Ref. 1. Furthermore, a similar effect was predicted[2,3] for surface positron states at the interface metal – vacuum. As shown by esti-mations,[2] in the case of QMFs dimensional effects in ADAP appear, when their thickness is on the order of the wavelengths of electrons in the QMF. This circumstance causes troubles for both, the preparation of QMFs for ADAP experiments and their realization with monochromatic positron

---

*adxam_payziev@rambler.ru

J.P. Reithmaier et al. (eds.), *Nanostructured Materials for Advanced Technological Applications,*
© Springer Science + Business Media B.V. 2009

beams (necessity of high vacuum, cleaning of the surface, maintenance of the uniformity of the film thickness, etc.). To overcome these troubles we started to study monocrystalline HTSC materials with a layered electronic structure, where conducting layers of Cu–O(1) alternate with non-conducting layers in the direction of the crystal **c** axis. As the conducting layers have strictly identical thicknesses (about the inter-plane distance of 0.24 nm) such substances are very convenient to study dimensional effects by the widely accessible positron isotope methods. Some classes of HTSC s consists of such materials, synthesized on the basis of the systems La–Ba–Cu–O, Y–Ba–Cu–O, Bi–Sr–Ca–Cu–O, or Ti–Ba–(Ca)–Cu–O. These systems have a perovskite-like layered structure connected by Cu–O(1) planes, and show strongly anisotropic characteristics for some physical properties which are connected with the movement of conduction electrons within or across these layers (electrical resistance, the maximum critical field, light reflection).[4]

In the present work the effect of anisotropy of ADAP in monocrystalline HTSC samples such as $La_{2-x}M_xCuO_4$ (M = Ba, Sr) is theoretically investigated.

2. Materials and Methods

We considered a layered La–Ba–Cu–O HTSC crystal consisting of alternating conducting Cu–O(1) and non-conducting La–O(2) layers. As is well-known,[5] for the superconducting properties of HTSCs the conducting Cu–O(1) layers are responsible. Since the anisotropic part of ADAP is caused by positron annihilation in the conducting layers, we shall concentrate our attention on the 2D system of conduction electrons. Here we use the model of free electrons within the Cu–O(1) layers, while in perpendicular direction 1D states of electrons were calculated in the framework of one-dimensional potential wells.

3. Results and Discussion

According to Ref. 6, the probability of positron annihilation with emission of the two annihilation quanta is:

$$\omega(\overline{p}) = \sum_{n-} \int \frac{Sd\overline{K}_{11}}{(2\pi)^2} \rho(k)$$

$$\rho(k) = \left| \iint \exp(ipr)\phi_-(r)\phi_+(r)dr \right|^2,$$

where p is the sum of the momenta of the two annihilation photons, $\bar{k}_{II}$ its parallel component, S the surface area of the Cu–O(1) layer, ϕ_- and ϕ_+ the known wave functions of an electron and a positron in conducting layer, V the volume of this layer, and r the electron radius vector. Here the summation will be carried out only over the occupied electronic states in the conducting layer. In the three-dimensional jelly model of electrons the form of the ADAP curve is isotropic in all directions of the momentum space. In the case of thin conducting HTSC ceramics layer, where quantum dimensional effects become essential, it is necessary to orient the monocrystalline HTSC in two directions relative to the axis of the angular distribution annihilation photons detection system: along the conducting layer and in a perpendicular direction. As was indicated above, the thickness of the conducting layer in monocrystalline $La_{2-x}M_xCuO_4$ compounds is the distance d between the conducting Cu–O(1) plane and the insulating La–O(2) layer. Estimations show[1] that for d = 0.24 nm there is one allowed, filled electron subzone in the spectrum of electronic states. In this case the expression for the longitudinal and perpendicular components of the ADAP spectrum for the positron captured on the lowest level (n = 1) are:

$$R_\perp = 4(2\pi)^4 \frac{\sin^2 P_\perp d/2}{(P_\perp d)^2 \left[(P_\perp d)^2 - (2\pi)^2\right]^2} \quad R_{II} = \left(1 - \frac{P_y^2}{K_F^2 - \pi^2/d^2}\right)^{1/2} \quad (1)$$

It is shown that the basic contribution to the ADAP spectra is given by the first quantum state of electrons in the Cu–O(1) layer (n = 1), while the contribution of higher states have a maximum for annihilation quantum emission angles corresponding to the position of discrete electronic levels in the conducting layer. In the total ADAP spectrum, the position of separate film levels corresponds to the position of a plateau: in R_{11} also a basic contribution is given by the first subzone with a cutting angle of $(k_F^2 - \pi^2/d^2)$. With increasing number of a subzone its contribution to R_{11} decreases. The cutting angle of the total ADAP spectrum corresponds to the number of the subzone, in which the positron is located at the moment of its annihilation (in this case n = 1). According to (1), the ADAP for R_1 shows the presence of weak oscillations for greater momenta and periodically repeated features in the form of plateaus, the position of which corresponds to the positions of the subzones in momentum space.

For the parallel component of the ADAP, spectra cutting of the R_{II} component takes place (1). It shows that the FWHM of curve $R_\perp$ for the captured states of positrons is more appropriate than that for the free positron state. This circumstance is probably connected with the restricted movement of the positron in cross direction in the potential well, which leads to an increase of the frequency of collisions of the positron with the walls, and to an additional widening of the annihilation lines in comparison with the case of free annihilation. Note that in both, the first and the second case weak oscillation of R_1 on the tail of the ADAP spectrum with a periodicity of 4π for the dimensionless parameter $\alpha = p\perp d$ are observed. For captured positrons the first maximum of the oscillation takes place at $\alpha = 5\pi$ with a intensity much smaller than that of the main maximum at $\alpha = 0$ $(10^{-2}\%)$. The same behavior of the ADAP curve takes place in case of a positron annihilation from free (not bounded) states in the potential well. It is shown that the FWHM for $R_\parallel$ is narrower than that for $R\perp$. From (1) it is easy to see that for the FWHM of $R_\parallel$ holds:

$$\Gamma_{11} = \frac{1}{2}\sqrt{3(k_F^2 - \pi^2/d^2)}$$

This FWHM depends on the thickness of the superconducting HTSC layer d and the Fermi-momentum of the electrons. For superconductivity valence electrons of oxygen O(1) ions in the Cu–O(1) layer are responsible, the concentration of which is, anyway, less than the electronic density of bulk metals. On the other hand the FWHM of $R_\perp$ does not depend on a parameter of the superconducting layer.

References

1. P.U. Arifov, A.A. Paiziev, Proceeding of Symposium on interaction of nuclear particles with a surface of a solids, Fan, Tashkent (1999), p. 10.
2. B. Rosenfeld, K. Jetie, W. Swiatkowski, Acta Phys. Pol. A 64, 93 (1983).
3. R. Ewertowski, A. Baranowsci, W. Swiatkowsci, Abstract of Paper 20 of the Polish Seminar on Positron Annihilation, 15–21 May 1983, p. 11.
4. S.N. Burmistrov, L.B. Dubrovskij, in: V. Ginzburg (Ed.), *High-temperature super-conductivity*, Nauka, Moscow (1988), p. 82.
5. C.P. Enz, Helv. Phys. Acta, 61, 741 (1988).
6. V.I. Grafutin, E.P. Prokop'ev, Phys.-Usp. 45, 59 (2002).

3. CHARACTERIZATION TECHNIQUES

EXAMPLES OF CURRENT INDUSTRIAL MICRO- AND NANOMATERIALS AND TECHNIQUES FOR THEIR CHARACTERIZATION

P. MORGEN[*1], B.O. HENRIKSEN[1], D. KYRPING[1],
T. SØRENSEN[1], Y.X. YONG[1], J. HOLSTM[1], J.R. AELSEN[2],
P.-E. HANSEN[3]
[1]Institut for fysik og kemi, SDU, Campusvej 55, DK-5230
Odense M, Denmark; South West Institute of Magnetism,
China (SIAM, www.siam.cn)
[2]Institute of Physics and Nanotechnology, Aalborg University,
DK-9220 Aalborg East, Denmark
[3]Danish Fundamental Metrology Ltd., Matematiktorvet 307,
DK-2800 Kgs. Lyngby, Denmark

Abstract. New functional materials and material structures with nanoscale features need new techniques for their characterization. In this condensed follow-up to the ASI presentation we will emphasize improvements of traditional microscopic methods, such as the combination of microscopy and spectroscopy with ion etching methods, and indicate some new corresponding trends in electron microscopy.

Keywords: micro- and nanostructured materials; characterization techniques; electron microscopy; ion etching

1. Introduction

A Danish innovation consortium has been set up under the leadership of the Danish Fundamental Metrology center, to propose standard procedures for the analysis of industrially relevant micro- and nanostructured materials in Denmark. This consortium has industrial partners, partners from the Danish technological service institute sector, and university institute partners.

[*]per@fysik.sdu.dk

It actually covers only a handful of relevant industrial products, so it is obvious, on the occasion of this NATO-ASI, to look to the rest of the (Danish) industry, and beyond, for good, relevant examples.

Included in our present activities are components for optical communications, imprint gratings, hard tool surface coatings, and nanoparticle-based coatings for photo-catalytic surface activations, to render surfaces super hydrophilic and self-cleaning. Danish industry is widely engaged in producing new materials based on the use of nanomaterials and nanocomposites, such as in catalysts and in (gigantic) windmill rotor blades; pharmaceutical companies are, in collaboration with universities, engaged in research around activated liposome (self-forming lipid bilayer) nanocapsules for targeted drug delivery.

In the energy and environmental sectors, besides wind energy components, new materials and elements are under close scrutiny for use in batteries and fuel cells, as well as for new photovoltaic applications, and for hydrogen storage. Finally, like in other parts of the World, sensors and actuators are developed for biomedical, environmental and human applications.

However, the major nanotechnological production in the world is still happening in the sector of Si-based electronic devices, following the Semiconductor Industry Road Map, and this industry also leads the quest for new "nanometric" standards and tools.

2. Examples of Industrial Micro- and Nanostructures

The deployment of new nanostructured materials is subject to careful tests of potential health hazards, and the press is well aware of possible mishaps, which could occur by the hazard of too early exposure to nanomaterials. There have also already been cases of direct fraud, where advertising campaigns have claimed nanoproperties of new products, which actually did not demonstrate the advertised properties. The belief that it is not possible to detect such fraud was toppled by top university researchers who exposed the fraud in television prime time programs.

Through series of discussions between researchers, industrialists, technology scouts, health authorities and the press, priorities of Danish nano-technology research and development have been set. Among the top priorities surface engineering and thin film based technologies were brought into focus. The Danish High Technology Foundation has recently generously awarded projects to manufacture fuel cells and improve the quality of fiber composite materials for use in windmill rotor blades, among other projects, to strongly support sustainable energy technology.

2.1. ACTUAL EXAMPLES OF NANOSTRUCTURED MATERIALS AND THEIR ANALYSES

Thin films on substrates are a very important class of micro- and nanostructured materials, which can be used for many different purposes. Many sensors, semiconductor lasers, and magnetic storage elements depend on a complex stacking of thin uniform layers or of layers where a graded composition through the layer is obtained as a result of the deposition process. In this contribution we have chosen to focus on hard disks and magnetic sensors based on a layered metallic stack on a ferrite substrate. These systems have important properties depending on the composition in depths below the normal information depths of surface sensitive spectroscopies. Thus a combination of ion etching, imaging and spectroscopic tools were employed to study these samples. We have further used a focused ion beam method to make a cross-section of a thin oxide grown on Si to determine its thickness from an image of the cross-sectioned sample obtained by a scanning electron microscope (SEM). These measurements will also demonstrate the capabilities for analysis of non-uniform samples with imbedded micro- or nanostructures.

2.2. EXPERIMENTAL TECHNIQUES

The nano-era can be seen as having begun with the invention of the scanning tunneling microscope (STM) as the forerunner of several other scanning probe instruments. The STM was found to be able to obtain images of a surface and adsorbed atoms, molecules or complex clusters on a metallic surface, with "atomic" resolution.[1] However, on semiconducting surfaces at the same resolution, we detect the shapes of the local density of electronic states with spectroscopic techniques by varying the bias between tip and surface. Thus for example a Si surface can be imaged to display the spatial distribution of empty and filled, localized bonds.[2]

However, the more classic tools as the scanning electron microscope (SEM) and of course the transmission electron microscope (TEM), have gradually been improved with new electron emitters, optics, and detectors, to offer magnifications of 10^6 times, and several imaging techniques, including the detection of x-rays emitted after electron impact ionization, to enhance the contrast caused by topography, crystalline phases, and elemental composition. SEM machines of today are compact instruments with quick access and the possibility of studying samples in a near atmospheric ambient (environmental SEM) and also samples of low conductivity. Other instruments use a focused ion beam as a probe and may detect species and clusters desorbed and fractured under ion impact (SIMS or TOF-SIMS[3]), or secondary electrons as in SEM, but with an even better imaging resolution due to a significantly smaller impact footprint at the surface from the penetration of the (swift) ions than for an electron beam.

3. Experiments

3.1. HARD DISK

We have looked at a hard disk to detect its composition through the layers.[4] To do this we have combined Auger electron spectroscopy and Auger electron microscopy with ion etching. The experiments were done in a modified PHI 560 system, in which the original computer has been replaced with a modern PC and new software written in LabView™. It performs x-ray induced photoemission, Auger electron spectroscopy with a focused electron beam (at up to 8 keV energy), and imaging with the detection of secondary- and backscattered electrons. Many sequences of experiments can be automated, such as the recording of the sputter profile shown in Figure 1. The experimental results in Figure 1 are converted to a concentration versus depth diagram in Figure 2.

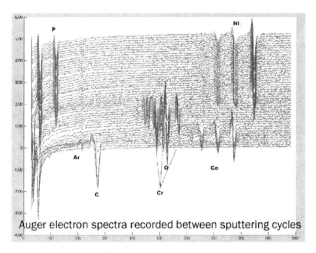

Auger electron spectra recorded between sputtering cycles

Figure 1. Auger spectra of a hard disk recorded at regular intervals of sputtering with 1 keV Ar+ ions. The front (lower) spectrum is from the top surface. The elements found from the Auger spectra are indicated: P, Ar, C, Cr, O, Co, and Ni. The electron energy was 3 keV.

The results of Figure 2 were obtained using standard Auger electron sensitivity factors for the analyzer used here.[5] The depth scale is obtained by using the SRIM[6] code, with the sputter rate determined from the ion beam flux measured with a Faraday cup, which is inserted at the position of the sample before the experiment, and using the effective density of the sample determined from the concentrations. The ion intensity and time of sputtering was controlled by a software interfacing the controlling supply of a motorized valve setting the pressure of Ar in the ionization volume of the ion gun, and the sputter gun high voltage. The formula to determine the *relative atomic concentrations* of element i, C_i, $i:1,...,$ is

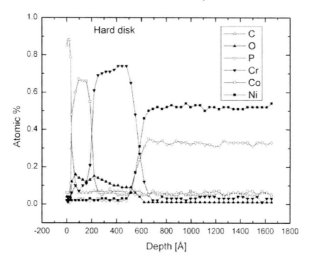

Figure 2. Sputter profile of a hard disk: concentrations and depths are determined from the experimental data of Figure 1.

$$C_i = \frac{I_i / S_i}{\sum_j I_j / S_j}$$

where I_i is the measured intensity and S_i are the relative sensitivity factor.[5] The effective sputter rate (*thickness/time*) in every interval is calculated by SRIM using the average concentration and density, obtained with the concentrations determined from the Auger signals:

$$\frac{dz}{dt}(t) = \frac{j_P Y_M(t)}{eN(t)} = \frac{j_P Y_M (\rho(\sum_j c_j(t)n_j)}{e \sum_j c_j(t)n_j}$$

Here, $Y_M(...)$ is the effective sputter yield for the composition with the average density $\rho(...)$ determined by the actual concentrations of the elements in the sampled volume.

In Figure 3 the evolving sputter crater is clearly visible; its edge shows the cross section of the disk, due to the less intense ion beam at the rim. The different elements in the layers reflect the electron beam with different strengths, thus the different grey levels. Figure 4 gives a less enlarged view of the entire crater.

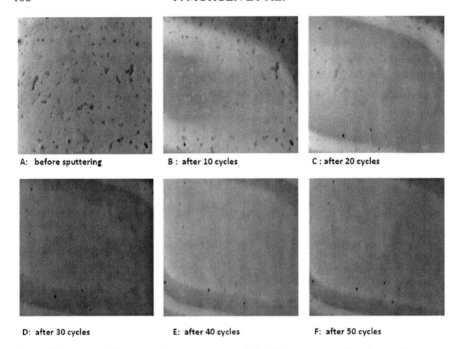

A: before sputtering   B : after 10 cycles   C : after 20 cycles

D: after 30 cycles   E: after 40 cycles   F: after 50 cycles

Figure 3. Images of the sputtering process recorded with backscattered and secondary electrons at 60 times magnification. The crater dimension is 2 × 2 mm.

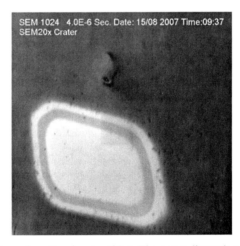

Figure 4. View of sputtered crater (20×). The crater dimension is 2 × 2 mm.

At the upper right corner (seen as a darker square in Figure 4) we have performed scanning Auger microscopy, as shown in Figure 5a, b. Figure 6 is a corresponding backscatter/secondary electron image. All images were recorded at 300 times magnification. The Auger images are recorded using the negative maximum intensity in the differentiated spectra, which appear with different levels of black for the element selected to form the image.

These images further show that the dark spots in Figure 3 are protrusions originally on the top surface, which are propagating with the sputtering process; in Figure 5a, b, these protrusions are seen to contain the elements (colored black) from the layer above. These are examples of nanostructures, which are exposed through the sputtering process, and therefore analyzable with this method.

Figure 5. (a) Carbon layer black (left); (b) Co layer black (middle).

Figure 6. Backscatter/secondary electron image (right).

3.2. MAGNETIC SENSOR CHIP

In a similar study of a (relatively) thick, layered structure of a magnetic sensor chip, the sputter crater reaches hundreds of microns in depth, as shown in Figure 7. The chip is formed with a top Au layer, two differently deposited Cu layers, a Cr layer, a thin Ta layer, while the substrate is a ferrite material. In this case the thin Ta-layer at around 80 μm depth was the object of interest, which is difficult to reveal in a single experiment. X-ray dot maps, which cannot be reproduced here due to lack of contrast could not easily discern the geometry of the Ta layer as part of the sputter crater; but looking at the crater edge with two different techniques, truly backscattered- and secondary electrons, like shown in Figure 8, permits us to detect the presence of Ta, due to both its reflection power, similar to the top layer of Au, and its different coefficient of secondary electron emission. After making these images it was possible to find the Ta layer among the layers seen in the Auger sputter profile, as shown in Figures 9 and 10. For a more detailed analysis it will be necessary to perform a new sputter profile, dwelling around the Ta layer with shorter sputter cycles. In the process of growing the sensor structure the nitriding of Ta is required, but in the present case at the relevant depth, no signal of N was detected. Again a more thorough study is needed with a special emphasis on providing more detailed spectra at the relevant depth. Needless to say that such procedures are very demanding concerning their reproducibility, as these sputter profiles have been collected over a period of several days.

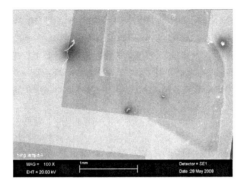

Figure 7. Secondary electron image of the sputtered area of the magnetic sensor chip.

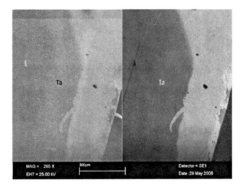

Figure 8. Images of edge of sputter crater: electron energy 25 keV; left image: backscattered electrons; right image: secondary electrons (SEI detector). The Ta-layer is indicated in the well of the sputter crater from its grey level in the two images.

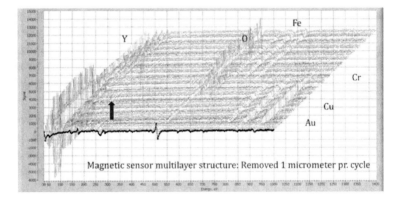

Figure 9. Auger sputter profile of magnetic sensor chip. The arrow points to the expected location of Ta Auger peaks.

Thus even relatively thick samples may be studied with this technique to obtain important compositional information with nanoscale depth resolution.

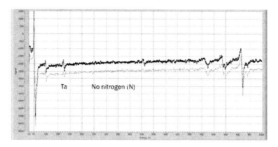

Figure 10. Spectra from the Ta containing layer (Figure 9).

3.3. THIN OXIDE LAYERS GROWN THERMALLY ON SILICON

Previously we have applied the sputter profiling method to study the thickness and interface structures of thermally grown oxides on Si.[7] Recently we got access to focused ion beam methods,[8] which are powerful add-ons to SEM. With this technique a narrowly focused ion beam with high energy using Ga ions at a high flux are used to cut sharply walled cross-sections perpendicular to the layers of complex structure, subsequently allowing the imaging from the side of the cross-section. In the process a thick Pt layer is deposited on top of the section of the sample to be cut. This preserves the original surface and allows a cut with a sharp edge. It is being done by dosing the sample with a metal-organic vapor while cracking this vapor with the electron beam. The resulting structure is depicted in Figure 11, including a measure of the thickness of the oxide based on the parameters of the image.

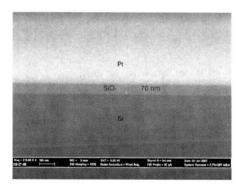

Figure 11. Cross section of oxide grown on Si, obtained with the FIB method.

4. Summary of Experimental Procedures and Non-uniform Samples

In the above examples the samples are primarily uniform, but impurities and nanoscaled topographical features develop, which are resolvable with the focused electron beam. In principle therefore also non-uniform samples can be analyzed, but the accumulation of data from such samples may be too complicated or too time consuming. A new kind of instrumentation is appearing, however, to be able to cope with such cases. This instrument combines the features of a normal SEM with ultrahigh vacuum conditions for detailed surface analyses, offering combinations of scanning electron microscopy and all the other surface analytical spectroscopies, including STM, where the STM tip and the region imaged by it may be seen by the electron beam at up to a million times magnification.

5. Observations and Recommendations

In this contribution emphasis has been on the potential of "classical" surface analysis techniques combined with imaging for the analysis of nanoscaled features, especially complex layered structures, which are and will be an important class of nanotechnological devices. As in this example the power of other techniques like XPS (x-ray induced photoemission) for nanostructure analysis is also expected to improve, due to developments in instrument sensitivities and capabilities and to new theoretical approaches.

ACKNOWLEDGEMENTS

P.M. acknowledges the organizers of this NATO ASI for having, once again, provided a vigorous and stimulating scientific and social ambience for the exchange of ideas and knowledge, and the participants for their spirits and engagement in reporting about their current activities.

References

1. G. Binnig, H. Rohrer, IBM J. Res. Dev. 30, 4 (1986).
2. R.J. Hamers et al., Phys. Rev. Lett. 59, 2071 (1987).
3. D. Briggs and M.P. Seah, *Practical Surface Analysis, Vol. 1* (Wiley, Chi-chester, 1997).
4. P. Morgen et al., Vacuum 82, 922 (2008).
5. L.E. Davis, N.C. MacDonald, P.W. Palmberg, G.E. Riach, R.E. Weber, *Handbook of Auger Electron Spectroscopy* (Physical Electronics Industries, Eden Prairie, MN, 1976).

6. (SRIM2008) For a general discussion, see J. F. Ziegler, J. P, Biersack, U. Littmark, The *Stopping and Range of Ions in Solid, Vol. 1* (Pergamon Press, New York, 1984). The program is available at the web address http://www.srim.org/index.htm#HOMETOP
7. P. Morgen et al., J. Vac. Sci. Technol. A 23, 201 (2005).
8. D. Brown et al., Web presentation http://www.davidjbrown.org.uk/presentations/lon don%20env%20min%20compressed.pdf

RAMAN SPECTROSCOPY OF UNCD GRAIN BOUNDARIES

M. VERES[*1], S. TÓTH[1], E. PEREVEDENTSEVA[2],
A. KARMENYAN[3], M. KOÓS[1]

[1]Research Institute for Solid State Physics and Optics of the Hungarian Academy of Sciences, H-1525 Budapest, Hungary
[2]Department of Physics, National Dong-Hwa University Taiwan, ROC
[3]Institute of Biophotonics Engineering, National Yang-Ming University, Taipei, Taiwan

Abstract. Near-infrared excited Raman spectroscopy was used to investigate the bonding configuration of nanocrystalline and ultra-nanocrystalline diamond thin films. It was found that by limiting the excitation volume it is possible to detect characteristic vibrations arising from the grain boundary region. A statistical analysis of the spectra measured in different points of the film surface allows the determination of dominant structural units present in this region.

Keywords: ultra-nanocrystalline diamond; grain boundary; Raman spectroscopy; surface-enhanced Raman spectroscopy

1. Introduction

Due to their unique, advantageous physical and chemical properties[1-3] nanocrystalline (NCD) and ultra-nanocrystalline diamond (UNCD) are promising materials for many fields of modern technology. These layers have a composite structure where diamond crystallites are embedded in an amorphous carbon matrix. The interface between the two phases (grain boundary region) is one of the most important structural regions determining the properties of the films, thus it is extremely important to characterize this interfacial region in detail, preferably by easy to use, non-destructive methods.

---
[*]vm@szfki.hu

J.P. Reithmaier et al. (eds.), *Nanostructured Materials for Advanced Technological Applications*,   115
© Springer Science + Business Media B.V. 2009

In many respects Raman spectroscopy, which is widely used for the characterization of different carbon-based materials,[4-6] could be an ideal tool for the characterization of the grain boundaries of NCD and UNCD films. The typical spectrum of an UNCD thin film excited by visible light consists of the narrow diamond peak at 1,332 cm^{-1}, two broad bands (D and G peaks) around 1,350 and 1,580 cm^{-1} related to vibrations of the amorphous carbon matrix and two additional peaks around 1,150 and 1,480 cm^{-1} assigned to vibrations of sp^2 hydrocarbon chains (polyacetylene-type structural units) formed at the grain boundaries and interfaces.[7] However this method has several drawbacks when using visible excitation. Raman spectra of UNCD films provide averaged information on the bonding configuration of all carbon atoms in the excitation volume, belonging to the diamond crystallites, the surrounding amorphous carbon matrix and the interface region. Because of the resonant enhancement of the scattering in the sp^2 clusters of the matrix[8,9] the intense bands of this phase will overlap the scattering from the other parts of the structure, especially if those are such minor regions like grain boundaries. Therefore, to make Raman spectroscopy more sensitive to vibrations of this interfacial region, the scattering contribution of the amorphous phase have to be eliminated or lowered reasonably.

One way to realize this is to change the excitation wavelength. By increasing the excitation energy the difference between the scattering cross-sections of the sp^2 and sp^3 structural units decreases. The use of UV excitation allows even a direct detection of the characteristic peaks of the sp^3 matrix of different amorphous carbon materials.[4] However the UV light could induce structural changes in the sample, so the measurements require extreme care. Earlier we have found[6] that for amorphous carbon thin films the use of near-infrared excitation could also provide additional information on the structure compared to the visible light probing. Due to the different dispersion behavior of the component peaks we were able to detect four peaks in the spectra instead of the well-known two (the D and G peaks).

This work was aimed to investigate the structure of NCD and UNCD thin films by near-infrared excited Raman spectroscopy.

2. Experimental

Raman spectra were recorded on a Renishaw 1000 Raman spectrometer attached to a Leica microscope. The excitation beam was focused into spots having diameters between 1–20 μm. A diode laser operating at 785 and the 488 nm line of an Ar-ion laser served as excitation sources. Commercial NCD and UNCD thin films of different grain size (UNCD Aqua Series of

Advanced Diamond Technologies), prepared by MW-CVD method were used for the experiments.

3. Results and Discussion

Figure 1a shows the near-infrared and visible excited Raman spectra of an NCD film recorded with 20.0 and 1.0 μm spot diameters in the same point of the sample surface. The "conventional" near-infrared and visible excited Raman spectra of the NCD film (those taken with 20.0 μm spot size in Figure 1a) look very different, but both constitute of the broad bands mentioned earlier. The use of a smaller excitation spot has almost no effect on the spectrum taken with 488 nm excitation. However, the character of the spectrum excited at 785 nm changes remarkably. In the 800–2,000 cm^{-1} region a number of narrow peaks appear on the background of the broad bands. The small width of these peaks implies that they arise from well-defined, molecular-like structural units. Such units can form in the interface region between the diamond and amorphous carbon phases where the crystalline diamond structure constrains some ordering on the neighboring atoms in the distance of a few nanometers from its surface. To detect these vibrations the excitation volume has to be lowered to a level where the amount of the amorphous phase is small enough, so that its Raman bands will not overlap the small intensity peaks in the spectrum. Additionally the

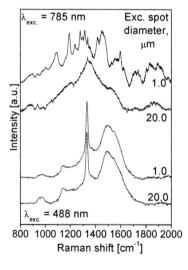

Figure 1. (a) 488 nm (lower curves) and 785 nm (upper curves) excited Raman spectra of a NCD film taken with different excitation spot diameters. (b) SEM image of the sample surface with an illustration of the 1 μm excitation spot. The average grain size is 250 nm.

excitation energy has to be tuned in order to minimize the efficiency of the resonant scattering in the amorphous phase. As can be seen in Figure 1b, only a few diamond crystallites fall into the excitation spot when using a 1 μm spot diameter, but the resonant scattering on the sp^2 carbon atoms is still too high for the observation of other peaks with visible excitation.

Near-infrared excited measurements performed on different points of the sample surface showed that there is some variation in the features of the spectra. Figure 2 shows a series of spectra taken along a line with 0.5 μm steps. It can be seen that a number peaks is present in each spectrum, but some other vary in intensity and can even disappear. From Figure 2 it can be concluded that the narrow peaks arise from very definite regions of the structure. As the excitation spot moves on the surface, it excites a given region (crystallite) in different manners, so the vibrations of the latter will be excited with different intensities. E.g. the peak at 1,230 cm^{-1} (marked by an arrow on Figure 2) has a low intensity in the first spectrum, but as the region (crystallite) it belongs to enters more into the excitation volume, the band will be excited more intensely till reaching the point of maximum overlapping. After this the intensity of the peak will decrease. This way it is possible to study the characteristic vibrations of individual grains of NCD films.

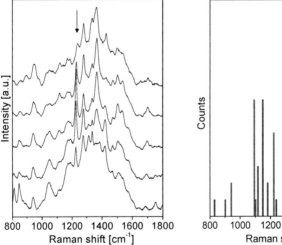

Figure 2. (a) Series of near-infrared excited Raman spectra recorded along a line on an NCD film surface with steps of 0.5 μm. (b) Most frequent Raman peaks in a set of spectra taken in different points of the NCD surface.

To obtain general information on the bonding configuration of the film the most frequent peaks were counted and the number of their occurrence in a series of 50 spectra was determined. Figure 2b shows the positions of the most frequent peaks found. They fall mainly into the 1,000–1,600 cm^{-1} region characteristic for carbon based materials including NCD. It is interesting to note that the 1,334 cm^{-1} diamond peak is not the most frequent one, which indicates that there are regions in the NCD film lacking this peak at all. The highest occurrence have bands in the 1,400–1,550 cm^{-1} region. A number of peaks can be found around 1,420 cm^{-1} where the band of sp^2 hydrocarbon chains (trans-polyacetylene) is located in the Raman spectrum of NCD excited at 488 nm. The region of the other sp^2 hydrocarbon peak around 1,100 cm^{-1} also contains several peaks (the dispersion causes the shift of these peaks to lower wavenumbers when using near-infrared excitation[7]). The presence of more than one peak is probably due to the length differences of sp^2 hydrocarbon chains formed on the diamond surface.

The identification of the observed peaks is not simple since the peak positions are affected by a number of different parameters. We were able to identify some peaks earlier[10] as vibrations of sp^3 CH$_x$ structural units. For instance the 1,298, 1,313 and 1,415 cm^{-1} bands are presumably related to different vibrations of the sp^3 CH$_2$ group[11] and similar to those observed for polyethylene.[12,13] The peak at 1,573 cm^{-1} is related to C–C vibrations of sp^2 rings while the bands around 1,510 cm^{-1} arise from sp^2 chains.

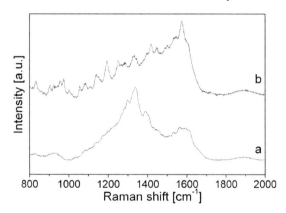

Figure 3. Near-infrared excited SERS spectra of (a) UNCD film (average grain size is 25 nm) and (b) individual diamond crystallites put on the top of a gold film (grain size is 50 nm).

The use of the microscope and of the near-infrared excitation led to excellent results in the case of the NCD films with average grain sizes above 150 nm, but the detection of the above mentioned features was not possible for layers with grain size below 100 nm, since the number of grains excited simultaneously is too large. Their vibrations overlap resulting in broad bands in the spectra.[10] The detection of the grain boundary peaks would require the decrease of the excitation volume further, below 100 nm, which is impossible with traditional instruments due to the diffraction limit. However, there is a possibility to enhance scattering in small regions of the sample by using surface enhanced Raman scattering (SERS). Putting gold nanoparticles in contact with the surface of an UNCD film of 25 nm average grain size, the vibrations of structural units from the surrounding of the particles can be enhanced selectively (Figure 3, lower curve). The enhancement takes place only in a small region around the 10 nm sized gold particles, and only a few crystallites are involved in the process. As it can be seen on Figure 3, narrow peaks appear on the background of the broad "normal" Raman bands of the UNCD film. The latter have a high intensity since the normal Raman scattering from the whole excitation volume is superposed onto the SERS peaks. Similar features were observed in the SERS spectrum measured on individual diamond crystallites put on the top of a gold film (Figure 3, upper curve), which could help to identify the peaks observed in the near-infrared Raman spectra of NCD and UNCD films.

4. Conclusions

By limiting the excitation volume (by a microscope and/or by SERS) it is possible to detect characteristic peaks of molecular structural units from the grain boundaries in near-infrared excited Raman spectra of NCD and UNCD thin films. The method can be used to determine the bonding configuration of individual grains, as well as the dominant structural units of the film by statistical methods.

ACKNOWLEDGEMENTS

M. Veres is grateful for the support of the Bolyai János Research Scholarship of the Hungarian Academy of Sciences.

References

1. C. Zuiker, A.R. Krauss, D.M. Gruen, X. Pan, J. C. Li, R. Csencsits, A. Erdemir, C. Bindal and G. Fenske, *Thin Solid Films* 270, 154 (1995).
2. M. Hupert, A. Muck, J. Wang, J. Stotter, Z. Cvackova, S. Haymond, Y. Show and G.M. Swain, *Diamond Relat. Mater.* 12, 1940 (2003).
3. O.A. Williams, S. Curat, J.E. Gerbi, D.M. Gruen and R.B. Jackman, *Appl. Phys. Lett.* 85, 1680 (2004).
4. A.C. Ferrari and J. Robertson, *Phys. Rev. B* 64, 075414 (2001).
5. A.C. Ferrari and J. Robertson, *Phys. Rev. B* 61, 14095 (2000).
6. M. Koós, M. Veres, S. Tóth and M. Füle, in: *Carbon: the future material for advanced technology applications*, ed. by G. Messina and S. Santangelo, Springer Series Topics in Applied Physics, v. 100 (Springer, Berlin, 2005) p. 415.

7. A.C. Ferrari and J. Robertson, *Phys. Rev. B* 63, 121405 (2001).
8. N. Wada and S.A. Solin, *Physica B* 105, 353 (1981).
9. A.C. Ferrari and J. Robertson, *Phys. Rev. B* 61, 14095 (2000).
10. M. Veres, S. Tóth and M. Koós, *Appl. Phys. Lett.* 91, 031913 (2007).
11. N.B. Colthup, L.H. Daly and S.E. Wiberley, *Introduction to infrared and Raman spectroscopy* (Academic Press, New York, 1990).
12. R.P. Wool, R.S. Bretzlaff, B.Y. Li, C.H. Wang and R.H. Boyd, *J. Polym. Sci. B: Polym. Phys.* 24, 1039 (1986).
13. M. Veres, M. Füle, S. Tóth, I. Pócsik, M. Koós, A. Tóth, M. Mohai and I. Bertóti, *Thin Solid Films* 482, 211 (2005).

STRUCTURAL STUDY OF MULTI-COMPONENT GLASSES BY THE REVERSE MONTE CARLO SIMULATION TECHNIQUE

P. JÓVÁRI[*][1], I. KABAN[2]
[1]*Research Institute for Solid State Physics and Optics, H-1525 Budapest, POB 49, Hungary*
[2]*Institute of Physics, Chemnitz University of Technology, D-09107 Chemnitz, Germany*

Abstract. Using the example of amorphous $Ge_2Sb_2Te_5$ it is shown on how the local order at the level of pair distribution functions, coordination numbers and most probable interatomic distances can be revealed by combining the information obtained by different experimental techniques, when the measured data are modeled simultaneously by the reverse Monte-Carlo simulation technique (RMC). Special attention is paid to the information content of individual datasets. The capability of the new RMC implementation to assess the reliability of model structures is demonstrated.

1. Introduction

Elucidating the structure of disordered materials is one of the oldest challenges of condensed matter physics. The theory of diffraction of X-rays by non-crystalline materials was initiated by Debye in 1915.[1] The earliest speculative model that accounts for the non-periodicity of a covalent solid was given most likely by Zachariassen in 1932.[2] Although the formalism of the radial distribution function connected experimental data with microscopic structural quantities in a simple and pictorial way, the diffraction study of disordered materials remained an exotic technique with only a few applications to simpler systems (e.g. carbon black[3] or liquid Ar[4]). This area was given a new impetus only by the appearance of the first research reactors. These sources provided experimentalists with neutron beams that could be used for diffraction measurements. Though the available intensities were quite low in the early days of neutron scattering, the possibility of isotopic

---

[*]jovari@lxserv.kfki.hu

J.P. Reithmaier et al. (eds.), *Nanostructured Materials for Advanced Technological Applications*, 123
© Springer Science + Business Media B.V. 2009

substitution[5] meant that neutron diffraction was the most important tool of liquid and amorphous structure determination from the seventies to the nineties.

This situation has been rapidly changing since synchrotron based experimental techniques such as high energy X-ray diffraction[6] and various X-ray absorption methods became available for the scientific community. Synchrotron radiation has several advantages over neutrons. These techniques meet better the requirements of present day materials science (study of 'libraries', i.e. several slightly different compositions usually prepared in small quantities). High intensities result in shorter scan times and smaller sample quantities (often in the mg range). It should also be mentioned that technologically relevant materials are often prepared by evaporation or sputtering; thus it may be difficult to obtain samples in 'neutron quantities' (0.1–1 g, depending on sample composition and source intensities). Isotopic substitution may also become prohibitively expensive in such cases. For the above reasons, neutron diffraction cannot be regarded as a privileged technique. Experimentalists involved in structural studies are in most cases forced to rely on pieces of information obtained by different methods such as diffraction techniques and extended X-ray absorption fine structure spectroscopy (EXAFS).

2. Overview of Experimental Techniques

In a diffraction experiment one can determine (within certain limitations) $S(Q)$, the so-called total structure factor (TSF). For X-ray diffraction, the TSF can be obtained from the coherent scattering intensity in the following way:

$$S(Q) = \frac{I_{e.u.}^{coh}(Q) - \left\{ \left\langle f^2(Q) \right\rangle - \left\langle f(Q) \right\rangle^2 \right\}}{\left\langle f(Q) \right\rangle^2} \qquad (1)$$

Here $Q = 4\pi \sin\theta/\lambda$ (2θ is the scattering angle and λ is the radiation wavelength), and

$$\left\langle f^2(Q) \right\rangle = \sum_i c_i f_i^2(Q), \qquad \left\langle f(Q) \right\rangle = \sum_i c_i f_i(Q), \qquad (2-3)$$

c_i is the concentration of the ith atomic species and $f_i(Q)$ is the atomic form factor. The e.u. subscript refers to 'electron units', which means in practice that the measured intensities should be rescaled to obtain a curve oscillating around $\left\langle f^2(Q) \right\rangle$.

In the Faber-Ziman formalism,[7] $S(Q)$ is related to the partial structure factors of the sample investigated through

$$S(Q) = \sum_{i \leq j} w_{ij}(Q) S_{ij}(Q) , \tag{4}$$

where i and j run over the atomic species. The weights w_{ij} depend on the composition of the sample and on the type of radiation. For X-ray scattering they are given by

$$w_{ij}(Q) = \frac{(2 - \delta_{ij}) c_i c_j f_i(Q) f_j(Q)}{\sum_{ij} c_i c_j f_i(Q) f_j(Q)} . \tag{5}$$

For (nuclear) neutron scattering $f_i(Q)$ should be replaced by the Q-independent b_i coherent neutron scattering lengths.

Two remarks should be made here:

(i) The above defined TSF oscillates around unity, while the sum of the weights w_{ij} (when they exist) is also 1. Some authors use different normalization. Interested readers are referred to Ref. 8 for an overview of the terminology.

(ii) The w_{ij} weights can always be calculated for X-ray diffraction once the form factors have been chosen. For neutron diffraction, however, the denominator of Eq. (5) can be zero as for some elements or isotopes b_i is negative. The most well-known examples are Mn, Ti, H. In such cases, the Faber-Ziman formalism cannot be applied.

The partial structure factors are related to the partial pair correlation $g_{ij}(r)$ functions through

$$S_{ij}(Q) = 1 + \frac{4\pi\rho_0}{Q} \int r \sin Qr (g_{ij}(r) - 1) dr . \tag{6}$$

Here ρ_0 means the number density of the sample. The information given by $g_{ij}(r)$ is only one-dimensional. A further problem is that an n-component material is characterized by $n(n + 1)/2$ partial pair correlation functions, which may all contribute to the total structure factor. In the ideal case, $n(n + 1)/2$ partial pair correlation functions can be separated if we have the same number of independent diffraction measurements. However, as the number of partial pair correlation functions is already six for a three-component sample while the number of independent diffraction experiments is usually two (X-ray and neutron diffraction), the most cases can be considered as highly non-ideal. Alternative sources of information are thus needed if one wants to understand the structure of multi-component glasses. Beyond doubts, the most useful technique is extended X-ray absorption fine structure spectroscopy (EXAFS). For an overview of EXAFS experiment and theory we refer to Refs. 9 and 10. Here we just mention that $\chi(k)$, the

'EXAFS signal' extracted from absorption data, can be related to the partial pair correlation functions in a rather straightforward way:

$$\chi_i(k) = \sum_j \chi_i^j = \sum_j 4\pi\rho_0 c_j \int_0^R r^2 \gamma_{ij}(r,k) g_{ij}(r) dr \qquad (7)$$

Here the subscript i refers to the absorber atom; $\gamma_{ij}(r,k)$ is the photoelectron backscattering matrix that gives the k-space contribution of a j-type back-scatterer at a distance r from the absorber atom. As $\gamma_{ij}(r,k)$ depends on both the phase and amplitude of the backscattered waves, in favorable cases it is possible to distinguish between different types of backscatterers. This is an additional source of information in comparison with diffraction techniques. The upper limit of the integration R is usually below 4 Å, in line with the fact that EXAFS is a short-range technique giving information mostly on the first coordination shell. Elements of the backscattering matrix $\gamma_{ij}(r,k)$ should be calculated for each i–j pair by dedicated programs (e.g. FEFF[11,12]).

Providing information only about the environment of absorbing atoms, EXAFS is element sensitive. Further advantages are the low sample quantity needed and the high sensitivity. In favorable cases EXAFS can give useful structural information even if the concentration of absorbers is well below 10%. Nearest neighbor distances can be extracted rather accurately (usually with an uncertainty of ±0.01–0.02 Å) from an EXAFS data-set, while the error of coordination numbers is about 10–20% and can be considerably higher if the concentration of absorber or backscatterer atoms is too low.

3. RMC-Modeling

The reverse Monte Carlo simulation technique (RMC)[13] is a framework for the simultaneous interpretation of diffraction and EXAFS data. As there are some more recent studies discussing the method in detail,[14,15] we give here only a very brief summary of RMC.

Simulation starts from an atomic configuration (either random or crystalline) having the number density and composition of the alloy modeled. An initial set of pair correlation functions is calculated and refined through several millions iterations. In each elementary step an atom or a group of atoms is moved around randomly, while model structure factors and EXAFS spectra of the resulting atomic configuration are calculated via Eqs. (6), (4) and (7) and compared with the experimental data. The simulations are performed until a satisfactory agreement is achieved. In most cases diffraction and EXAFS data are fitted, but in principle any technique can be modeled if the signal can be calculated from atomic coordinates. The idea that the atomic structure can be revealed by simulating diffraction data by randomly moving around atoms in a simulation box was certainly not new in 1988. The first of such studies was – most likely – reported by Rechtin et al.[16] who modeled amorphous selenium by fitting a laboratory X-ray diffraction measurement. The principal difference between their implementation and present day RMC versions is that they accepted only moves improving the agreement between experimental and model curves. This way each simulation was trapped into the local minimum nearest to the starting configuration. In order to avoid this, McGreevy and Pusztai[13] suggested that some moves worsening the fits should also be accepted. The difference is rather small from the point of view of computing, but the consequences are crucial for obtaining reliable structural models.

4. Application to Amorphous $Ge_2Sb_2Te_5$

$Ge_2Sb_2Te_5$ and $GeSb_2Te_4$ are important for practical applications as they show the phase-change memory effect.[17,18] A number of studies have been devoted to elucidate the structure of the amorphous phases and to understand the mechanism of the fast phase change. However, the similarity of Sb and Te for many experimental techniques, including XRD, ND and EXAFS, makes it extremely difficult to set up consistent structural models. The structure of $Ge_2Sb_2Te_5$ was investigated by diffraction techniques and EXAFS; the five datasets were modeled simultaneously by RMC in Ref. 19. It was established that besides Te–Ge and Te–Sb bonds characterizing the crystalline modifications of these materials, a significant amount of Ge–Sb and Ge–Ge bonds can also be found in the amorphous state. In the present contribution we would like to show, using the example of this alloy, how the lack of experimental information can lead to misleading conclusions about the structure of multi-component glasses, while simultaneous exploitation of different techniques may help to avoid such traps. This will be done by comparing simulation results obtained by modeling different datasets. In line with the results of Ref. 19, Ge–Te, Sb–Te, Sb–Ge and Ge–Ge bonds will be allowed in all simulations.

At first we consider a model based only on the fit of X-ray and neutron diffraction data (Figure 1). It is clear that it is not possible to draw any conclusion about the local order in the alloy by considering only total structure factors or pair distribution functions. Also, modeling of the structure with RMC cannot help to resolve the six partial pair correlations. This is well exemplified by the coincidence of the positions of the first peaks of $g_{GeTe}(r)$ and $g_{GeSb}(r)$ (Figure 2).

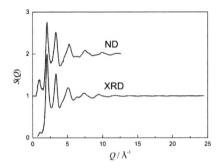

Figure 1. X-ray and neutron diffraction structure factors of amorphous $Ge_2Sb_2Te_5$. The neutron diffraction curve is shifted by 1 for clarity.

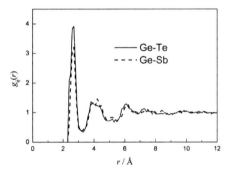

Figure 2. Ge–Sb and Ge–Te partial pair correlation functions obtained by simulating XRD and ND data.

In the next step, Ge K-edge data were added to the simulation. One may assume that this is the most direct probe of the environment of Ge. While this is certainly true in general, in our case, due to the similarity of Sb and Te, the information content of Ge absorption edge data is insufficient. This is illustrated in Figure 3, where χ_{Ge}^{Sb} and χ_{Ge}^{Te}, the contributions of Ge–Sb and Ge–Te pairs to the model Ge–K edge signal (see Eq. (7)) are compared. These 'partial EXAFS curves' can be readily obtained from the *.pek* output files of the new RMC implementation. The two curves oscillate in phase up to 12 Å^{-1} showing that these contributions are not resolved.

Addition of Te K-edge data may 'sharpen' the models by giving a more reliable estimate of the Ge–Te coordination number and distance. However, this still cannot distinguish between Sb and Te atoms. The situation can be significantly improved only when Sb–K-edge data are included in the modeling. As the electron numbers of Ge and Te/Sb are sufficiently dissimilar, phase shifts of photoelectrons backscattered by Ge and Te are sufficiently different to resolve Sb–Ge and Sb–Te/Sb correlations.

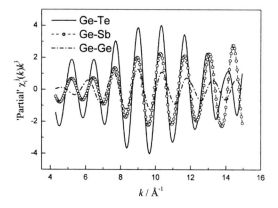

Figure 3. Ge–Te, Ge–Sb and Ge–Ge contributions to the model Ge K-edge EXAFS curve obtained by simulating XRD, ND and Ge K-edge EXAFS data.

We considered a bond as 'real' if the following two criteria are met: (i) the fit should get better upon the formation of the peak of the corresponding pair correlation function, and (ii) the position of newly formed peak should be close to the sum of the covalent radii of the participating atoms. When both Sb–Te and Sb–Sb pairs were formed, the Sb–Sb peak position could be freely varied between 2.8 and 3.2 Å by changing the cutoff. On the other hand, the Sb–Ge and Sb–Te distances were clearly determined by the simulations, and not by the constraints. Thus, according to the above classification, Sb–Te and Sb–Ge bonds are real features while Sb–Sb pairs are not. The Sb–Ge coordination number is about 0.6. Though it is small, the comparison of χ_{Sb}^{Ge} and χ_{Sb}^{Te} curves clearly shows that the two curves oscillate in antiphase in the low k region (Figure 4), and Sb–Ge bonds should be included in the simulation to get a satisfactory fit.[19]

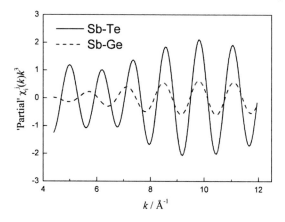

Figure 4. Sb–Te and Sb–Ge contributions to the model Sb K-edge EXAFS curve.

5. Summary

In this study it was demonstrated how the short range order of a ternary amorphous alloy can be determined by fitting simultaneously several diffraction and EXAFS measurements by the reverse Monte Carlo simulation technique. In the frame of this method, it is possible to identify 'correlations' (i.e. bonds or nearest neighbor pairs) which most likely do not exist in the system investigated, and to find those that are needed for satisfactory data interpretation. It was shown for the example of amorphous $Ge_2Sb_2Te_5$, how subtle details of the microscopic structure can be revealed by combining the information of different experimental techniques which complement each other. Though the method is not new, the range of sensibly addressable systems is rapidly increasing due to the recent development of synchrotron based experimental techniques (high energy X-ray diffraction and EXAFS) and theoretical methods of computing X-ray absorption cross section.

References

1. P. Debye, Annalen der Physik IV. 46, 809 (1915).
2. W. H. Zachariassen, J. Am. Chem. Soc. 54, 3841 (1932).
3. B. E. Warren, J. Chem. Phys. 2, 551 (1934).
4. A. Eisenstein and N. S. Gingrich, Phys. Rev. 58, 307 (1940).
5. J. E. Enderby, D. M. North, and P. A. Egelstaff, Phil. Mag. 14, 961 (1966).
6. C. Kunz, J. Phys.: Condens. Matter 13, 7499 (2001).
7. T. E. Faber and J. M. Ziman, Phil. Mag. 11, 153 (1965).
8. D. A. Keen, J. Appl. Cryst. 34, 172 (2001).
9. B. -K. Teo, *EXAFS: Basic Principles and Data Analysis* (Springer, New York, 1986).
10. J. J. Rehr and R. C. Albers, Rev. Mod. Phys. 72, 621 (2000).
11. A. L. Ankudinov, B. Ravel, J. J. Rehr, and S. D. Conradson, Phys. Rev. B 58, 7565 (1998).
12. FEFF project homepage: http://leonardo.phys.washington.edu/feff/
13. R. L. McGreevy and L. Pusztai, Mol. Simulat. 1, 359 (1988).
14. R. L. McGreevy J. Phys.: Condens. Matter 13, R877 (2001).
15. O. Gereben, P. Jóvári, L. Temleitner, L. Pusztai, J. Optoel. Adv. Mater. 9, 3021–3027 (2007). The RMC++ code is available at http://www.szfki.hu/~nphys/rmc++/downloads.html
16. M. D. Rechtin, A. L. Renninger, B. L. Averbach, J. Non-Cryst. Solids 15, 74 (1974).
17. J. Feinleib, J. de Neufville, S. C. Moss, and S. R. Ovshinsky, Appl. Phys. Lett. 18, 254 (1971).
18. N. Yamada and T. Matsunaga, J. Appl. Phys. 88, 7020 (2000).
19. P. Jóvári, I. Kaban, J. Steiner, B. Beuneu, A. Schöps, A. Webb, Phys. Rev. B 77, 035202 (2008).

DYNAMIC LIGHT SCATTERING AS A PROBE OF NANOSIZED ENTITIES: APPLICATIONS IN MATERIALS AND LIFE SCIENCES

S.N. YANNOPOULOS[*]

Foundation for Research and Technology Hellas, Institute of Chemical Engineering and High Temperature Chemical Processes, FORTH/ICE-HT, P.O. Box 1414, GR-26 504 Patras, Greece

Abstract. The ability of dynamic light scattering (DLS) to probe dynamics in non-crystalline media and to provide by virtue of this rather accurate estimations of dimensions of nanosized entities is discussed in this contribution. Selected applications of DLS in materials science (supercooled liquids) and life sciences (use of DLS as a diagnostic tool for early cataract detection in the ocular lens) are briefly discussed.

Keywords: dynamic light scattering; supercooled liquids; diagnosis of cataract

1. Introductory Remarks: Structure vs. Dynamics

A great body of experimental work in materials and life science is driven nowadays by the imperative need to characterize fast and accurately nanosized entities. Electron microscopies offer a convenient way to directly visualize the size and morphology of objects in immobile environments at the nanoscale. However, several important aspects in physics, materials and life sciences occur in liquid-like environments where the mobility of particles is important. In most cases an *in situ* study of the properties of such systems is desirable because "drying" procedures to isolate the nanoscale objects onto a substrate usually alter their morphological properties. A powerful method that can be used to determine non-invasively particle sizes in mobile environments is dynamic light scattering.

---

[*]sny@iceht.forth.gr

J.P. Reithmaier et al. (eds.), *Nanostructured Materials for Advanced Technological Applications*, 131
© Springer Science + Business Media B.V. 2009

By definition, DLS can provide information on the *dynamics* of a medium. This has to be contrasted to structure-probing techniques dealing with *static* properties. In solids, atoms are situated at rather fixed positions executing small periodic oscillations around their equilibrium positions. Structural probes such as Raman scattering, X-ray and neutron diffraction, etc. can furnish information on static (time independent) parameters such as bond lengths, bond angles, force constants, short range order (i.e. local bonding geometry), etc. On the contrary, in liquids atoms undergo diffusive motion in combination with periodic oscillations around their equilibrium positions. In this case time is important and dynamical properties must be determined.

To understand dynamics in liquids one has to introduce the concept of fluctuations. When an external perturbation is applied to a fluid, the disturbance will be damped by so-called *relaxation* processes in the medium. In a macroscopic description, dissipation phenomena include diffusion, viscous flow and thermal conduction. However, even in the absence of an external perturbation, spontaneous microscopic fluctuations always occur in the medium at any finite temperature above absolute zero. These fluctuations are dissipated by the same mechanism as those induced by external perturbations. The determination of the time scale of these fluctuations is important for understanding the dynamics of a fluid.[1] The theory of Fluctuation was devised mainly by Einstein and Smoluchowski almost a 100 years ago.[1,2] It was suggested that thermal fluctuations are an infinite Fourier series with a distribution of wavevectors (q) and frequencies (ω) propagating along all directions in the medium. These fluctuations induce local inhomogeneities in the local density or equivalently in the refractive index of the medium. Light is scattered by these local inhomogeneities or density fluctuations.

2. Dynamic Light Scattering: Particle Sizing Issues

DLS (or equivalently Photon Correlation Spectroscopy or Quasi-Elastic Scattering) is a technique used to measure dynamics in liquids over a broad time scale spanning almost 10 decades in time, i.e. 10^{-7} to 10^3 s. In the simple case of illuminating a particle suspension using laser light (i.e. monochromatic and coherent), light is scattered to all directions (Rayleigh scattering) if the particles are small enough compared to the light wavelength ($\lambda \approx 500$ nm in the visible spectrum). The light scattered by the surrounding particles undergoes either constructive or destructive interference giving rise to the speckle pattern shown in Figure 1. This is due to the fluctuations occurring in the fluid medium as a result of the *Brownian motion* executed by the suspended particles. The scattered intensity measured by a detector placed at some angle θ from the axis of the transmitted beam will fluctuate in time resembling a noisy pattern in Figure 1.

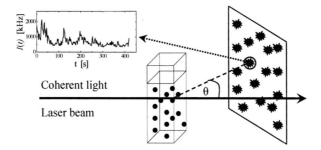

Figure 1. A typical DLS experiment. Light is scattered from a particle suspension. Spots on the screen represent speckles which blink on and off as time proceeds. The time dependence of the scattered intensity at the encircled speckle is depicted by the intensity trace $I(t)$ vs. t.

This noisy pattern, known as the intensity trace, contains information about the dynamics of the moving particles. This information can be obtained by constructing the intensity-time correlation function (TCF). This function gives an account of the similarity of the scattered signals as a function of time. At two times that do not differ appreciably, particles do not move considerably and hence the light intensities at the same speckle at these two times are practically identical or correlated. At long delay times the Brownian motion of the particles causes a loss of the correlation, and the two intensities become quite dissimilar or uncorrelated. The normalized intensity TCF that describes quantitatively the dynamics of moving particles is defined as $g^{(2)}(q,t)=<I(q,0)\ I(q,t)>/<I(q,0)>^2$, where q is the wavevector which depends on the scattering angle and the light wavelength: $q=4\pi n/\lambda_0 \sin(\theta/2)$. $g^{(2)}(q,t)$ is related to the electric field TCF $g^{(1)}(q,t)$, which is the physically significant function, through the relation $g^{(2)}(q,t)=B\ [1+f^* |g^{(1)}(q,t)|^2]$, where f^* is a coherence factor depending on instrumental parameters.

By its definition, a normalized TCF decays monotonically from 1 to 0 at long delay times. In simple cases the decay is simple exponential. In most cases of practical significance $g^{(1)}(q,t)$ deviates from the exponential form because its decay starts earlier and finishes later than the exponential case. Then $g^{(1)}(q,t)$ is considered as a *stretched exponential* (SE) decay. There are two main ways to model a SE decay. The first approach is empirical and accounts for stretching via the stretching exponent β ($0 < \beta \leq 1$) in the modified simple exponential function, i.e. $g^{(1)}(q,t) = A \exp[-(t/\tau)^\beta]$ where A is the zero-time intercept and τ is the characteristic decay or *relaxation time*. Alternatively, the stretching exhibited by $g^{(1)}(q,t)$ can be accounted for by a superposition of simple exponential decays, namely,

$g^{(1)}(t) = \int L(\ln \tau) \exp(-t/\tau) \, d \ln \tau$. In this case, L(lnτ) stands for the distribution of the relaxation times and its determination is usually accomplished with the aid of specific inverse Laplace transformation (ILT) techniques such as the CONTIN code. The estimation of the relaxation time τ leads to the determination of the diffusion coefficient of the particle $D=1/(\tau q^2)$. Finally, an *apparent hydrodynamic radius* can be estimated through the Stokes–Einstein relation, $R_h{}^{app}=k_BT/6\pi\eta D$ where η is the solvent viscosity. The above arguments show that DLS is a very useful technique for the *in situ* determination of nanoscale "particle" sizes. It should be noted that the term "apparent" is used to account for the fact that the estimated R_h is the radius of an ideal sphere whose D is equal to that of the "particle" in the suspension. Particle sizes determined by DLS fall within the submicron scale, i.e. $1 < R_h < 10^3$ nm.

3. Applications of DLS in Material Sciences

DLS has amply been used to study dynamics and interparticle interactions in macromolecular solutions,[2] and colloidal systems.[3] Only few DLS studies have been focused on neat liquids, i.e. systems that do not fall within the two above mentioned classes. Since neat liquids are composed of small molecules, a high viscosity is needed in order to slow-down their dynamics so as to fall within the time window of DLS. Therefore, most of these studies deal with glass-forming liquids which exhibit a strong temperature dependence of their viscosity while approaching the glass transition temperature, T_g, cooled down from high temperatures.[4–6] This temperature regime corresponds to the so-called *supercooled liquid* range. In view of the fact that the concept of a suspended "particle" is not valid in this case the question arises: what kind of information can be gained from studies of bulk atomic liquids? In these supercooled liquids atomic/molecular diffusion and reorientational motions give rise to the *structural relaxation* which is actually probed by DLS. The corresponding structural relaxation time is a strongly temperature dependent parameter that is eventually proportional to the liquids shear viscosity; the two quantities are related via the Maxwell equation, i.e. $\eta_s \approx G_\infty \tau$, where G_∞ is the high-frequency shear modulus of the liquid.

In brief, the main information that can be gained from DLS studies in supercooled liquids is twofold. First, a supercooled liquid can be categorized in the strong/fragile classification.[†] *Strong* liquids exhibit Arrhenius temperature dependence of their relaxation time over a broad time scale

---

[†]The terms strong and fragile are not related to the mechanical properties of the material.

from T_g up to the melting point. *Fragile* liquids are characterized by a temperature dependent activation energy exhibiting a strong increase while approaching T_g. This increase has been associated with the formation of cooperatively rearranging regions whose compositions are not different from that of the surrounded environment but their dynamical properties differ; typical sizes are within 2–10 nm. Second, there are several glasses (mostly chalcogenides) which exhibit nanoscale phase separations as revealed by structural studies. In such glasses, which highly absorb visible light, a recently developed DLS technique operating at infrared frequencies, IR-DLS, has been developed and used to study several glasses from the binary systems As_xS_{100-x} and As_xSe_{100-x}. Two-step relaxation decays were observed associated with two local environments, a chalcogen-rich one and one close to stoichiometry. Finally, dynamical aspects of λ-transitions of elemental sulfur have been recently revealed by IR-DLS.[7]

4. Applications of DLS in Life Sciences

Abnormal protein condensation (aggregation) has been related to more than 20 diseases; among these are lens cataracts and the sickle cell disease as well as the neurodegenerative Alzheimer's and Parkinson's diseases. The condensation process can lead to various forms including fibers, amorphous aggregates, gels, and crystalline products, which can impede cellular functions. All these products entail a change in the "particle" size and can in principle be studied by DLS. The ocular lens is the most characteristic example of a mammalian tissue that exhibits complex colloidal behavior. The lens can be considered as a dense aqueous suspension of globular proteins. Lens opacification, or *cataract*, reflects changes of the interactions between the proteins. Cataract is considered today as the most important cause of preventable blindness worldwide; its diagnosis is currently made clinically at the mature level of the disease. As the life expectancy is gradually increasing, age-related diseases like cataract will become more prominent among the population, with a high societal impact. The need for the development of a tool for early diagnosis of cataract onset is therefore obvious. The advantage of DLS is that it can provide a sensitive and reliable methodology for non-invasive, early diagnosis. The transparency of the lens and the other ocular tissues, which are situated externally to it, as well as the dependence of the scattered intensity on the sixth power of the "growing particle" radius add to the advantages of using DLS.

We have recently in our laboratory undertaken systematic DLS studies of various mammalian lenses in order to determine the proper spectral parameters that can be used as reliable indicators of cataract onset. To demonstrate the difficulty of this task we show in Figure 2 the TCF from a cataract-free porcine lens, taken at the physiological body temperature. The TCF is very broad due to the high volume fraction of proteins and can be fitted with at least four decay steps as shown by the ILT curve. The main difficulty is to reliably and uniquely rationalize these four modes in terms of the diffusive motions of the constituent protein "particles". To tackle this issue we followed a hierarchical approach where the lens proteins were isolated; studies of dilute, semidilute, and dense suspensions were undertaken.[8] Having identified the various relaxation steps of the broad TCF of the lens nucleus, we employed an easily realizable cataract model – the *cold cataract* – where a cataract develops gradually to the desired extent by cooling the lens. Systematic studies on several lenses led us to the conclusion that the stretching exponents of the first decays step can be used as a good and reliable indicator of cataract onset.[9]

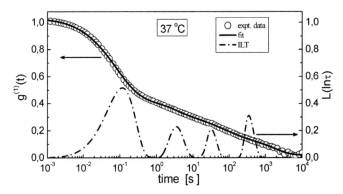

Figure 2. A typical electric-field time correlation function of the porcine lens nucleus. The solid line is the best-fit curve obtained by SE analysis.

References

1. J.P. Boon and S. Yip, *Molecular Hydrodynamics* (Dover, New York, 1991).
2. B.J. Berne and R. Pecora, *Dynamic Light Scattering* (Wiley, New York, 1976).
3. J.K.G. Dhont, *An Introduction to Dynamics of Colloids* (Elsevier, Amsterdam, 1996).
4. S.N. Yannopoulos, G.N. Papatheodorou, G. Fytas, Phys. Rev. E 53, R1328 (1996).
5. S.N. Yannopoulos, G.N. Papatheodorou, G. Fytas, Phys. Rev. B 60, 15131 (1999).
6. E.A. Pavlatou, S.N. Yannopoulos, G.N. Papatheodorou, G. Fytas, J. Phys. Chem. 101, 8748 (1997).
7. T. Scopigno, S.N. Yannopoulos, F. Scarponi, K.S. Andrikopoulos, D. Fioretto, G. Ruocco, Phys. Rev. Lett. 99, 025701 (2007).
8. A. Giannopoulou, A.J. Aletras, N. Pharmakakis, G.N. Papatheodorou, S.N. Yannopoulos, J. Chem. Phys. 127, 205101 (2007).
9. V. Petta, G.N. Papatheodorou, N. Pharmakakis, S.N. Yannopoulos, Phys. Rev. E 77, 061904 (2008).

THE COLLAPSE OF HYDRODYNAMIC RADII IN PLURONIC PE6400 MICELLES IN VICINITY OF SUPRAMOLECULAR TRANSITION: DYNAMIC LIGHT SCATTERING, HEAT CAPACITY AND SOUND VELOCITY MEASUREMENTS

Z. BAKAEVA[*1], K. IGAMBERDIEV[1], P. KHABIBULAEV[1],
P. STEPANEK[2], P. CERNOCH[2]
[1]*Heat Physics Department, Tashkent, Uzbekistan Academy
of Sciences, 700130 Tashkent, Uzbekistan*
[2]*Institute of Macromolecular Chemistry, Prague 6,
Czech Republic*

Abstract. This work reports the results of a study of the microstructural and dynamical properties of self-assembled triblock copolymers. It was performed using adiabatic calorimetry, ultrasonic spectroscopy and dynamic light scattering. In the vicinity of supramolecular transitions, first aggregates occur; with further increasing temperature a spontaneous forming of compact structures was observed in the system. These results are supported by data of the shear viscosity and free internal volume parameters calculated in the framework of the Frenkel-Andrade approach.

Keywords: triblock copolymer; supramolecular structures; sound velocity; heat capacity; hydrodynamic radius

1. Introduction

The anomalous micellization process, which accompanies association processes in block copolymers, has remained an attractive problem for chemists, physicists, and engineers during the last decade. The construction and characterization of nanosized thermosensitive biocompatible objects in aqueous solutions is a very promising approach for solving important tasks in nanorobotics, drug encapsulation techniques, and as solving agent for hydrophobic molecules. Very interesting of view of both, fundamental and

---

[*]zulonok@yandex.ru

J.P. Reithmaier et al. (eds.), *Nanostructured Materials for Advanced Technological Applications,*   137
© Springer Science + Business Media B.V. 2009

applied research, are Pluronics, amphiphilic triblock copolymers of poly-ethylene oxide and polypropylene oxide (PEO–PPO–PEO); this class of high-molecular compounds demonstrates an attractive ability to form supra-molecular disperse phases.[1] A distinctive property of this class of polymers is the thermosensibility: upon small changes of the temperature, the poly-mer molecules form micelles which are subsequently linked together to form oriented structures. These triblock copolymers show thermo-reversible transitions that are similar to the reactions of biomolecules to external influ-ences in nature. The understanding of the physical and chemical nature, structure and behaviour of these systems still remains a subject of intensive research. While static and structural properties of PEO–PPO–PEO polymers are investigated more or less in detail, the dynamic properties and changes of the topological structure of these polymers under external influences are not yet well understood.

2. Materials and Methods

For our study the commercially available Pluronic PE 6400 (BASF, Germany), was used without further purification. The water for preparing solutions was twice distilled and additionally deionized. The microstructure of the Pluronic supramolecular structure was investigated using adiabatic calorimetry and sound velocity measurements.[3] In addition, density measure-ments were performed. For studying the dynamics and hierarchical structures, dynamic light scattering was used.[4] The solutions were prepared by weight; before the measurements they were filtered trough 0.05 μm filters to elimi-nate dust.

3. Results and Discussion

It is known from the literature that solutions of commercial samples demon-strating the anomalous association contain a small amount of molecules with longer chains of the polypropylene oxide block caused by composition polydispersity.[2] For our aim, namely the characterization of nanostructured particles formed in the vicinity of physiological temperatures, this is a favorable aspect. Pluronic L64 (analogous of Pluronic PE6400) shows detectable aggregates at 298 K only at concentrations above ~6%, probably indicating a multiple association process.[5] At T ~ 308 K, however, essen-tially invariant values for the hydrodynamic radius were found over a wide concentration range, and the micelles were monodisperse: these systems are more likely represented by a closed association model. This micellar state, described by a distribution of micelles homogeneous in size and aggregation number, was considered as supramolecular equilibrium. However, a few reports focus on the investigation of processes, which occur in the vicinity of the critical point of association resulting in supramolecular equilibrium. The data of heat capacity and sound velocity[3] indicate that a micelle formation process occurs in a wide range of temperatures. The amplitude of the heat capacity peak, which is observed at association temperatures, grows with increasing polymer concentration, while at the same time its position is shifted to lower temperatures. To describe this process we investigated microstructural and rheological properties using classical theories.[6] On the basis of the measured heat capacity and sound velocity, a calculation of complex calorimetric parameters was performed. The isochoric heat capacity C_v, the isobaric heat capacity C_p and the internal free volume values v_f were estimated. The temperature dependence of the ratio C_p/C_v is given in Figure 1. It demonstrates that in the temperature range of 295–323 K the system reaches the supramolecular equilibrium when the C_p/C_v ratio tends to unity, which is caused by a correlation between the internal and external degrees of freedom of the copolymer molecules.

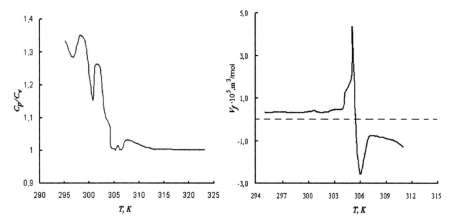

Figure 1. Temperature dependence of the isobaric and isochoric heat capacity ratio of a Pluronic PE6400 copolymer in 6% aqueous solution (left) and internal free volume at the same conditions (right).

Before reaching the supramolecular equilibrium, the ratio C_p/C_v is >1. This can be explained in the framework of classical approaches by the dynamical prevalence of external degrees of freedom over the internal ones; this description can be applied to the molecularly dispersed state of macromolecules. The trend of C_p/C_v ratio to unity has a nonlinear character; it exhibits two minima at 297 and 300 K. This behaviour can be explained by the formation of metastable intermediate complexes which dissociate with increasing temperature. Thus we can conclude that the supramolecular association in the Pluronic PE6400/water system has a step-like character with increasing temperature.

The internal free volume of Pluronic PE6400 in water solution exhibits a very interesting behaviour (see the right side of Figure 1). At the first steps of association processes it demonstrates a sharp increase in close vicinity of the supramolecular equilibrium of the system. The association in Pluronic in aqueous solutions is usually driven by hydrophobic interactions between the middle blocks of the polymer chains; it is accompanied by a conformational rearrangement related to the movement of the polypropylene block to the front of the Pluronic molecule which requires a significant value of the system energy as was established in Ref. 3. It is known also from theoretical predictions[6] that the water density close to hydrophobic surfaces is smaller than in the water bulk. The depletion layer near to hydrophobic surfaces is relatively higher than that near hydrophilic ones and is increasing with temperature. Thus we can suppose that at temperatures close to the association temperature polymer molecules exist in the associated state, but these structures have local regions with significant depletion thicknesses; this state is registered about 1 K before the transition temperature. With further increasing temperature, the free volume of the system becomes negative at 305.7 K which corresponds to the process of spontaneous dehydration of micelles which occurs during the formation of compact structures in the course of micellisation.

In further investigation of the dynamics of triblock copolymers we extracted the shear viscosity from the calorimetric parameters on the basis of the Frenkel theory[7]:

$$\eta_s = \frac{A_0 \alpha_p^2}{\beta_T}$$

where α_p is the volume expansion coefficient, β_T the isothermal compressibility, and A_0 is a pressure independent temperature coefficient. On the basis of the calculation of η_S from our experimental data it is possible to consider the behaviour of this parameter in the vicinity of the supramolecular transition where η_S decreases and shows a fluctuation character (Figure 2).

The changing of shear viscosity can be interpreted as fluctuations due to a diffusion of polymer molecules. At further increasing temperatures, η_S decreases up to a critical value at 320 K, caused by a collapse of micelles during clouding processes.[8] This result is confirmed by the dynamic light scattering measurements in Figure 3, which show the temperature dependence of the hydrodynamic radii of molecularly dispersed polymer and larger particles.

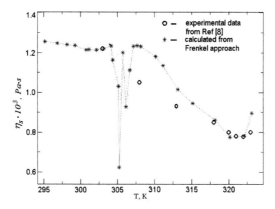

Figure 2. The shear viscosity calculated for a Pluronic PE6400 copolymer in 6% aqueous solution vs. temperature.

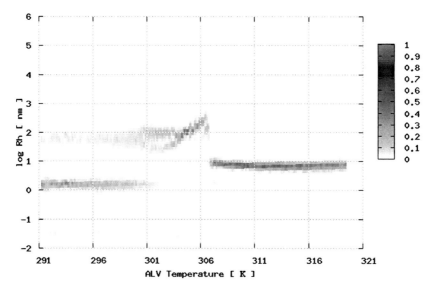

Figure 3. Temperature dependence of the hydrodynamic radii for a Pluronic PE6400 copolymer in 6% aqueous solution.

As has been shown in the 3D-matrix of dynamic modes at temperatures below the transition temperature, the size of dispersed polymer molecules is 2 ± 0.4 nm. In the same temperature region the second dynamical mode appears, which corresponds to big clusters existing in the solution and that have a wide size distribution with an average of 130 ± 30 nm. After filtering the solutions trough 0.05 μm filters at room temperature, the big clusters remain in the dispersions which proves that they are flexible and noncompact. The big clusters are in dynamical equilibrium with the free polymer molecules. In the second temperature region above ~301 K the system demonstrates anomalous association, when a major amount of polymer molecules is involved to form large structures, which are growing with increasing temperature up to critical value of ~307 K. In the third region the micellar state is established in the system when the hydrodynamic radii of the structures decrease from 850 to 10 nm within 0.4 K. At temperatures above 307 K the micellar size does not depend on temperature.

This interesting behaviour of the hydrodynamic radii of associates occurs in course of anomalous associations due to the presence of compositional hydrophobic impurities.[9] We mean here by the term "impurity" the Pluronic molecules containing longer chains of the polypropylene block. We suggest that hydrophobic impurities are responsible also for the big premicellar clusters. By DLS experiments it was shown that impurities encourage the association processes to form large clusters as more or less thermodynamically stable hierarchical structures even at low temperatures. The concentration of these clusters grows in the critical region, while a dynamic competition between the structures occur: the system is converging to the most stable association numbers.

ACKNOWLEDGMENTS

The authors acknowledges support of this work by the INTAS Program for young scientists No. 06-1000023-6276, the Grant Agency of the Czech Academy of Sciences No. 4050403, and partial support by the NATO Reintegration grant EAP.RIG.982545.

References

1. L. Tianbo et al., Prog. Polym. Sci. 28, 5 (2003).
2. M.J. Kositza et al., Langmuir 15, 322 (1999).
3. Z.S. Bakaeva et al., J. Eng. Phys. Thermophys. 79, 153 (2006).
4. P. Stepanek et al., Int. J. Polym. Anal. Charact. 12, 3 (2007).
5. P. Alexandris and R.A. Hatton, Collods Surf. A. 96, 1 (1995).
6. Mamatkulov S.I. et al., Langmuir, 20, 4756–4763 (2004).
7. Ya. I. Frenkel, *Theory of Kinetics in Liquids* (in Russian), (NAUKA, Leningrad, 1975), p. 589.
8. M.M. Jebari et al., Polym. Int. 55, 176 (2006).
9. Z. Tuzar and P. Kratochvíl. In: E. Matijevic (Ed.), *Surface and Colloid Science 15*, (Plenum Press, New York, 1993), p. 1.

4. PREPARATION TECHNIQUES

SOME STRUCTURAL MODIFICATIONS AT THE NANOMETRIC SCALE INDUCED BY SWIFT HEAVY ION IRRADIATION

J.-C. PIVIN[*1], F. SINGH[2], A. KUMAR[2], M.K. PATEL[3], D.K. AVASTHI[2], D. DIMOVA-MALINOVSKA[4]

[1]*Centre de Spectrométrie Nucléaire et de Spectrométrie de Masse, CSNSM-IN2P3,bâtioment 108, 91405 Orsay Campus, France*
[2]*IUAC, P.O. Box 10502, Aruana Asaf ali Marg, New Delhi 110067, India*
[3]*High Pressure Physics Division, BARC, Purnima Tromblay, Mumbai 400085, India*
[4]*SENES, Bulgarian Academy of Sciences, 72 Tsarigradsko Chaussee, 1784 Sofia, Bulgaria*

Abstract. The effects of thermal spikes occurring in the wake of swift heavy ions on the structure of oxides, oxide-metal composites and carbon based targets are reported. Linear strings of metal nanoparticles are formed along the ion tracks in silicate glasses but nanowires of amorphous carbon in fullerenes or polymers. Changes of the size, shape or stress state of the particles induced by swift heavy ion irradiation in composites affect their optical and magnetic properties.

Keywords: ion irradiation; metal nanoparticles; magnetic films; plasmon resonance

1. Introduction

The applications of ion implantation to the doping of semiconductors or the synthesis of phases in superficial layers of solids received great attention during the past 40 years. The structural modifications induced by swift ions passing through targets without undergoing atomic collisions are less known by the community of material scientists. However, they offer many possibilities due to the nanometric diameter of the cylindrical volume perturbed

---

[*]pivin@csnsm.in2p3.fr

J.P. Reithmaier et al. (eds.), *Nanostructured Materials for Advanced Technological Applications,* 145
© Springer Science + Business Media B.V. 2009

along their path and to the out-of-equilibrium nature of these transformations.[1] The ion energy is transferred to the electronic system in the form of a
high density of ionizations and electronic excitations (the transferred energy
per unit length of ion path is called electronic energy loss S_e). The repulsion
between ionized atoms causes Coulomb explosions in insulating targets. On
the other hand, electronic excitations lead to an intense electronic heating,
which is transferred to the atomic network by electron–phonon coupling. It
results in a strong agitation similar to that of a thermal spike, lasting 10^{-11} to
10^{-9} s depending on the thermal properties of the target and the electron–
phonon coupling constant. Calculations of thermal exchanges account for
temperature increases of several thousand degrees (often above the vapori-
zation temperature) in the wake of heavy ions with energies above 500
keV/atomic mass unit.[2] High temperature phases or glasses are formed
during the freezing stage of targets possessing initially the structure stable
at room temperature. The precipitation of a secondary phase may occur or,
on the contrary, preexisting particles may be dissolved, depending on the
parameters enumerated thereafter. This paper focuses on structural changes
induced by swift heavy ions (SHI) in oxides and oxide-metal composites
and their consequences for the optical or magnetic properties of systems of
technological interest. SHI irradiation effects in polymers and fullerenes are
also mentioned for their numerous applications in electronics, chemistry
and biomaterials.

2. Modifications of Structures

The effects of Coulomb explosions are generally disregarded, because in
many materials they could hardly be distinguished from those of thermal
spikes, which last a much longer time, and they are probably eclipsed by the
latter. The strong directionality of covalent bonds compared to ionic ones
causes that the structural order is less easily restored in covalent compounds, after atoms have been displaced in the collisional regime.[3] This rule
applies also in the electronic regime of ion stopping, as shown by numerous
experimental investigations.[4] Oxides in which M–O bonds have a strong
covalent character, such as TiO_2 or SiO_2, get amorphous whenever the density of electronic excitations is high enough for inducing a local melting,
whereas more ionic oxides (MgO, ZnO, Al_2O_3 and all oxides with fluorite
structure such as ZrO_2, UO_2) re-crystallize. The radiation resistance of
spinels, perovskites, pyrochlores, or other mixed compounds depends on the
weakest bonds. However, a strong correlation is also observed between the
stability of $A_xB_yO_z$ compounds and the ability of A and B cations to exchange
sites by recombination of Frenkel pairs, created during collision cascades or
thermal spikes.[5] For instance, pyrochlores formed from sesquioxides A_2O_3

and dioxides BO_2 undergo a transformation into a chemically disordered crystal with a fluorite structure when the ratio of the A and B cation radii does not exceed 1.6. On the contrary they get amorphized when the sizes of A and B are more dissimilar.[6] These laws have important implications in the choice of matrices for the storage or transmutation of nuclear wastes and also for the magnetic properties of oxide compounds.

Another consequence of thermal spikes is that the fluid cylinder, since its density is generally lower than that of the solid matrix, is under pressure and consequently exercises a strong radial compressive stress on the surrounding matrix. The latter effect accounts for instance for the decrease of the lattice parameter, which is observed in many oxides resistant to amorphization, and for their polygonization.[4] The compression of the matrix by the hot core of tracks affects the properties of the target until the tracks become close to each other and overlap. It induces a magnetostriction in magnetic oxides and is probably also responsible for the polymerization of fullerenes.[7]

Polymers and fullerenes constitute a good example for discussing the effects of the electronic stopping magnitude on the continuity of the so-called latent tracks remaining after the quenching of the spikes. The name refers to their generally higher sensitivity to chemical etching compared to that of the undamaged matrix, which is used for the fabricating of porous membranes, especially of polymers.[1] It is worth to note first that the formation of amorphous carbon in the track core in organic polymers and fullerenes does not require an intense heating, because these molecular solids are highly sensitive to electronic excitations. In fact they get also amorphous, more progressively, under the cumulative effects of ion impacts even under bombardment with H or He ions of a few MeV, or under electron or UV irradiation.[1] However, the formation of isolated nanowires of amorphous C and the polymerization of fullerenes in the track halos occur only under SHI irradiation producing spikes.[7] The formation of conducting nanowires is evidenced by AFM imaging in the conducting mode as shown in Figure 1. The influence of the magnitude of S_e on the continuity of the latent tracks is well illustrated by the electrical properties of these conducting paths in fullerene or polymer films.[7,8] The threshold voltage for conduction decreases, and the areal density of efficiently conducting paths for a given voltage V increases with S_e. When S_e is low, the paths are discontinuous and their I(V) characteristics is not ohmic. In the case of fullerene films, an explanation of this behavior is that the polymerized track halos contribute to the transport only for higher voltages, in contrast to metallic islands of amorphous C formed in the track cores. Such conducting wires in otherwise insulating films find applications as ohmic contacts of nanometric size or field emission tips, the areal density of which is easily controlled by the choice of the ion fluence.

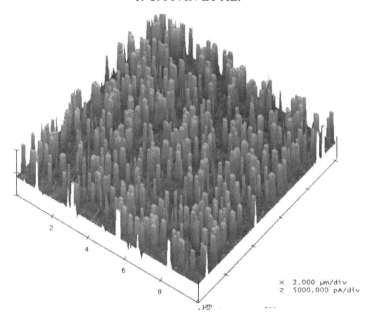

Figure 1. AFM image in the conducting mode under an applied voltage of 2 V of a 500 nm thick C_{60} film irradiated with $5 \times 10^9/cm^2$ Au ions of 120 MeV.

Just as for the amorphization of fullerenes or polymers, the occurrence of spikes is not necessary for inducing the nucleation of clusters of noble metals in glasses or even of transition metals in inorganic gels.[1] Low densities of electronic excitations are sufficient for provoking oxido-reduction reactions, and clusters are formed when the local concentration of reduced metal atoms is large enough after a given number of track overlaps. But the nucleation of clusters of transition metals in silicates is observed only under SHI irradiation because they are more inert than gels.[9] The requirement of a strong heating is proven indirectly by the much lower efficiency of annealing treatments in reducing atmospheres for obtaining the same effect. Another interesting feature of the precipitation produced by SHI irradiation is the narrow distribution of sizes of the obtained particles (limited by the track size) and their alignment along the tracks.[9,10]

Now, when irradiating an oxide matrix already containing metal particles, spikes may result in different effects, depending on the metal solubility, the diffusion coefficient of the metal atoms in the melt, the particle size and the volume density. Particles smaller than the tracks tend to be dissolved and metal atoms to evaporate, when the temperature reaches the sublimation point, as for instance for Ag in silica for linear energy densities S_e of about 10 keV/nm (Ag or Au ions of 100 MeV). This electronic sputtering (evaporation), which occurs as a general rule in any material above a S_e threshold, finds some applications in the patterning of thin films.[1] Beside that, when a

composite exhibits a noticeable distribution of particles sizes and a sufficient volume fraction of particles to permit the transport of metal atoms
from dissolved particles towards others which have not been completely
melted because their size exceeds that of the track, a ripening of the largest
particles is observed. They also become elongated along the molten zone
(Figure 2). Another effect can also induce an elongation, but it is efficient
only in systems with low volume fractions of particles. It results from the
compressive stress exercised by the matrix, whenever the latter undergoes
an anisotropic densification named hammering. This process is completely
different from the lattice compaction of crystalline oxides mentioned above
and is efficient only in amorphous targets such a silica.[11]

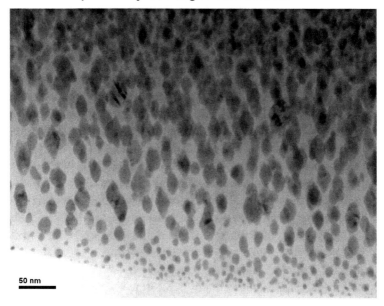

Figure 2. Elongation of FePt particles in a silica film containing 5 at % Fe and 5 at % Pt,
deposited by magnetron sputtering, annealed at 700°C for growing particles of the quadratic
intermetallic compound, then irradiated with $10^{13}/cm^2$ Au ions of 120 MeV.

3. Modifications of Optical and Magnetic Properties

Significant changes of the band gap of oxides (for instance a decrease from
3.5 to 2.8 eV for SnO_2) are observed as a result of either the lattice compression, the modification of the grain size, or the amorphization of the
target. Such effects have of course strong implications for the transfer of
charges between localized levels of doping elements in the gap and the conduction or valence bands of the host, with applications in the conversion of
solar energy, waveguides or spintronics. A stronger overlap between the 3d
level of magnetic transition metals and the O2p band in TiO_2, ZnO or SnO_2

is for instance expected to enhance the correlation between magnetic spins and consequently the ferromagnetism in these dilute magnetic semiconductors (DMS).

The lattice disordering or amorphization of magnetic oxides can as well induce a decrease or an increase of the permanent magnetization and the Curie temperature. For instance, the site exchange between M^{2+} and Fe^{3+} cations in ferrimagnetic ferrites increases the interaction between magnetic cations,[12] whereas the amorphization of $RFeO_3$ perovskites (R = Er, Ho or Y) decreases their magnetization.[13] When the ion fluence is limited such that the tracks don't overlap, the compressive radial stress exercised by the disordered core of the tracks on the surrounding matrix sometimes provokes a tilt of the easy magnetization axis from in-plane to out-of-plane, which is of interest for the magnetic recording at high density.[14]

The dissolution of metal particles is generally not intended, with the noticeable exception of magnetic metal particles in DMS, since they cannot contribute to the magnetically driven conduction or optical polarization, and in addition are paramagnetic at room temperature when of small size.[15] The elongation of nanoparticles parallel to the ion beam direction modifies their depolarization or demagnetization factors and consequently the optical or magnetic anisotropy of thin films embedding these particles. The surface plasmon resonance of noble metal particles is red-shifted when the electric field is polarized parallel to the elongation axis and vice versa, which results in a dichroism of the system.[10] On the other hand, the easy magnetization axis of ferromagnetic particles, such as those shown in Figure 2, becomes perpendicular to the surface when elongated under the effect of SHI irradiation.[9,10] Note, however, that a similar rotation of the easy axis is observed for Fe particles, which remain spherical under the same conditions of irradiation, as long as their volume fraction is limited to a few percentage.[10] In this case, the perpendicular magnetization is ascribed to the in-plane compression of the particles as a consequence of the matrix hammering. When the volume fraction of the Fe particles is increased, the magnetostriction results only in an increase of the coercive field and the remnant magnetization perpendicular to the surface, nevertheless interesting for the recording stability.

A last possible application of SHI in magnetism which is worth to note is the patterning of thin films, induced either by electronic sputtering or by a periodic cracking under the effect of hammering.[16] The parallel grooves formed at the surface of oxide films can be used as templates, for instance for growing wires of Co with a forced in-plane [001] orientation.

4. Conclusions

SHI irradiation induces the formation of columnar defects with a nanometric diameter and the nucleation of nanoparticles aligned along these tracks, which are appealing for their optical or magnetic properties. In addition, the local melting of tracks during the thermal spikes provokes changes of the size, shape or spatial distribution of nanoparticles in composites, which are liable also to modify the optical or magnetic anisotropy of thin films. SHI irradiation effects in composite systems need to be investigated more extensively for modeling, as they have been explored mainly in silica-metal systems.

References

1. J.C. Pivin, in: A. Vaseashta, D. Dimova-Malinovska, J.M. Marshall (Eds.). *Nanostructured and Advanced Materials* (Springer, Berlin, NATO Sciences Series 204, 2005), p. 155.
2. M. Toulemonde, C. Trautmann, E. Balanzat, K. Hjoirnt, A. Weidinger, Nucl. Instrum. Meth. Phys. Res. B 216, 1 (2004).
3. K. Trachenko, J. Phys.: Condens. Mat. 16, R1491 (2004).
4. A. Turos, L. Nowicki, F. Garrido, L. Thomé, R. Fromknecht, J. Domagala, Acta Phys. Pol. B 30, 1611 (1999).
5. K.E. Sickafus, R.W. Grimes, J.M. Valdez, A. Cleave, M. Tang, M. Ishimaru, S.M. Corish, C.R. Stanek, B.P. Uberuaga, Nat. Mater. 6, 217 (2007).
6. M.K. Patel, V. Viyayakumar, D.K. Avasthi, S. Kailas, A.K. Tyagi, Nucl. Instrum. Meth. Phys. Res. B 266, 2898 (2008).
7. A. Kumar, D.K. Avasthi, A. Tripathi, D. Kabiraj, F. Singh, J.C. Pivin, J. Appl. Phys. 100, 1 (2006).
8. A. Kumar, F. Singh, D.K. Avasthi, J.C. Pivin, Nucl. Instrum. Meth. Phys. Res. B, 266 (14), 3257 (2008).
9. F. Singh, D.K. Avasthi, O. Angelov, P. Berthet, J.C. Pivin, Nucl. Instrum. Meth. Phys. Res. B 245, 214 (2006).
10. J.C. Pivin, F. Singh, Y.K. Mishra, D.K. Avasthi J.P. Stoquert, Surf. Coat. Technol., 2009 (in press).
11. S. Klaumünzer, C.L. Li, S. Löffler, M. Rammensee, G. Schumacher, H.C. Neitzer, Rad. Eff. Def. Sol. 108, 131 (1989).
12. F. Studer, Ch. Houpert, D. Groult, J. Yun Fan, A. Meftah, M. Toulemonde, Nucl. Instrum. Meth. Phys. Res. B 82, 91 (1993).
13. M. Bhat, B. Kaur, R. Kumar, P.A. Joy, S.D. Kulkarni, K.K. Bamzai, P.N. Kotru, Nucl. Instrum. Meth. Phys. Res. B 243, 134 (2006).
14. F. Studer, C. Houpert, H. Pascard, R. Spohr, J. Vetter, J. Yun Fan, M. Toulemonde, Rad. Eff. Def. Sol. 116, 59 (1991).
15. W.K. Choi, B. Angali, H.C. Park, J.H. Lee, J.H. Song, R. Kumar, Adv. Sci. Technol. 52, 42 (2006).
16. R.S. Chauhan, D.C. Agarwal, S. Kumar, S.A. Khan, I. Sulania, W. Bolse, D.K. Avasthi, J. Nanosci. Nanotechnol., 2009 (in press).

FABRICATION OF METAL NANOPARTICLES IN POLYMERS BY ION IMPLANTATION

A.L. STEPANOV[*]
*Kazan Physical-Technical Institute, Russian Academy
of Sciences, Sibirsky Ttrakt 10/7, 420029 Kazan, Russia
Laser Zentrum Hannover, Hollerithallee 8, 30419 Hannover,
Germany*

Abstract. This article considers some features of metal nanoparticle fabrication in polymer matrices by ion implantation. This technological approach is very promising for the development of optical composite materials. The polymer layers were irradiated by silver ions with high doses up to 10^{17} ion/cm^2. Optical density spectra of these composites demonstrate that the silver nanoparticles exhibit unusually weak and wide plasmon resonance spectra. The formation of silver nanoparticles in carbonized polymers by ion irradiation is reviewed. Based on the Mie theory, optical extinction spectra for metal particles in polymer and carbon matrices are simulated and compared with optical spectra for complex silver core–carbon shell nanoparticles.

Keywords: metal nanoparticles; ion implantation; radiation; polymers; surface plasmon resonance; Mie theory

1. Introduction

The task of designing new polymer-based composite materials containing metal nanoparticles (MNPs) is of current interest particularly for the fabrication of nonlinear optical switches, magneto-optic data storages, directional connectors, etc. The nonlinear optical properties of these composites stem from the dependence of their refractive index on the intensity of the incident light. This effect is associated with the MNPs, which exhibit a high nonlinear third order susceptibility when exposed to ultra-short (ps or fs)

[*] a.stepanov@lzh.de

J.P. Reithmaier et al. (eds.), *Nanostructured Materials for Advanced Technological Applications,* 153
© Springer Science + Business Media B.V. 2009

laser pulses.[1] Light-induced electron excitation in MNPs (the so-called surface plasma resonance SPR),[2] which shows up most vividly in the range of
linear absorption, gives rise to nonlinear optical effects in the same spectral
range. In practice, the SPR effect can be enhanced by increasing the
nanoparticle concentration in the host matrix, i.e. by increasing the volume
fraction of the metal phase (filling factor). Systems with a higher filling factor offer a higher nonlinear third-order susceptibility.[1]

   Metal nanoparticles can be embedded in a polymer matrix by a variety
of techniques such as chemical synthesis in an organic solvent,[2] vacuum
deposition on viscous polymers,[3] plasma polymerization combined with
metal evaporation,[4] etc. However, they all suffer from disadvantages, such
as a low filling factor or a large distribution in size and shape of the
nanoparticles, which offsets the good optical properties of the composites.
One promising method is ion implantation,[5] which provides a controllable
synthesis of MNPs at various depths under the surface and unlimitedly high
impurity doses. Despite intensive studies of MNP synthesis by ion implantation in dielectrics, such as non-organic glasses and crystals, which was
started in 1973 by Davenas et al. with Na and K ions[6] and in 1975 Arnold
with Au ions,[7] the formation of nanoparticles in organic matrices was realized only at the beginning of the 1980s by Koon et al. in their experiments
on the implantation of Fe ions into polymers.[8] By ion implantation, it is possible to produce almost any metal–dielectric (specifically, metal–polymer)
composite.

   Noble metals exhibit the most pronounced SPR effect and, hence, the
highest nonlinearity of the optical properties of MNPs in dielectrics. Quite
recently such materials have been fabricated by Ag ion implantation into
epoxy resins,[9] polymethylmethacrylate (PMMA)[10] and polyethylene tere-
phthalate.[11] The aim of this work was to study the SPR-related linear optical
properties of MNPs introduced into a polymer matrix by implantation.

2. Experimental Procedure

1.2-mm thick PMMA plates, which are optically transparent in a wide spectral range (400–1,000 nm), were used as substrates. They were implanted by
30 keV Ag^+ ions with doses in the range from $3.1 \cdot 10^{15}$ to $7.5 \cdot 10^{16}$ ion/cm^2 at
an ion current density of 4 $\mu A/cm^2$ in a vacuum of 10^{-4} Pa using an ILU-3
ion implanter. In a control experiment, Xe-ion implantation into PMMA at
the same conditions was performed. Spectra of the optical density were
measured from 300 to 900 nm by a Hitachi 330 spectrophotometer. All
spectra were recorded in a standard differential mode in order to normalise
substrate effects. The samples obtained were also examined by transmittance electron microscopy (TEM).

The optical spectra of spherical MNPs embedded in various dielectric media were simulated in terms of the Mie electromagnetic theory,[12] which allows one to estimate the extinction cross section σ_{ext} for a light wave incident on a particle. This value is related to the intensity loss ΔI_{ext} of an incident light beam of intensity I_0 passing through a transparent particle-containing dielectric medium due to absorption σ_{abs} and elastic scattering σ_{sca}, where $\sigma_{ext} = \sigma_{abs} + \sigma_{sca}$. From the Lambert–Beer law it follows

$$\Delta I_{ext} = I_0 \left(1 - e^{-\#\sigma_{ext} h}\right), \tag{1}$$

where h is the thickness of the optical layer and # the density of the nanoparticles in the sample. The extinction cross section is connected to the extinction constant γ as $\gamma = \#\sigma_{ext}$.

The experimental spectral dependencies of the optical density OD are given by

$$OD = -\lg(I/I_0) = \gamma\lg(e)h; \tag{2}$$

hence, for samples with electromagnetically non-interacting nanoparticles, it possible to put $OD \sim \sigma_{ext}$. Therefore, experimental OD spectra are compared with modeled spectral dependencies which are expressed through σ_{ext} calculated by the Mie theory.

3.   Experimental Results

As follows from TEM, Ag ion implantation results in the formation of silver nanoparticles. As an example, the micrograph in Figure 1 shows spherical nanoparticles synthesized in PMMA with a dose of $5.0 \cdot 10^{16}$ ion/cm^2.

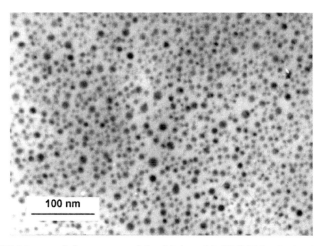

Figure 1. TEM image of silver nanoparticles fabricated in PMMA by Ag ion implantation.

Microdiffraction patterns of the composite samples demonstrate that the MNPs have the fcc structure of metallic silver. The diffraction image consists of very thin rings (corresponding to polycrystalline nanoparticles) imposed on wide diffuse faint rings from the amorphous polymer matrix. By comparing the experimental diffraction patterns with standard ASTM data, it possible to conclude that the implantation does not form any crystalline compounds involving silver ions.

Optical absorption spectra of PMMA irradiated by xenon and silver ions at various doses are shown in Figure 2. It can be seen in Figure 2a that with increasing xenon ion dose the absorption of the polymer in the visible (especially in the close-to-UV) range increases monotonically. This indicates the presence of radiation-induced structural defects in the PMMA. The implantation by silver ions not only generates such radiation defects but also causes the nucleation and growth of MNPs. Therefore, along with the absorption intensity variation as in Figure 2a, a selected absorption band associated with silver nanoparticles is observed (Figure 2b). For the lowest ion dose, the maximum of this band is near 420 nm; it shifts to the red spectral region (up to ~600 nm) with increasing dose, simultaneously with a band

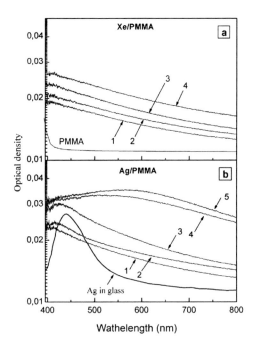

Figure 2. Optical density spectra of PMMA irradiated by (a) xenon and (b) silver ions with doses of (1) $0.3 \cdot 10^{16}$, (2) $0.6 \cdot 10^{16}$, (3) $2.5 \cdot 10^{16}$, (4) $5.0 \cdot 10^{16}$, and (5) $7.5 \cdot 10^{16}$ ion/cm². Also shown is a spectrum of SiO_2 implanted with silver ions ($5.0 \cdot 10^{16}$ ion/cm²).[14]

broadening. The maximum of this band is not sharp, although it is definitely related to the SPR effect in the silver nanoparticles. Such broad SPR absorption is untypical for silver nanoparticles in PMMA. When silver particles are synthesized in PMMA by the convection melting technique,[13] the SPR band is very sharp, unlike our experiment. Figure 2b also shows the *OD* spectrum for inorganic silica glass irradiated by silver ions under the same implantation conditions as.[14] The particle size distributions in SiO$_2$ and PMMA are nearly the same. SiO$_2$ has a refractive index close to that of PMMA. However, the absorption of Ag nanoparticles in the glass (Figure 2b) is much more narrow and intense than the absorption of the MNPs in the polymer.

4. Simulation of the Optical Extinction

The attenuation (extinction) of an optical wave propagating in a medium with MNPs depends on the SPR absorption and the light scattering efficiency. The wavelength of the optical radiation, the particle size, and the properties of the environment are governing factors in this process. Within the framework of classical electrodynamics (the Maxwell equations), the problem of the interaction between a plane electromagnetic wave and a single spherical particle was exactly solved in terms of the optical constants of the selected materials by Mie.[2,12] The complex values of the optical constants ε_{Ag}[15] and ε_{PMMA}[16] in the visible range were used The extinction was calculated for particles of sizes between 1 and 10 nm to be in consistence with the experimental sizes (Figure 1).

As a first step of simulation, consider the simplest case where Ag nanoparticles are incorporated into the PMMA. Simulated extinction spectra of Ag nanoparticles embedded in a polymer matrix are shown in Figure 3. The extinction feature is a very wide band, which covers the entire spectral range. In the given range of particle sizes, the position of the SPR absorption maximum (near 440 nm) is almost independent of the particle size. At the same time, the intensity of the extinction band grows while the band itself somewhat narrows with increasing particle size. Comparing the modeled and experimental spectra, it can be seen that Figure 3 refers to the situation where PMMA is implanted by silver ions with doses between $0.33 \cdot 10^{16}$ and $2.5 \cdot 10^{16}$ ion/cm^2 (Figure 2b, curves 1–3). This range of doses corresponds to the early stage of MNP nucleation and growth in the *OD* spectral band with a maximum between 420 and 440 nm. Thus, it is possible to conclude that ion implantation in this range of doses results in the formation of Ag nanoparticles, as also revealed by TEM. However, at higher implantation doses, the measured *OD* spectra and the modeled spectra shown in Figure 3 diverge.

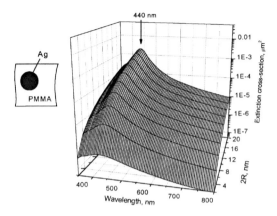

Figure 3. Simulated optical extinction spectra for silver nanoparticles embedded in PMMA vs. particle size.

To explain the experimental dependences corresponding to high-dose silver implantation into PMMA, the difference between implantation into polymers and inorganic materials (silicate glasses, crystals, etc.) should be considered. The most important distinction is that, as the dose increases, so does the number of dangling chemical bonds in the polymer along the track of an accelerated ion. As a consequence, gaseous hydrogen, low-molecular hydrocarbons (e.g. acetylene), CO, and CO_2 evolve from the matrix.[17] In particular, ion-irradiated PMMA loses methoxy groups $(HCOOCH_3)$.[18] The evolution of several organic fractions leads to the accumulation of carbon in the irradiated polymer layer, and radiation-induced chemical processes may cause chain linking. Eventually, an amorphous hydrogenated carbon layer is produced.

To take into account this specific phase structure of the irradiated polymer, it is of interest to analyse the optical properties (extinction) of Ag nanoparticles embedded in an amorphous carbon matrix (C-matrix). For this system, the extinction cross section spectra vs. particle size dependence was simulated (Figure 4) in the same way as for the MNP–PMMA system using the complex optical constant ε_C for amorphous carbon.[19] As before (Figure 3), throughout the entire particle size interval, the extinction spectra exhibit a single broad band, which covers the visible range, but with a maximum at a longer wavelength (~510 nm). This wavelength position of the maximum, which is observed upon changing the matrix, may be assigned to the longer wavelength *OD* band in the experimental spectra for PMMA, which arises when the Ag ion dose exceeds $2.5 \cdot 10^{16}$ ion/cm^2 (Figure 3b; curves 3, 4). It seems that this spectral shift can be associated with the fact that the pure polymeric environment of the Ag nanoparticles turns into amorphous carbon as the implantation dose rises. The broader extinction bands of the C-matrix (Figure 4), compared with those of PMMA (Figure 3) also count in favor of this supposition, since a broadening of the extinction bands is observed in the experiments as well (Figure 2b). In a number of experiments, however, the carbonization of the polymer surface layer depended on the type of the organic material and the accelerated ions, as well as on the implantation parameters, and was completed at doses of $(0.5-5.0)\cdot10^{16}$ ion/cm^2 but the entire material was not carbonized. The carbon fragments may reach several tens of nanometers in size.[17] Thus, the assumption that the polymer irradiated is completely carbonized, which was used in the simulation (Figure 4), does not completely correspond to the real situation. Therefore, extinction spectra for nanoparticles represented as a silver core covered by a carbon shell in an insulating matrix (PMMA) will be analyzed in terms of the Mie relationships for shelled cores.[20]

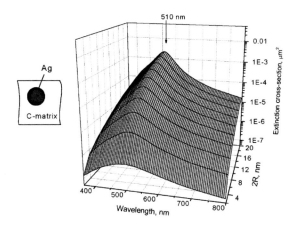

Figure 4. Simulated optical extinction spectra for silver nanoparticles embedded a the C-matrix vs. particle size.

Optical extinction spectra for Ag nanoparticles with a fixed size of the core (4 nm) and a varying thickness of the carbon shell (from 0 to 5 nm) are shown in Figure 5. The maximum of the SPR bands of the particles is seen to shift from 410 nm (uncovered particle, Figure 5) to approximately 510 nm. Simultaneously, the SPR band intensity decreases, while the UV absorption increases, so that for a shell thickness of 5 nm the absorption intensity at 300 nm exceeds the SPR absorption of the particles. Both effects (namely, the shift of the SPR band to longer wavelengths and the increased absorption in the near ultraviolet) agree qualitatively with the variation of the experimental *OD* spectra (Figure 4b) when the implantation dose exceeds $2.5\cdot10^{16}$ ion/cm^2. Thus, our assumption that the increase of the carbonized phase fraction with implantation dose is responsible for the variation of the *OD* spectra (Figure 2b), is sustained by the simulation of the optical extinction for complex particles (Figure 5).

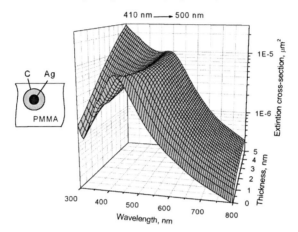

Figure 5. Simulated optical extinction spectra for 4-nm silver nanoparticles with a carbon shell that are placed in a PMMA matrix vs. sheath thickness.

When analyzing the optical properties of nanoparticles embedded in a medium, effects arising at the particle–matrix interface, such as static and dynamic redistributions of charges between electronic states in the particles and the environment in view of their chemical constitution should be taken into account.[21] Consider first the static charge redistribution. When an atom is deposited (adsorbed) on the MNP surface, the energy levels of this atom change their positions compared with those of the free state.[22] When the number of adsorbed matrix atoms becomes significant, their contact generates a wide distribution of density of states. Additionally, the adsorbed atoms are separated from surface atoms of the metal by a tunnel barrier. The gap between the energy positions of the adsorbed atoms and the Fermi level of the particles depends on the type of the adsorbate. The overlap between the energy positions of the matrix atoms and the energy positions of the silver surface atoms depends on the rate with which the electrons tunnel through the barrier. Accordingly, the conduction electron density in the particles embedded will change compared with that in the particles placed in vacuum (without adsorbates): it decreases if the electrons tunnel toward the adsorbed atoms but increases when the electrons tunnel in the reverse direction. Eventually, equilibrium between the particle and the matrix is reached; i.e. a constant electrical charge (Coulomb barrier) forms at the nanoparticle surface.

Such a static charge redistribution due to the deposition of an adsorbate on the particle surface and the respective change in the electron concentration in the MNPs can also be observed in the SPR absorption spectra.[2] The incorporation of Ag nanoparticles into a carbon matrix of C_{60} fullerenes (or the deposition of carbon on the nanoparticle surface) reduces the concentration of $5sp$ electrons in the particle roughly by 20%, since they are trapped by matrix molecules.[21] It was shown that the decrease of the electron concentration shifts the MNP extinction spectrum toward longer wavelengths. The shift of the SPR extinction band to the longer wavelengths with increasing implantation dose in the present experiment (Figure 2) may also be explained by the formation of carbon shells around the silver nanoparticles, which trap conduction electrons.

The dynamic charge variation with time at the particle–matrix interface causes the electron concentration in the particle to fluctuate. This fluctuation influences directly the SPR relaxation. The lifetime of excited conduction electrons in the particle defines the SPR spectral width. Here, the contribution from electrons scattered at the interface (because of restrictions imposed on the electron free path[2]) adds up with the dynamic charge variation at the interface. Thus, the temporal capture of conduction electrons from the particle broadens the SPR-related extinction spectra. Such an effect was demonstrated with silver nanoparticles embedded in a C_{60} matrix.[21] Silver nanoparticles in a carbon matrix exhibit a much broader SPR band than those in free space. We may therefore suppose that, as the dose rises, the dynamic charge redistribution may broaden the SPR spectra of silver nanoparticles synthesized by ion implantation in PMMA. The implantation carbonizes the irradiated layer with increasing dose and raises the amount of acceptor levels on the MNP surface, which change the relaxation time of excited electrons.

ACKNOWLEDGEMENTS

The author is grateful to the Alexander Humboldt Foundation (Germany) and the Austrian Scientific Foundation in the frame of the Lise Meitner program for financial support. This study was also sponsored in part by the Russian Foundation for Basic Research (grant no. 04-02-97505-p) and the OFN project of the Russian Academy of Sciences "Advanced Materials and Structures".

References

1. C. Flytzanis, F. Hache, M.C. Klein, D. Ricard, and P. Rousignol, *Nonlinear Optics in Composite Materials* (Elsevier Science, Amsterdam, 1991).
2. U. Kreibig and M. Vollmer, *Optical Properties of Metal Clusters* (Springer, Berlin, 1995).

3. S.N. Abdullin, A.L. Stepanov, Yu.N. Osin, and I.B. Khaibullin, Surf. Sci. 395, L242 (1998).
4. A. Heilmann, *Polymer Films with Embedded Metal Nanoparticles* (Springer, Berlin, 2003).
5. P.T. Townsend, P.J. Chandler, and L. Zhang, *Optical Effects of Ion Implantation*, (Cambridge University Press, Cambridge, 1994).
6. J. Davenas, A. Perez, P. Thevenard, and C.H.S. Dupuy, Phys. Status Solidi A 19, 679 (1973).
7. G.W. Arnold, J. Appl. Phys. 46, 4466 (1975).
8. N.C. Koon, D. Weber, P. Pehrsson, and A.I. Sindler, Mater. Res. Soc. Symp. Proc. 27, 445 (1984).
9. A.L. Stepanov, S.N. Abdullin, R.I. Khaibullin, V.F. Valeev, Yu.N. Osin, V.V. Bazarov, and I.B. Khaibullin, Mater. Res. Soc. Symp. Proc. 392, 267 (1995).
10. A.L. Stepanov, S.N. Abdullin, V.Yu. Petukhov, Yu.N. Osin, R.I. Khaibullin, and I.B. Khaibullin, Phil. Mag. B 80, 23 (2000).
11. Y. Wu, T. Zhang, A. Liu, and G. Zhou, Surf. Coat. Technol. 157, 262 (2002).
12. G. Mie, Ann. Phys. (Leipzig) 25, 377 (1908).
13. W. Scheunemann, and H. Jäger, Z. Phys. 265, 441 (1973).
14. A.L. Stepanov and D.E. Hole, Recent Res. Dev. Appl. Phys. 5, 1 (2002).
15. M. Quinten, Z. Phys. B 101, 211 (1996).
16. M.A. Khashan, and A.Y. Nassif, Opt. Commun. 188, 129 (2001).
17. D.V. Sviridov, Russ. Chem. Rev. 71, 315 (2002).
18. B. Pignataro, M.E. Fragala, and O. Puglisi, Nucl. Instrum. Meth. B 131, 141 (1997).
19. E.D. Palik, *Handbook of Optical Constants of Solids* (Academic, London, 1997).
20. J. Sinzig and M. Quinten, Appl. Phys. A 58, 157 (1994).
21. U. Kreibig, M. Gartz, and A. Hilger, Ber. Bunssenges. Phys. Chem. 101, 1593 (1997).
22. J. Hölzl, F. Schulte, and H. Wagner, *Solid Surface Physics* (Springer, Berlin, 1979).

RAPID LASER PROTOTYPING OF POLYMER-BASED NANOPLASMONIC COMPONENTS

A.L. STEPANOV[*1], R. KIYAN[2], C. REINHARDT[2],
A. SEIDEL[2], S. PAS-SINGER[2], B.N. CHICHKOV[2]
[1]*Kazan Physical-Technical Institute, Russian Academy of Sciences, Sibirsky trakt 10/7, 420029 Kazan, Russia*
[2]*Laser Zentrum Hannover, Hollerithallee 8, 30419 Hannover, Germany*

Abstract. Renewed and growing interest in the field of surface plasmon polaritons (SPPs) comes from a rapid advance of nanostructuring technologies. The application of two-photon polymerization technique for the fabrication of dielectric and metallic SPP-structures, which can be used for localization, guiding, and manipulation of SPPs waves on a subwavelength scale, is studied. This technology is based on nonlinear absorption of near-infrared femtosecond laser pulses. Excitation, propagation, and interaction of SPP waves with nanostructures are controlled and studied by leakage radiation imaging. It is demonstrated that created nanostructures on metal film are very efficient for the excitation and focusing of SPPs. Examples of passive and active SPP components are presented and discussed.

Keywords: surface plasmon polaritons; femtosecond laser; nanoparticles; two-photon polymerization; waveguides; plasmon leakage radiation

1. Introduction

Miniaturization is a central guideline in modern science and technology. Semiconductor electronic devices, which are fabricated today in industry, are characterized by low heat dissipation and limited propagation speed of electrical signals.[1,2] An alternative, allowing a solution of these problems, is to use optical data transmission. Therefore, miniaturized light sources and detectors are required as key elements for modern telecommunication and

---

[*]a.stepanov@lzh.de

J.P. Reithmaier et al. (eds.), *Nanostructured Materials for Advanced Technological Applications*, 163
© Springer Science + Business Media B.V. 2009

interconnector systems to combine with or to replace conventional electronic devices, sensors, and actuators. Nowadays, both improved preparation and analytical methods allow the realization of structures in the range of *nanometers*. Many physical properties of material structures are dramatically modified when their sizes are reduced below 1 μm and approach the length scales of the effective light wavelengths. It is expected that the understanding and application of these novel effects will allow the development of a wealth of new methods and devices for applications in medicine, electronics, photonics, and material science.

Research at the nanoscale level has already been conducted for many years, whereas in optics only first attempts towards nanostructures have been undertaken recently. The miniaturization of conventional optical elements such as, for example, waveguides as well as the resolving power of optical microscopes is restricted by the diffraction limit. In the visible spectral range, this limit corresponds to about half a micron. A promising solution to overcome the diffraction limit is to use optics based on light coupled to collective electron oscillations at the interface of a metal and a dielectric.[3,4] Such mixed photon/electron surface modes are known as surface plasmon polaritons (SPPs).[5] When SPP waves are excited on metal nanostructures, the dimensions of the surface structures rather than the exciting light wavelength determine the special SPP electromagnetic field profile. This allows to overcome the diffraction limit and the realization of optical devices with nanoscale dimensions.

As a propagating electromagnetic waves, SPPs in an extended metal structure can be used for the guiding of light fields, corresponding to *signal transfer* or *optical addressing*.[6,7] A dispersion relation characterizes SPPs waves and predicts that the plasmon wavelength is smaller than the vacuum light wavelength for any given light frequency.[5,7] The metals of choice are usually gold or silver, as these metals possess high concentrations of free conducting electrons and show SPP wave modes in the visible or near-infrared spectral range.

In summary, propagating SPPs in surface nanostructures can be used for electromagnetic signal propagation and nanoscale light field manipulation. Propagating plasmon effects can lead to the realization of nanoscale plasmon-based (*plasmonic*) optical devices. SPP waves can provide an interface between macrooptics and plasmonic nanocomponents. The following examples demonstrating the broad prospects of SPP nanodevices:

➢ In analogy to electronics higher integration densities and easier fabrication techniques allow the realization of 2D and 3D devices using SPPs modes.

➢ The ongoing increase of computer processor frequencies can soon lead to a situation when the processor will be working faster than the communication speed via electrical signals in metal wires. Faster intra- and inter-chip connections are required. Conventional optical waveguides cannot be miniaturized to meet the demand of present and future integration densities. Instead of that, SPP waveguides could serve as optical interconnects with nanoscale sections.

➢ As mixed photon/electron oscillations, SPP waves could offer an elegant solution for the transformation of optical and electrical signals in a new generation of electro-optical devices.

The aim of this article is to present a short overview of resent progress in the field of SPPs with nanostructures fabricated by a state-of-the-art femtosecond laser technology, namely two-photon polymerization (2PP).[8,9]

2. Sample Preparation

The optical SPP elements were produced by 2PP technique. Dielectric structures were fabricated upon a glass substrate by 2PP of inorganic-organic hybrid polymer ORMOCER. This polymer can be polymerized using a radical photoinitiator. For 2PP a femtosecond laser (Spectra-Physics Model Tsunami) was used. This system delivers pulses at the wavelength of 780 nm with a duration of 80 fs and a repetition rate of 80 MHz. The average power of 40 mW was applied. A schematic of the experimental set-up used for the sample fabrication is illustrated in Figure 1.

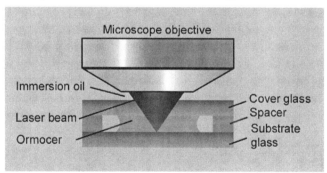

Figure 1. Schematic of the experimental set-up for sample preparation by 2PP.

Femtosecond laser pulses were focused by an immersion oil objective (Nikon, 100×, numerical aperture 1.3). A liquid polymer droplet was sandwiched between the glass substrate and a cover glass. A plastic frame fixed their separation. For the fabrication of the structures, the laser beam was focused through the cover glass and the ORMOCER layer on the substrate glass surface. During the structuring, the laser beam was scanned over the sample surface by a galvo scanner. Samples were washed in isobutyl-methylketon (4-metheyl-2-pentanone) to remove liquid non-irradiated polymer. Then a 50 nm thin gold film was deposited on the dried samples with surface nanostructures by electron sputtering (Figure 2).

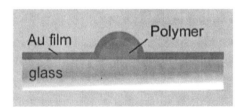

Figure 2. Schematic cross-section of the fabricated sample.

3. Leakage Radiation Imaging

The intensity decay length of a plane SPP wave in a metal film located between two dielectric media defines its intrinsic decay length $L_{int} = 1/2k''$, where k'' is defined as the imaginary part of the complex surface plasmon wave vector $k_{SPP} = k' + ik''$. Intrinsic losses are caused by inelastic scattering of conduction electrons, scattering of electrons at interfaces, and leakage radiation (LR). Leakage radiation is emitted from the interface between the metal thin film and a higher-refractive-index dielectric medium (substrate), e.g. glass.[5] When the electromagnetic plasmon field crosses the metal film and reaches the substrate, leakage radiation appears at a characteristic angle with respect to the interface normal. This radiation permits the detection of SPPs in the far-field[7,10,11]; this approach was used in our study.

The experimental set-up used for leakage radiation imaging is shown in Figure 3. For the local excitation of SPPs, the linearly polarized light from a Ti:Sapphire laser with wavelengths between 760 and 900 nm is focused through a microscope objective onto the surface of nanostructured gold or silver films. LR is detected by means of an oil immersion objective in optical

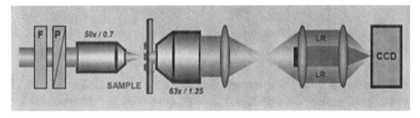

Figure 3. Experimental scheme of a leakage radiation microscope. SPPs are excited by laser light on a structured metal film on a glass substrate. F (gray) filter, P polarizer.

contact with the glass substrate. By focusing this objective on the metal/
glass interface, this radiation can be collected and either imaged or focused
onto a photo-detector. The distribution of SPPs is photo- or videotaped by
a charge-coupled-device (CCD) camera. Each pixel element of an image
captures the flux of LR emerging from the corresponding position at the
metal-air interface. The latter is a direct quantitative measure of the SPP
intensity at this position.

4. Nanoplasmonic Components

SPP wave focusing on the flat metal surface was studied with a curved (circular) line of nanoparticles. Scanning electron microscope (SEM) images of
these structures are presented in Figure 4.

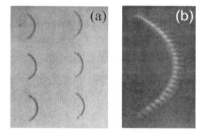

Figure 4. SEM image of (a) a sample with several surface structures, (b) an individual structure. The radius of the curved chains of nanoparticles is equal to 10 μm. The particle in-plane
size (diameter) and inter-particle distance are 350 and 850 nm, respectively.

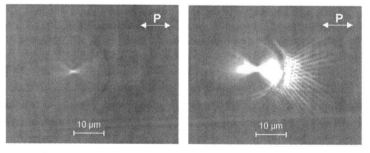

Figure 5. Experimental LR images of a SPP focused by the surface structure. The arrow
indicates the incident light polarization. The laser intensity is higher on the left image.

Focusing of SPPs by curved chains of nanoparticles was observed with
by LR for moderate sizes of the incident laser beam (Figure 5, left). The
SPP excitation was obtained by a focusing laser beam (at normal incidence)
directed to the centre of the chain. The exciting laser spot was in this case
(Figure 5, right) equal to approximately 10 μm in diameter. Interference of

divergent SPPs excited by different particles results in SPP focusing with a focal point located at the centre of curvature of the nanoparticle chain. As can be seen, the focal SPP waist is very well localized and has relatively small width. In addition to the SPP focusing one can see that there is a system of SPP rays on the other side of the nanoparticle chain. The origin of this effect is again the interference between the SPPs produced by the individual particles. Details of a numerical modelling of plasmon focusing efficiency are presented in Ref. 12.

When the incident laser spot is decreased down to 3 μm, the SPP focusing effect becomes less pronounced (Figure 6), since only a few scattering centers are excited by the laser beam. The position where the laser beam is focused on the structure is indicated in the images by the dashed circles. The SPP focal waist broadens resulting in a nearly parallel and relatively narrow beam. In such case it is possible to control the propagation direction of the narrow SPP beam (always directed perpendicular to a local tangent of the particle chain) simply by changing the position of the incident laser spot along the chain (Figure 6). This effect is useful for SPP manipulations in complex micro-optical elements utilizing chains of separated scatterers.

Figure 6. Experimental LR images of the SPPs excited by the surface structure. The diameter of the incident light spot for the SPP excitation (dashed circles) is equal to 3 μm. (a) The exciting laser spot is at the centre of the structure, (b) and (c) the laser spot is shifted away from the centre. The arrow indicates the incident light polarization.

SPP device technology makes use of the properties of surface waves; electromagnetic excitations propagating along and bound to the interface between a metal and a dielectric. The intrinsically two-dimensional nature of SPPs is ideally suited for the realization of a planar waveguide, which can be implemented from polymers. The guiding properties of dielectric SPP structures were studied with straight 60 μm long ridges fabricated by 2PP. The SPPs are launched at the gold–air interface by a focused laser beam at a wavelength tip of 1,520 nm in the Kretschmann excitation configuration. The topography of the polymer ridges and the SPP near-field distribution imaged with a scanning near-field optical microscope are shown in Figure 7. It can be seen that a relatively wide SPP beam is partially coupled into, propagating along, and coupled out of the polymer ridge. At the same time, the rest of the (diverging) SPP beam continues to propagate along the air–gold interface and is reflected by the neighboring ridge.

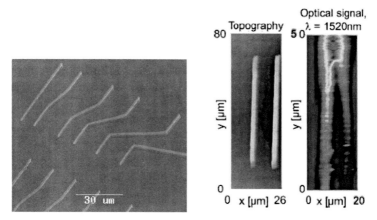

Figure 7. SEM images of dielectric stripes and bends fabricated on a gold-covered surface (left and centre) and SPP near-field intensity distribution (right) of waveguide demonstrating guiding of SPP waves.

2PP provides the possibility to realize and investigate plasmon propagation in 3D surface structures. The schematic of a first experiment is shown in Figure 8, where plasmons are excited on the left defect structure and propagate over a ramp structure. In this case LR from the plasmons on the ramp structure is less visible compared to the leakage radiation from SPPs on plane surfaces.

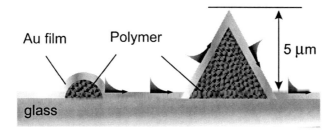

Figure 8. Schematic of a 3D surface structure for SPP propagation.

The LR image together with the corresponding SEM image of the test structure are shown in Figure 9. This result clearly indicates that SPPs can propagate on smoothly curved surfaces with curvatures larger than the SPP wavelength.

In conclusion, laser fabrication of dielectric 2D and 3D surface plasmon polariton (SPP) structures on metal surfaces by two-photon induced poly-merization of a high refractive index inorganic–organic hybrid polymer has been studied. Optical properties of the fabricated dielectric SPP structures have been investigated by leakage radiation microscopy of the propagating SPPs. Effective excitation and focusing of the SPPs have been demonstrated with dielectric structures. The demonstrated results on excitation and manipulation of SPP fields and the simplicity of laser fabrication technique provide interesting prospects for the realization of future plasmonic devices.

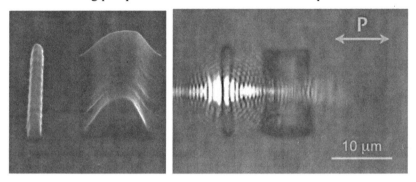

Figure 9. SEM image of the 3D structure fabricated by 2PP (left) and LR image of SPPs excited on the ridge and propagating over the 3D structure (left).

ACKNOWLEDGEMENTS

Author is grateful to the Alexander Humboldt Foundation (Germany) and the Austrian Scientific Foundation in the frame of the Lise Meitner program for financial support. The authors would like to thank the European Network of Excellence "Plasmo-Nano-Devices" and the "Plasmocom" projects for financial support. This study was also sponsored in part by the Russian Foundation for Basic Research (grant no. 04-02-97505-p) and the OFN project of the Russian Academy of Sciences "Advanced Materials and Structures".

References

1. H. Wong, and H. Ieai, Phys. World 18, 40 (2005).
2. B. Hitz, Photon. Spectra 39, 52 (2005).
3. W.L. Barnes, A. Dereux, and T.W. Ebbesen, Nature 424, 824 (2003).
4. E. Ozbay, Science 311, 189 (2006).
5. H. Raether, *Surface Plasmons* (Springer Tracts in Modern Physics, Berlin, 1988).
6. J.-C. Weeber, Y. Lacroute, and A. Dereux, Phys. Rev. B 68, 115401, 1 (2003).

7. A. Drezet, A. Hohenau, D. Koller, A.L. Stepanov, H. Ditlbacher, B. Steinberger, F.R. Aussnegg, A. Leitner, and J.R. Krenn, Mater. Sci. Eng. B 149, 220 (2008).
8. J. Serbin, A. Egbert, A. Ostendorf, B. Chichkov, R. Houbertz, G. Domann, J. Schulz, C. Cronauer, L. Frohlich, and M. Popall, Opt. Lett. 28, 301 (2003).
9. J. Serbin, A. Ovsianikov, B. Chichkov, Opt. Express 12, 5221 (2008).
10. A.L. Stepanov, J.R. Krenn, H. Ditlbacher, A. Honenau, A. Drezet, B. Steinberger, A. Leitner, and F.R. Aussenneg, Opt. Lett. 30, 1524 (2005).
11. R. Kiyan, C. Reinhardt, S. Passinger, A.L. Stepanov, A. Hohenau, J.R. Krenn, and Chichkov, B.N., Opt. Express 15, 4205 (2007).
12. A.B. Evlyukhin, A.L. Stepanov, R. Kiyan, and B.N. Chichkov, J. Opt. Soc. Am. B 25, 1011 (2008).

DESIGN OF CERAMIC MICROSTRUCTURE OF NANOSCALED TRANSITIONAL ALUMINA

E. FIDANCEVSKA[*1], J. BOSSERT[2], V. VASSILEV[3],
R. ADZISKI[1], M. MILOSEVSKI[1]
[1]Faculty of Technology and Metallurgy, Ruger Boskovic 16,
University St. Cyril and Methodius, 1000 Skopje,
FYR Macedonia, Republic of Macedonia
[2]Institute of Materials and Technoogy.Friedrich Schiller
University, Lobdergraben 32, Jena, Germany
[3]University of Chemical Technology and Metallurgy,
Kl.Ohridski Blvd 8, Sofia, Bulgaria

Abstract. Nanoscaled transitional alumina (γ-, δ-, θ- $\rightarrow$ α-phase) consolidated at P = 500 MPa and T = 1,500°C for 30 min gave compacts with a density of 75% of theoretical density (TD) and a vermicular microstructure. The temperature of the transition to α-Al_2O_3 was 1,304°C. After mechanical activation by attriting, the transformation temperature to α-Al_2O_3 was reduced to 1,080°C. The density of the compacts consolidated at P = 500 MPa and T = 1,500°C for 30 min was 95% TD. The grain size distribution reached from 100 to 600 nm.

Keywords: nanoscaled alumina; mechanical activation; microstructure

1. Introduction

Fabrication of nanostructured alumina from transitional alumina powders with a high specific surface area and ultra fine particle sizes is characterized by rapid grain growth. Generally, the grain growth is accompanied by a vermicular microstructure consisting of a network with large pores. It is also known that the transformation of alumina proceeds by nucleation and

---
[*]emilijaf@tmf.ukim.edu.mk

J.P. Reithmaier et al. (eds.), *Nanostructured Materials for Advanced Technological Applications*,   173
© Springer Science + Business Media B.V. 2009

growth processes. Kumagai et al.[1] showed that the nucleation step may effectively be eliminated by supplying nuclei to the system. They showed that 5×10^{13} seeds/cm^3 or 1.5 wt % 0.1 μm α-Al$_2$O$_3$ particles represent an optimum concentration for the θ- to α-Al$_2$O$_3$ transformation. Also, structural defects formed by mechanical activation act as heterogenous nucleation sites in the system, controlling the transformation to α-Al$_2$O$_3$ and enhancing the densification. The effect of milling on the appearance of surface defects in alumina is discussed in Refs. 1–5.

The present investigation is focused on the fabrication of dense α-Al$_2$O$_3$ nanoceramics, starting from nanocrystalline transitional alumina. Investigation of the role of mechanical activation on the transformation to α-Al$_2$O$_3$ and the design of the microstructure of nanoscaled alumina were the final goals of this paper.

2. Experimental

The nano aluminium oxide powder (99.88%) consisting of δ-, γ-, θ-, and α-Al$_2$O$_3$ used in this study was purched from IBU-tech, Germany. The powder morphology was observed via transmission electron microscopy (TEM). Differential thermal analysis and thermogravimetry (DTA/TG) investigations were performed by a NETZSCH STA 409C in air in the temperature range from 20–1,500°C using a heating rate of 10 Kmin^{-1}. Mechanical activation (wet milling) of the starting powder was performed by a attrition mill (NETZSCH) using alumina balls at pH 5 for 30, 60 and 90 min. The dried powder was consolidated by cold isostatical pressing (CIP) at 500 MPa (WEBER PRESSEN KIP 500 E). Sintering was realized with a constant heating rate of 10 Kmin^{-1} in the interval from RT-1,500°C/30 min. A high-resolution scanning electron microscope (Leica IS 110, Germany) was employed to observe the microstructure of sintered samples.

3. Results and Discussion

The morphology of the powder presented in Figure 1 shows that the size of primary alumina particles was 10–60 nm. The specific surface area was 111.5 m^2/g.

A part of these nanosized primary particles in the as-received powder was aggregated; the aggregate size ranged from 100–600 nm.

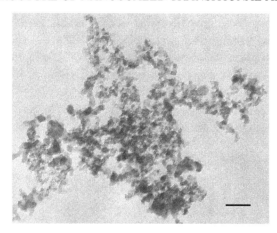

Figure 1. TEM photograph of the starting transitional alumina powder (the bar represents 8 nm).

A DTA analysis of the starting powder showed one exo peak at 1,304°C which corresponds to the α-Al₂O₃ transformation (Figure 2, left, lower curve). Cold isostatic pressing at 500 MPa shifted the exo peak to 1,126.4°C (Figure 2, left, upper curve). The transformation temperature is reduced by 177.4°C. After sintering in the interval RT-1,500°C for 30 min, the density was 2.9 g/cm³, which is 0.73 TD. The open porosity was 26%.

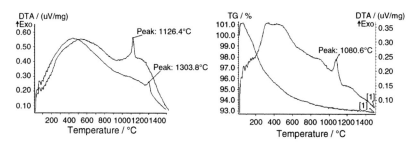

Figure 2. Left: DTA analysis of as received powder (lower curve) and cold isostatically pressed powder (upper); right: DTA/TG of mechanically activated and pressed powder at 500 MPa.

The specific surface area of the mechanically activated powders was 114.5 m²/g after 30 min milling, 108.5 m²/g after 60 min and 80.6 m²/g after 90 min. Mechanical activation causes changes in the structure of a material, which have a direct influence on properties depending on the structure, e.g. transport and reactive properties.[2] According to Dynys and Halloran,[6] after milling the nucleation frequency increased by four orders of magnitude. After 30 min milling a uniform distribution of α-Al₂O₃ is present, besides the metastable δ-, γ- and θ-Al₂O₃.

DTA of the powder mechanically activated and pressed at 500 MPa showed the phase transformation to α-Al$_2$O$_3$ at 1,080.6°C (Figure 2, right).

The microstructures of sintered samples of the starting and the mechanically activated powders are shown in Figure 3. Figure 3 (right) shows a vermicular network with elongated pores with sizes of 0.5–1.0 μm. The density of this system was 0.73 TD. Figure 3 (left) shows a compact of Al$_2$O$_3$ with a density 0.95 TD and grain sizes from 100–600 nm.

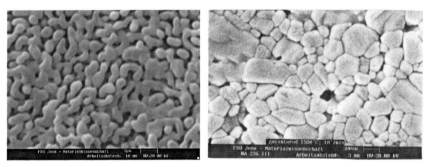

Figure 3. SEM micrographs of a fractured sample sintered at 1,500°C; left: no mechanical activation (bar 1 μm); right: after mechanical activation (bar 300 nm).

4. Conclusion

Nanosized transitional alumina powder consolidated by an isostatic pressure of 500 MPa and sintered at RT-1,500°C/30 min yielded compacts with a vermicular structure and a density of 0.73 TD. The elongated pores have sizes of 0.5–1 μm. Using mechanical activation, realised by attriting for 30 min, an active nano-powder was obtained, in which the transformation to α-Al$_2$O$_3$ was taking place at 1,080.6°C. The consolidated (RT-1,500°C for 30 min) system has a density of 0.95 TD and a grain size of 100–600 nm. The obtained alumina represents a potential bioinert material.

References

1. M. Kumagai and G. Messing, J. Am. Ceram. Soc. 68, 500 (1985).
2. S. Kirchuchi, T. Ban, K. Okada, N. Otsuka, J. Mat. Sci. Lett. 11, 471 (1992).
3. J. Bossert and E. Fidancevska, Sci. Sintering 39, 117 (2007).
4. R. Oberacker, S. Poehmitzsh, H. Hofius, Cfi/Ber. DKG 78, 45 (2001).
5. M. Nikolic, N. Nikolic, S. Radic, Sci. Sintering 32, 149 (2000).
6. W. Dynys and J.W. Halloran, J. Am. Ceram. Soc. 65, 442 (1982).

5. PREPARATION AND CHARACTERIZATION OF NANOSTRUCTURED MATERIALS

5.1. NANOTUBES

CARBON NANOTUBE COMPOSITE MATERIALS:

OPPORTUNITIES AND PROCESSING ISSUES

W.K. MASER[*1], A.M. BENITO[1], P. CASTELL[1], R. SAINZ[1],
M.T. MARTÍNEZ[1], M. NAFFAKH[2], C. MARCO[2],
G. ELLIS[2], M.A. GÓMEZ[2]
[1]Instituto de Carboquímica (CSIC), Department
of Nanotechnology, C/Miguel Luesma Castán 4, E-50018
Zaragoza, Spain
[2]Instituto de Ciencia y Tecnología de Polímeros (CSIC),
Department of Physics and Polymer Engineering, C/Juan
de la Cierva 3, E-28006 Madrid, Spain

Abstract. In this article we present a general introduction into the field of carbon nanotubes composites. The opportunities for achieving novel high performance materials with superior properties are highlighted and the challenges to be overcome are discussed. Here, the focus lies on the following key issues: Nanotubes' dispersion and aggregation, matrix compatibility, load-transfer and interface interactions. Processing strategies for different classes of polymers towards advanced functional nanomaterials are presented.

Keywords: nanocomposites; carbon nanotubes; thermoplastics; electroactive polymers

1. Introduction

Today, nanoscience and nanotechnology provide powerful concepts for the development of novel high performance materials. Here nanostructuring, i.e. the manipulation of matter at the molecular level, is an important bottom-up strategy to develop materials with improved and tailored properties. A nanostructuring concept of increased importance is the use of nanofillers yielding in highly functional nanocomposite materials, especially in the

---

[*]wmaser@icb.csic.es

J.P. Reithmaier et al. (eds.), *Nanostructured Materials for Advanced Technological Applications*, 181
© Springer Science + Business Media B.V. 2009

case of polymer matrices and the corresponding polymer nanocomposites. The idea of employing nanofillers is based on the following: Firstly, they should provide novel functionalities to the polymer without sacrifying the polymers inherent characteristics and adding excessive weight, and secondly, they also should induce changes of the arrangement of the polymer chains which in itself already could results in significant materials advantages.

Of particular interest here are carbon nanotubes (CNTs),[1,2] fundamental nanoscale objects of unique structure and very special mechanical, thermal, electrical, electronic and optical properties. Their dimensions and properties make them excellent candidates for achieving particular interactions with polymer chains and highly favorable synergetic effects. Thus, CNTs are highly attractive nanobuilding blocks for obtaining novel nanocomposites with superior properties.

With this article we provide the reader with the background knowledge on carbon nanotube composite materials to understand, on one hand, their unique opportunities, and on the other hand, the various challenges to be overcome and the typical processing approaches in this broad and fascinating field of research.

2.   Carbon Nanotubes: Fundamental Nanoscale Objects

2.1. STRUCTURE

A carbon nanotube (CNT) can be described as a seamlessly rolled-up sheet of graphene, resulting in an open tubular structure composed of carbon atoms arranged in a hexagonal network. The tubular structure is closed by the inclusion of six pentagonal defects into the hexagonal network at each end of the open cylinder to form the corresponding semi-fullerene[3] caps. Figure 1 explains the basic structures of carbon nanotubes.

There are three different ways in which a graphene sheet can be rolled up onto itself thus leading to the three basic forms of CNTs, i.e. the family of single-wall carbon nanotubes (SWNTs): (i) Armchair tubes: These are obtained when rolling-up the sheet in such a way that carbon–carbon bounds are perpendicular to the cylinder axis. Looking at the cross-section of such a tube one observes an "armchair"-like structure of the carbon atoms. (ii) Zig-zag tubes: Here, carbon–carbon bounds are parallel to the tube axis and the cross-section results in a "zig-zag"-like pattern of carbon atoms. (iii) Chiral nanotubes: Here, the hexagons of the original graphene sheet are winding-up around the tube axis in a helical way letting form the carbon atoms a helical (chiral) pattern along the length of the tube.

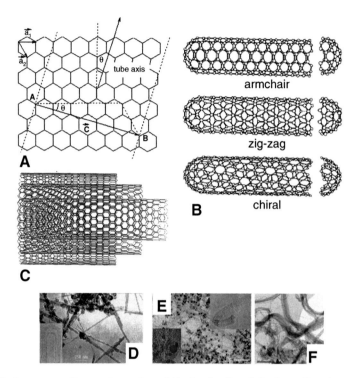

Figure 1. Structures of a graphene sheet (A), the family of single-wall carbon nanotubes (B), a multi-wall carbon nanotube (C). TEM images of as-produced CNT materials: straight and highly graphitized multi-wall carbon nanotubes (D) and single-wall carbon nanotubes (E), both arc-produced, bundles of highly bended multi-wall carbon nanotubes (F) produced by CVD process.

Additionally, there exists the family of multi-wall carbon nanotubes (MWNTs): They are composed of individual SWNTs concentrically placed inside each other and separated by an interlayer distance of about 0.34 nm, slightly larger than in graphite due to the curvature. While length and diameter of the most inner tubes are similar to SWNTs, the outer diameters, depending on the number of individual nanotubes, easily can reach values up to 20 or 30 nm.

2.2. PROPERTIES

Carbon nanotubes are characterized by a very close and unique structure–property relationship. Composed of only carbon atoms, carbon nanotubes are very light objects with a very low density of around 1.5 g/cm^3. Having all atoms at the surface, CNTs possess very high specific surface areas of about 1,400 m^2/g (outer surface) or even 2,800 m^2/g (if the inner surface is considered as well), comparable or even better than highly activated carbons. Furthermore, carbon–carbon bounds are one of the strongest in nature. Therefore, very small individual SWNTs may have Young's moduli up to 1,800 GPa (100 times stronger than steel), asymptotically reaching the values of graphite for very large diameter tubes.[4] The maximum tensile strength can reach values of up to 30 GPa. Apart of the high stiffness individual SWNTs show a surprisingly high mechanically flexibility.[5] Upon bending to very large angles, SWNTs will not break. Instead, they show a reversible buckling behaviour conferring SWNTs an extremely high toughness.

Having micrometer dimensions along their axis and nanometer-sized diameters, CNTs are nanoscale objects with aspect ratios (ratio length to diameter) as high as 1,000. This is an important property whenever it comes to applications where percolation issues play a critical role.

Furthermore, the structural arrangement of carbon nanotubes clearly defines the electronic properties.[5] One third of all SWNTs behave as metals, e.g. all armchair tubes, while two thirds of all SWNTs are semiconductors, with an energy gap inversely proportional to the tube diameter.[6–8] Additionally, the one-dimensional electronic character of SWNTs provides these nanoscale objects fundamental quantum properties.[5] Moreover, experimentally it has been found that SWNTs show ballistic transport[9] even at room temperature, have current densities[10] as high as 10^{10} A/cm^2 (copper and aluminium show values between 10^7 and 10^{10} A/cm^2), and behave as excellent electron emitters[11,12] with low turn-on fields of 1.5–5 V/μm at 1 mA/cm^2 and a low energy spread of 0.25 eV. Finally, it is worthwhile mentioning that CNTs possess an excellent thermal conductivity[13] of up to 3,000 W/mK, superior of diamond, one of the best thermal heat conductors.

For all this it becomes clear that carbon nanotubes with their structure and the combination of their unique properties – light, strong, flexible, metal, semiconductor, heat conductor, electron emitter – are of special interest for the development of novel high performance nanocomposite materials.

However, before profiting from CNTs' unique properties and using these for developing functional composite materials, one has to be aware that individual CNTs have a strong tendency to agglomerate and to form bundles (see the TEM images in Figure 1). The smaller their diameter, the larger are the corresponding Van der Waals forces and the higher their mutual attraction. Moreover, since CNT soot materials are not homogeneous at all but contain a more or less wide spread in their distribution of diameters and chiralities, the agglomeration behaviour usually varies between samples from different origins. Together with the fact that nanotubes with their graphite-like network are chemically quite inert and can thus not easily be dissolved this is a major point of concern for the development of carbon nanotubes composites as will be explained in more detail in the next section.

3. Carbon Nanotubes Composites

3.1. OPPORTUNITIES AND CRITICAL ISSUES

The development of high performance polymer/nanocomposite materials is one of the most promising field for CNT applications impacting on a broad range of technological sectors, such as aeronautics, automotive industry, telecommunications, textile industry, energy, and electronics. Being a multi-functional additive for fields with high market expectations thus is the driving force for industrial mass production of carbon nanotubes. The basic idea is to "simply" incorporate nanotubes into a matrix material and transfer their unique properties to the host system resulting in a material with enhanced functional, structural and processing properties.

One of the most prominent features of CNTs is their high aspect ratio. Combined with their low weight and conducting (electrical and thermal) properties CNTs present great opportunities whenever conducting filler networks in insulating matrices have to be developed. Ultra-low percolation thresholds in the range 0.001 and 0.01 wt % of CNT filler (a factor of 1,000 lower than for spherical carbon black particles) have been observed to reach conductivity values around 0.1–1 S/cm.[14] Therefore, the formation of perco-lated CNT networks in different kinds of polymer matrices for the develop-ment of lightweight materials able to dissipate electrostatic charge, to shield electromagnetic radiation, and to dissipate heat, is of increased technological interest for commercial applications in the above mentioned fields.

On the other side, the mechanical properties of carbon nanotubes are of great interest for the development of light structural materials with signifi-cantly increased modulus and strength. Thermoplastics, such as polypro-pylene,[15] polystyrene,[16] as well as thermosets[17] such as epoxies are widely used for this purpose. Here, for low CNT loadings, typically in the region between 0.1 and 2 wt %, moderate improvements are observed (elastic moduli increase by a factor of 3–4, thermal conductivity by 1%). However, at higher loading rates, usually agglomeration takes place and the developed materials get highly inhomogeneous.

A further important characteristic of CNTs is their extended π-electron system. This is highly compatible with the conjugated backbone structure and the delocalized π-electron system of electroactive polymers, such as polyaniline, polythiophene, polypyrrole and respective derivatives. There-fore favorable synergetic interactions leading to highly functional materials can be expected. Indeed, composites of CNT/poly(p-phenylenevinylene) (CNT/PPV)[18–20] and CNT/poly(3,4-ethylenedioxythiophene) (CNT/PE-DOT)[21] have proven their functionality in light-emitting diodes and photovoltaic cells, while CNT/polyaniline (CNT/PANI) has been used as printable conductor for thermal-imaging techniques.[22]

In general, research results can be summarized as follows:

- Polymer matrix systems of highest technological interest such as the prominent members of the families of thermosets, thermoplastics and electroactive polymers are under close scrutiny.
- At very low CNT loadings already significant increases in the conductivity are obtained, especially when compared to more traditional filler systems.
- At low CNT loading fractions, only moderate mechanical improvements are observed.
- At higher CNT loadings, agglomeration of CNTs takes place, and non-homogeneous composites are obtained leading to deteriorated materials properties.
- Improvement of materials properties lack behind expectations for CNTs. Full load transfer, i.e. the transfer of CNTs unique properties to the matrix, has not been achieved yet.
- Compatibility of CNTs electronic structure with that of the polymer matrix favors synergetic effects.

From these general observations one can derive two critical key issues of paramount importance for the development of highly functional CNT-based composite materials, independent of the matrix in question.

i) Need for homogeneous CNT dispersion in the host system.
    Maximum stable separation of CNTs is required to avoid CNT agglomerations and thus sample inhomogeneities. Preventing nanotubes bundling, especially at higher loading concentrations (i.e. usually beyond 1 wt %) and achieving a homogeneous nanotube network throughout the matrix is here the task to be tackled. Good compatibility between matrix and CNTs will assist in this issue. CNTs badly dispersed in a matrix can be encountered as *the* major reason for non-satisfactory composite characteristics.

ii) Need for efficient transfer of CNT properties to the host matrix.
    The development of nanocomposites clearly is an issue of the interface (Figure 2), where the formation of a proper interface layer acting as link between CNT and the rest of the host material determines the degree of CNT property transfer and thus the final characteristics of the whole composite material. The interface layer simply may lead to a reorganized

polymer structure in the vicinity of the CNTs,[23,24] enough to observe already some first significant changes in the overall materials characteristics, or, in the best cases, even allow highly favorable CNT property transfer leading to completely new materials characteristics.[25] The development of a favorable interface layer impacting on the whole material characteristics again is directly linked to a high degree of CNT dispersion.

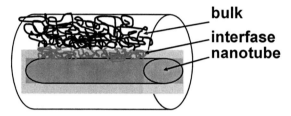

Figure 2. Cross section of a polymer/nanotubes composite material indicating nanotubes and polymer chains at the interface and bulk region.

Both, CNT dispersion and load-transfer are highly complex issues strongly depending on the type of CNT material, the CNT loading concentration and the host matrix system. All these parameters define the way towards successful compounding. Latest at this point, it should become clear, why a thorough knowledge of the type of CNTs and its tendency to form bundles is of great importance.

3.2. COMPOUNDING STRATEGIES

In the following typical concepts for developing polymer/CNT composites are presented:

i) Mechanical mixing of polymer and CNTs as solid components (pellets and powder) followed by shear-force melting (Figure 3A) and extrusion processes.

ii) Addition of CNTs to the molten state of a polymer followed by shear-force melting processes (Figure 3A), typical for thermoplastics, or by addition of cross-linking/curing agents, typical for thermosets.

iii) Dissolving the polymer and adding CNTs.

iv) Dissolving the polymer and adding a CNT dispersion prepared with the polymer solvent or/and a polymer compatible solvent (Figure 3B).

v) Dissolving the monomer, adding CNTs and initiate the polymerization process ("*in-situ* polymerization").

vi) Dissolving the monomer, adding a CNT dispersion prepared with the monomer solvent or/and a monomer compatible solvent and initiate the polymerization ("*in-situ* polymerization").

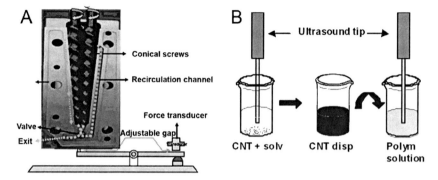

Figure 3. Typical approaches for polymer/CNT compounding: (A) Shear-force melting process using a twin screw compounder (DSM Xplore™ 15 cm³ twin-screw Micro-Compounder). Essential parameters are temperature, time, force, and rotating speed. (B) Processing in liquid phase. Use of an ultrasonic tip to disperse CNTs followed by mixing into the polymer solution under ultrasonication. Ultrasonication assists in dispersion and disrupting CNT bundles. Parameters of importance are time, power and frequency.

Mechanical mixing of polymer and CNTs and processing polymer or polymer/CNT melts as described in (i) and (ii) is favored for thermoplastics or thermosets since these polymers usually are not compatible with the addition of solvents. However, whenever polymers can be dissolved or have to be newly synthesized working with liquid CNT dispersions, as described in (iii)–(vi), is the most promising way towards functional composites. *In-situ* processes typically are applied when the polymer is not bought from shelf but can or has to be newly synthesized. In all cases, disrupting (exfoliating) CNT bundles and achieving stable and homogeneous dispersions of individual or at least very small bundles of CNTs is the key to success. For the mechanical compounding this usually is achieved in shear-force melting processes; for chemical compounding from the liquid phase, ultrasonication methods are employed. Therefore, for each type of polymer, CNT and solvent, a careful adjustment of process parameters for shear-melting (time, speed, force, temperature etc.) and ultrasonication (time, power, frequency, etc.) is required. Information on the dispersion state in liquid or solid form obtained from characterization techniques such as spectroscopy, electron microscopy, differential scanning calometry (DSC) etc. will provide the necessary feedback to optimize the dispersion approach. Eventually compatible agents have to be added in this process.

Concerning the issue of improving the interaction between CNTs and polymer two approaches usually are applied:

i) Covalent interactions between CNTs and polymer chains
   These are mediated by functionalized CNTs (carbonyl, carboxylic, amine groups, etc.) which can bind covalently to complementary groups

on the polymer chain. This favors strong connections and load transfer on one hand, but as well leads to deterioration of CNTs' original properties, on the other hand.

ii) Non-covalent interaction between CNTs and polymer chains ("polymer wrapping")

This is based on weak interactions (e.g. van der Waals interactions) between CNTs and polymer chains. It applies especially for polymers which have extended π-electron systems. Depending on the flexibility of the polymer backbone these polymers might form a more or less crystalline layer around the CNTs[26] or wrap around the CNTs.[27] This approach often results in changes of the conformation of the polymer chains at the CNT interface. These may lead to highly favorable interactions between both constituents which impact on the overall functional properties. Especially in the case of *in-situ* polymerization processing, where individual CNT can serve as molecular templates directing the growth and conformation of the polymer chains, highly functional composites can be achieved.[25,28–31]

In any case, at the end it will be the final application characteristics, which determines the approach to be taken. Once the final composite is fabricated, further processing towards a final end-product in form of films, coatings, fibers or any other shapes may be applied. Typical techniques for composites in liquids phase are drop-casting, spin-coating, impregnation, or spinning techniques while for composites in molten state extrusion or inject molding approaches are applied.

In the following examples from our latest research on polymer composites obtained by shear melting compounding as well as by *in-situ* polymerization are presented.

3.3. THERMOPLASTICS AND SHEAR-FORCE MELT MIXING

3.3.1. Polypropylene/CNT Composites

Polypropylene (PP) is a linearly structured thermoplastic polymer of great industrial relevance due to its excellent properties, namely light weight, excellent resistance to stress and high resistant to cracking, high operational temperatures with a melting point of 160°C, excellent dielectric properties, chemical resistance, non-toxicity, and easy production at relatively low price. However, processing difficulties due to low melt strength of PP, fracture at low temperatures and degradation upon long-term exposure to light demand further improvement. In this sense the incorporation of CNTs into PP is believed to significantly improve the performance of the matrix.

Compounding was performed as follows: Isotactic PP (ISPLEN™ 070 G2M, Repsol) was employed. The nanotubes were oxidized single-wall carbon nantoubes (SWNTs) from arc-discharge process[32] (see Figure 1E). Different concentrations of oxidized SWNTs (0.25, 0.5, 1.0, 1.5, 2 and 2.5 wt %) were blended with PP pellets (ca. 500 μm in size) before shear-force melting in the extruder (Haake Minilab Micro Compounder). The temperature of mixing was set at 180°C while the mixing time was 10 min. The rotation speed was set to 70 rpm. The obtained mixture was compressed at 180°C during 5 min to obtain moulded PP sheets of 10 × 10 × 1 mm in size.

Differential scanning calometry (DSC) was performed on PP and elaborated corresponding nanocomposites. The corresponding non-isothermal crystallization thermograms are displayed in Figure 4A; Table 1 summarizes the obtained crystallization peak temperature (T_c), the melting temperature (T_m), the crystallization enthalpy (ΔH) and the degree of crystallinity as a function of the SWNT content. It is remarkable that the incorporation of SWNTs had a strong effect on the observed crystallization temperature,

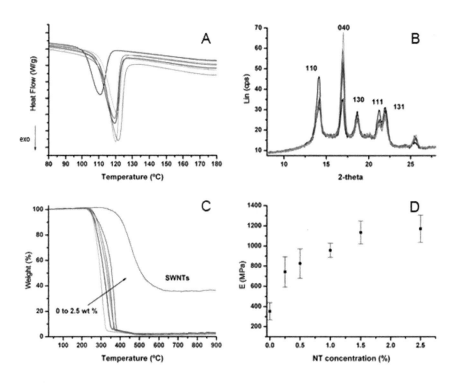

Figure 4. Characterization of PP and PP/SWNT nanocomposites at different SWNT concentrations. Non-isothermal crystallization thermograms (A), X-ray diffractograms with a characteristic reflection set for monoclinic (α-form) PP (B), thermogravimetric analysis TGA (C), and Young's modulus (D).

even at the lowest nanotube content. While neat PP showed an initial crystallinity of 65% the composite reaches values up to 78% at 2.5 wt % SWNTs loading. This increment in the crystallization corroborates that the addition of nanotubes enhances the nucleation process of the PP, as also previously observed by other authors.[33–36] On the other side, it is to note that the presence of SWNTs did not affect the crystalline polymorphism of PP. The X-ray diffractograms of the composites shown in Figure 4B unambiguously proved that PP preserves its original monoclinic (α-form).[37]

TABLE 1. Crystallization temperature (T_c), melting temperature (T_m), crystallization enthalpy (ΔH), crystallinity (%) and oxidation temperature (T_{oxid}) of neat PP and PP/SWNT nanocomposites.

SWNT	T_C	T_m	ΔH_f	Cryst	T_{oxid}	E-modul
(wt %)	(°C)	(°C)	(J/g)	(%)	(°C)	(MPa)
0	110.7	165.5	89.61	65.0	301	375
0.25	119.1	164.0	94.80	68.7	320	750
0.50	119.9	163.8	99.40	72.0	332	810
1.00	119.3	165.8	98.96	71.7	337	950
1.50	120.3	165.2	97.24	70.0	343	1,050
2.50	121.0	164.9	108.80	78.8	354	1,100

Since an orientation of PP crystallites induced by high aspect ratio fillers may improve the mechanical performance of the polymer,[38] in a next step the effects of orientation of PP crystallites in the PP/SWNT nanocomposites were studied. The diffractions related to the (110) and (040) planes are the most important as far as the α-crystal phase is concerned, since they are related with the a and b axes and provide information about the orientation of crystallites. The relationship between the orientations of PP in these two planes can be observed from the ratio of the diffraction intensities of the (110) plane to that of the (040) plane. If this ratio is below 1.3, the b axis predominates over the analyzed surface; if in contrast the relation is larger than 1.5 then the a axis is parallel to the surface. Meanwhile, if the relation lies between these two values, the mixture of all crystallites will be isotropic.[39] We observed that the intensities of the (110) and (040) diffractions significantly changed with the addition of CNTs (Figure 4). Neat PP had a ratio equal to 1.84, indicating that the a axis of the crystals were mainly parallel to the analyzed surface. The addition of only 0.25% of SWNTs decreased the previous value to 1.10 indicating that now the b plane was parallel to the surface and remains almost constant with increasing the content of CNTs up to 2.5 wt %. Despite of the orientation changes, it has to be mentioned that no changes in the crystalline phase were observed. In this sense, the orientation of the α-crystals parallel to the surface due to the anisotropy of the CNTs was a clear indication of the nucleation ability of the latter.

Furthermore, the thermal stability of the obtained nanocomposites was evaluated by means of thermogravimetrical analysis (TGA) (Figure 4C) under a 100 ml/min air atmosphere. While the neat PP decomposes in a single step at an oxidation temperature T_{oxid} of 301°C the degradation process of the nanocomposites systematically shifts to higher oxidation temperatures (also in a single step). For the nanocomposite containing 2.5 wt % the maximum oxidation temperature was centered at 354°C (Table 1). The amount of residue remaining at 900°C increased with higher SWNT loadings.

In order to evaluate the mechanical properties of the prepared PP/SWNT nanocomposites, uniaxial tensile test were carried out at room temperature according to ASTM D638M. Thus, bone-shaped samples were tested to failure at a crosshead speed of 10 mm/min in an Instron machine. In Figure 4D the obtained values for the Young's modulus for the nanocomposites are displayed as function of the SWNT content. The neat PP showed an initial Young's modulus of approximately 375 MPa. With higher nanotube loading fractions this value increased up to 1,100 MPa for the 1.5 wt % SWNT nanocomposite and remained almost constant independent of further nanotube loadings. Thus a clear reinforcement effect due to the presence of nanotubes has been observed. The presence of a plateau from 1.5 wt % on might point to problems of achieving homogenous SWNT dispersions in the PP host matrix at higher nanotube loading fractions. Thus the issue of obtaining a composite material in which nanotubes are homogeneously dispersed still remains a critical issue to be solved, at least for loading fractions beyond 1 wt %.

3.3.2. Nylon-6/CNT Composites

Polyamide-6 (nylon-6) is an important semicrystalline polymer which is used in many engineering products, such as fibres, films and various molding articles. Also here, the reinforcement of this technical polymer with nanotubes[40–43] is an issue of great interest. Current investigations are mainly focused on the mechanical properties of nylon/carbon nanotubes.[40,41] However, knowledge of the influence of carbon nanotubes on the crystallization and polymorphic behaviour of nylon nanocomposites is still scarce. Therefore our studies focus on studies on the effect of the concentration of SWNTs as well as MWNTs on nylon-6 crystalline phase transitions.

Nylon-6 was supplied in pellet form by La Seda de Barcelona S.A; SWNTs and MWNTs were synthesized by arc-discharge method.[32] Several concentrations of CNTs (0.1, 0.25, 0.5 and 1 wt %) were introduced in the nylon-6 matrix by melt-mixing using a micro-extruder (Thermo-Haake Minilab compounder) operated at 240°C and a rotor speed of 150 rpm for 10 min. The thermoplastic polymer used in this study was first dried in an oven at 120°C for 24 h before use, to minimize the effects of moisture. The shear-force parameters applied were adequate to minimize nanotube aggregation in the final nanocomposites as revealed by scanning electron microscopy (not shown).

Crystalline transformation of nylon-6 as a function of nanotube type and content were monitored by DSC thermograms. The DSC results (Figure 5) indicate that the presence of nanotubes systematically leads to an increase of the crystallization temperature of the nylon-6 matrix. As in the case of polypropylene, this again underlines the ability of nanotubes to act as efficient nucleating agent, independent of the type of nanotube used.

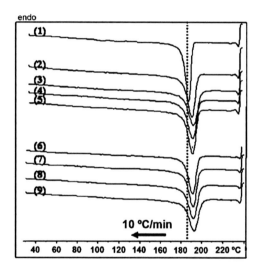

Figure 5. DSC thermograms of nylon-6/nanotube composites obtained during cooling from the melt to room temperature at 10°C/min. Neat nylon-6 (1), nylon-6/SWNT composites at SWNT concentrations of 0.1, 0.25, 0.5, 1 wt % (2–5), nylon-6/MWNT composites at MWNT concentrations of 0.1, 0.25, 0.5, 1 wt % (6–9).

Wide angel X-ray scattering (WAXS) of the nylon and the 1 wt % nylon/ nanotube composite samples (Figure 6) reveal that nylon-6 crystallizes first into a pseudo-hexagonal γ-phase from the molten state which then converts to the monoclinic α-form during cooling (Figure 6a). The presence of nanotubes favours the appearance of the more disordered γ-phase at higher temperatures. This reversible crystal-to-crystal phase transition is a gradual and continuous process typically observed in nylons upon heating from room temperature, known as Brill transition.[44,45] However, since for nylon/ MWNT nanocomposites prepared by in-situ polymerization the Brill transition was not found during the heating process,[40] its appearance also might be influenced by the processing conditions.

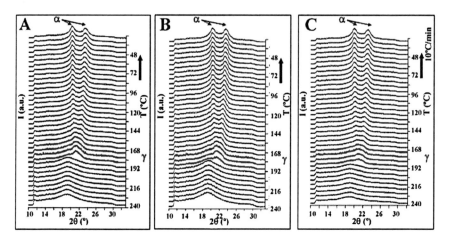

Figure 6. WAXS diffractograms of nylon-6 (A), nylon-6/SWNT (B) and nylon-6/MWNT (C) nanocomposites containing 1 wt % of nanotubes.

The presented studies consistently reveal that nanotubes act as nucleating agent which accelerate the crystallization rate of nylon-6 in nylon-6 nanocomposites and thus induce the appearance of structural transformations at higher temperature compared to neat nylon-6 and also increase the temperature of the α–γ transition of nylon 6. As in the case of polypropylene, a significant change of structural properties can be observed. However, this is strongly related to a good dispersion of the nanotubes in the host-matrix which finally determines the homogeneity of the overall composite material. Future efforts in compounding have to be done in order to overcome the dispersion/agglomeration problem.

3.4. ELECTROACTIVE POLYMERS AND IN-SITU POLYMERIZATION

3.4.1. *Polyaniline/Carbon Nanotube Composites*

"Electroactive polymers", also called "intrinsically conducting polymers" or "synthetic metals" form a class of polymers which possess the electrical, electronic, magnetic, and optical properties of a metal while retaining the mechanical properties (flexibility and toughness) and processibility commonly associated with a conventional polymer.[46,47] They are of special interest for the development of organic electronics.[47] The backbone of a conducting polymer consists of a truly conjugated and delocalized π electron system and thus is highly compatible with carbon nanotubes. Highly synergetic effects leading to improved or even novel properties can be expected by properly compounding electroactive polymers with carbon nanotubes.

As starting point we chose polyaniline (PANI) – a prominent member of the family of electroactive polymers – due its ease of synthesis and its tunable redox characteristics controlling the electrical and optical properties of interest for technological applications.[46] Our first attempts to directly mix nanotubes (as produced and oxidized) into PANI resulted in non-homogeneous material in which the nanotubes were strongly agglomerated.[48] To overcome this problem, we applied in-situ synthesis approaches, i.e. polymerization of PANI monomers in the presence of nanotubes[25,28,49] as shown in Figure 7. An aniline monomer solution in 1M HCl was added to nanotubes (arc produced MWNTs) suspended in 1M HCl; ammoniumperoxodisulphate (APS) was added as oxidant to initiate the polymerization which rapidly resulted in a dark green suspension. After several washing/filtering cycles we obtained a composite powder in which PANI was present in its primary doped form, i.e. as emeraldine salt (ES).[49] In a next step, we transformed ES into its soluble emeraldine base (EB) form.[25,28] Therefore, we de-doped the ES/MWNT composite with ammonium hydroxide. After several washing and filtering cycles EB/MWNT composites were obtained as powder materials in which MWNTs were homogeneously embedded.

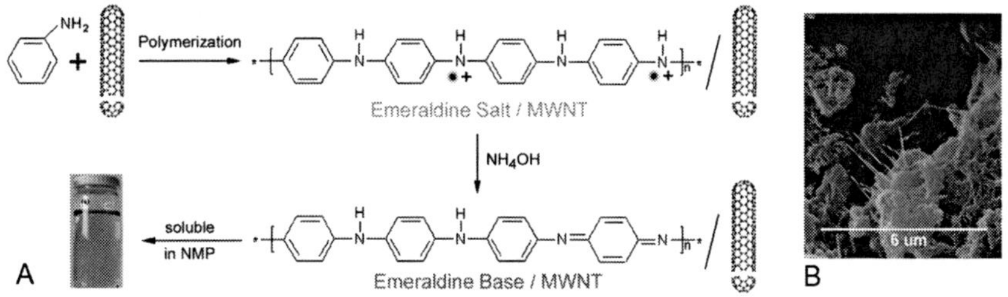

Figure 7. Synthesis of PANI/MWNT composite.[25,28,49] In-situ polymerization yields in the green emeraldine salt/MWNT composite which subsequently is transformed into the blue and soluble emeraldine base/MWNT composite (A); Scanning electron microscopy image of the emeraldine base/MWNT solid powder (B).

Most remarkably, the whole PANI/MWNT composites with nanotubes loadings as high as 50 wt % were completely "soluble" in organic solvents such as N-methylpyrrolidone (NMP) and dimethyl sulfoxide (DMSO). Straightforwardly, we could produce films, supported or free-standing by casting or coating methods as well as fibers (Figure 8). The composites films could be easily doped and de-doped switching between the EB and ES

states using common acids and bases.[28] Furthermore, PANI/MWNT composites also could be dissolved in camphorsulfonic acid introducing optical
activity to the corresponding dispersion.[29,30]

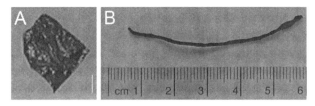

Figure 8. PANI/MWNT composites: free standing film (A), fiber (B).

We could show that the good processing properties are a direct consequence of the in-situ polymerization process in which straight arc-produced
MWNTs act as template for the growing PANI backbone. This resulted in
highly synergetic interactions leading on one side to improved processing
properties (solubility) as well as to enhanced physical properties, on the
other side. Here are to mention highly favorable charge transfer processes[49]
improved deprotonation stability,[29,30] a nine orders of magnitude higher
conductivity, and a significantly enhanced thermal stability by 200°C compared to neat PANI.[25] We thus could show that enhanced functionality and
processing properties of the obtained materials are directly linked to favorable interactions between carbon nanotubes and the polymer chains and
thus intimately related to the synthesis strategy. In a next step we currently
are developing a synthesis approach rendering PANI/MWNT soluble in
water or aqueous solutions, of great interest to printing technologies and the
fabrication of plastic electronics. The obtained knowledge then will be
transferred to other types of electroactive polymers.

4. Summary

We presented a general introduction into the field of carbon nanotube composite materials. Opportunities, challenges, critical issues and compounding
strategies were discussed in detail. Examples for successful compounding
strategies of nanotubes with thermoplastics, namely polypropylene and
nylon-6, by shear-force melting processes as well as synthesis approaches for
electroactive polymer/nanotubes composites, namely polyaniline, by in-situ
polymerization processes were given. On one hand, we showed that
enhanced materials properties and processing characteristics can be achieved.
On the other hand, we clearly pointed out that synthesis approaches for
highly functional and processible nanotube composite materials should be
designed in such a way that nanotubes' agglomerations should be avoided

while at the same time favourable interactions between nanotubes and the host matrix should be created. Taking this into account further progress towards novel high performance nanotubes-based composite materials can be expected.

ACKNOWLEDGEMENTS

Funding from the Spanish Ministry of Education and Science (MEC) and the European Regional Development Fund (ERDF) under projects MAT 2006-13167-C02-01 and -02, MAT2007-66927-C02-01, is gratefully acknowledged. M.N. acknowledges the CSIC for a postdoctoral contract (I3PDR-6-02), financed by the European Social Fund. We like to thank the European Commission for the synchrotron project at the Soft Condensed Matter A2 beamline at HASYLAB (DESY-Hamburg, I-20080056 EC).

References

1. S. Iijima, Nature 354, 56 (1991).
2. M.S. Dresselhaus, G. Dresselhaus, Ph. Avouris, (Eds.), *Carbon Nanotubes: Synthesis, Structure, Properties and Applications*, Topics in Applied Physics, Vol. 80 (Springer, Berlin, Germany, 2001).
3. H.W. Kroto, J.R. Heath, S.C. O'Brien, et al., Nature 318, 162 (1985).
4. J.W. Mintmire, D.H. Robertson, and C.T. White, Phys. Chem. Solids 54, 1835 (1993).
5. B.I. Yakobson, C.J. Brabec, and J. Bernholc, Phys. Rev. Lett. 76, 2511 (1996).
6. J.-C. Charlier, X. Blase, and S. Roche, Rev. Mod. Phys. 79, 677 (2007).
7. J.W. Mintmire and C.T. White, Carbon 33, 89 (1995).
8. N. Hamada and S. Sawada, Phys. Rev. Lett. 68, 1579 (1992).
9. C.T. White and T.N. Todorov, Nature 411, 649 (2001).
10. S. Frank, P. Poncharal, Z.L. Wang, and W.A. de Heer, Science 280, 1744 (1998).
11. W. Zhu, C. Bower, O. Zhou, G. Kochanski, and S. Jin, Appl. Phys. Lett. 75, 873 (1999).
12. G.Z. Yue, Q. Qiu, B. Gao, et al., Appl. Phys. Lett. 81, 355 (2002).
13. R.S. Ruoff and D.C. Lorents, Carbon 33, 925 (1995).
14. J.K.W. Sandler, J.E. Kirk, I.A. Kinloch, et al., Polymer 44, 5893 (2003).
15. P. Castell, W.K. Maser, A.M. Benito, et al., Polymer submitted (2007).
16. D. Qian, E.C. Dickey, R. Andrews, et al., Appl. Phys. Lett. 76, 2868 (2000).
17. M.J. Biercuk, M.C. Llaguno, M. Radosavljecvic, et al., Appl. Phys. Lett. 80, 2767 (2002).
18. S.A. Curran, P.M. Ajayan, W.J. Blau, et al., Adv. Mater. 10, 1091 (1998).
19. J.N. Coleman, S. Curran, A.B. Dalton, et al., Phys. Rev. B 58, R7492 (1998).
20. A. Star, J.F. Stoddart, D. Steurman, et al., Angew. Chem. Int. Edit. 40, 1721 (2001).
21. H.S. Woo, R. Czerw, S. Webster, et al., Synth. Met. 116, 369 (2001).
22. G.B. Blanchet, C.R. Fincher, and F. Gao, Appl. Phys. Lett. 82, 1290 (2003).
23. J. Sandler, G. Broza, M. Nolte, et al., J. Macromol. Sci. Phys. B 42, 479 (2003).
24. W. Chen and X.M. Tao, Macromol. Rapid Comm. 26, 1763 (2005).

25. R. Sainz, A.M. Benito, M.T. Martínez, et al., Adv. Mater. 17, 278 (2005).
26. K.P. Ryan, M. Cadek, V. Nicolosi, et al., Compos. Sci. Technol. 67, 1640 (2007).
27. M.I.H. Panhuis, A. Maiti, A.B. Dalton, et al., J. Phys. Chem. B 107 478 (2003).
28. R. Sainz, A.M. Benito, M.T. Martínez, et al., Nanotechnology 16, S150 (2005).
29. R. Sainz, W.R. Small, N.A. Young, et al., Macromolecules 39, 7324 (2006).
30. M.I.H. Panhuis, K.J. Doherty, R. Sainz, et al., J. Phys. Chem. C 112, 1441 (2008).
31. F. Picó, J.M. Rojo, M.L. Sanjuán, et al., J. Electrochem. Soc. 151, A831 (2004).
32. C. Journet, W.K. Maser, P. Bernier, et al., Nature 388, 756 (1997).
33. A.R. Bhattacharyya, T.V. Sreekumar, T. Liu, S. Kumar, L.M. Ericsson, R.H. Hauge, and R.E. Smalley, Polymer 44, 2373 (2003).
34. L. Valentini, J. Biagiotti, J.M. Kenny, and S. Santucci, Comp. Sci. Tech. 63, 1149 (2003).
35. L. Valentini, J. Biagiotti, M.A. Manchado, S. Samtucci, and J.M. Kenny, Polym. Eng. Sci. 44, 303 (2004).
36. M.A. Lopez Manchado, L. Valentini, J. Biagiotti, and J.M. Kenny, Carbon 43, 1499 (2005).
37. H. Xia, Q. Wang, K. Li ,and G.H. Hu, J. Appl. Polym. Sci. 93, 378 (2004).
38. E. Ferrage, F. Martin, A. Boudet, S. Petit, G. Fourty, F. Jouffret, P. Micoud, P. De Parseval, S. Salvi, C. Bourgerette, J. Ferret, Y.S. Gerard, S. Buratto, and J.P. Fortune, J. Mater. Sci. 37, 1561 (2002).
39. F. Rybnikar, J. Appl. Polym. Sci. 38, 1479 (1989).
40. H.L. Zheng, C. Cao, Y. Wang, P.C.P. Watts, H. Kong, X. Li, and D. Yan, Polymer 47, 113 (2006).
41. T. Liu, I.Y. Phang, L. Shen, S.Y. Chow, and W.D. Zhang, Macromolecules, 37, 7214 (2004).
42. J. Gao, M.E. Itkis, A. Yu, E. Bekyarova, B. Zhao, and R.C. Haddon, J. Am. Chem. Soc. 127, 3847 (2005).
43. R. Haggenmueller, F. Du, J.E. Fischer, and K.I. Winey, Polymer 47, 2381 (2006).
44. R. Brill, J. Prakt. Chem., 46, 161 (1946).
45. M. Naffakh, C. Marco, M.A. Gómez, G. Ellis, W.K. Maser, A.M. Benito, and M.T. Martínez, J. Nanosci. Nanotechnol. submitted (2008).
46. A.G. MacDiarmid, Angew. Chem. Int. Edit. 40, 2581 (2001).
47. T.A. Skotheim and J.R. Reynolds (Eds.), *Handbook of Conducting Polymers*, 3rd Edition (CRC Press, Boca Raton, FL, 2007, ISBN-10:1-4200-4360-9).
48. W.K. Maser, A.M. Benito, M.A. Callejas, et al., Mater. Sci. Eng. C 23 (2003).
49. M. Cochet, W.K. Maser, A.M. Benito, et al., Chem. Commun. 1450 (2001).

5.2. NANOPARTICLES

LASER ABLATION OF CdSe AND ZnO: ALKYLAMINE ASSISTED FORMATION OF MAGIC CLUSTERS

A. DMYTRUK[*,1,4], I. DMITRUK[2,4], R. BELOSLUDOV[3],
Y. KAWAZOE[3], A. KASUYA[4]

[1]*Institute of Physics of National Academy of Sciences
of Ukraine, prosp. Nauky 46, 03028 Kyiv, Ukraine*
[2]*Physics Department, National Taras Shevchenko University
of Kyiv, 2 Glushkov Prosp., Build. 1, 03680 Kyiv, Ukraine*
[3]*Institute for Materials Research, Tohoku University,
Katahira 2-1-1, Aoba-ku, Sendai, 980-8577, Japan*
[4]*Center for Interdisciplinary Research, Tohoku University,
Aramaki Aza Aoba, Aoba-ku, Sendai, 980-8578, Japan*

Abstract. CdSe and ZnO clusters have been produced by laser ablation of bulk powders of cadmium selenide and zinc peroxide, respectively, and studied by time-of-flight mass spectroscopy. Clusters of enhanced stability ("magic clusters") have been observed in both series of clusters, CdSe and ZnO. It has been found that addition of alkylamine to the bulk powder precursors significantly increases the abundance of the magic clusters in the series. It is suggested that alkylamine plays a catalyst role in the formation of magic clusters.

Keywords: cadmium selenide; zinc oxide; alkylamine; laser ablation; magic cluster

1. Introduction

Decreasing the size, drastic changes of material properties appear in the transition region between the solid state and atoms (molecules), i.e. in the range below about 200 monomers. Typically, clusters of this range are not stable enough to have a practical value, except they possess an enhanced stability ("magic clusters"), like the well-known carbon fullerenes,[1] or the recently discovered magic CdSe particles.[2] The other advantage of magic

---

*admytruk@gmail.com

J.P. Reithmaier et al. (eds.), *Nanostructured Materials for Advanced Technological Applications*, 201
© Springer Science + Business Media B.V. 2009

clusters is the atomic precision of size and shape distribution that considerably simplifies experimental research on nanomaterials and opens new applications, for example, where a regular arrangement of nanostuctures is necessary. As the structure of magic clusters may be significantly different from that of the bulk material (for example, graphite – fullerene,[1] wurtzite bulk CdSe – nested cage $(CdSe)_{33}$, $(CdSe)_{34}$ nanoparticles[2]), the laser ablation method is especially useful for production of such objects, because it produces clusters with structures inaccessible yet by other synthesis methods. Among others, II–VI group semiconductors, like ZnO and CdSe, are of interest as promising materials for photoelectrochemistry, photocatalysis, photovoltaics, optoelectronics, and others.

Several studies on laser ablation of bulk powders of II–VI group compounds have been done, in which magic clusters were observed.[2–6] Typically, the clusters were produced by laser ablation of solid targets in vacuum. A tendency for $A_{II}B_{VI}$ compounds (A_{II} = Cd, Zn; B_{VI} = S, Se, Te) to form stable clusters at certain magic numbers of monomers, namely 13, 19, 33, and 34, has been found.[2,6] However, no observation of ZnO and CdO magic clusters has been reported yet.

Here, we present time-of-flight mass spectroscopic studies on clusters of CdSe and ZnO produced by laser ablation of bulk powder precursors with and without assistance of alkylamine, and show the role of the latter in the cluster formation.

2. Experimental

CdSe and ZnO clusters were formed by laser ablation of bulk powder cadmium selenide and zinc peroxide (both from WAKO), respectively, in a Bruker Reflex III-T time-of-flight (TOF) mass spectrometer. The powder was mixed with a small amount of solvent (toluene or distilled water), with or without an alkylamine (decylamine (DA) and dodecylamine (DDA), respectively, both from WAKO). About 1 μl of the resulting suspension was dropped on a stainless steel target, dried in vacuum, and the target was inserted in the mass spectrometer. The chamber (flight tube) of the mass spectrometer was evacuated to about $5 \cdot 10^{-6}$ Pa. The spectrometer was equipped with a nitrogen laser, which produces 4 ns pulses of 337.1 nm light with an energy of 300 μJ at a repetition rate of 1–3 Hz. The laser beam was passed though a controlled attenuator (typically an attenuation level of 30–40% was used) and focused into a spot of about 20 μm on the target. Cooling by carrier gas injection was not used. The ions produced were accelerated by a potential of 19 kV, separated in the flight tube either in linear or in reflex mode, and registered by a microchannel plate detector.

Positive or negative cluster ions were measured, depending on the polarity of the accelerating potential. Typically the reflex mode was used, and the spectrum of positively charged clusters was acquired by accumulating the signal of 500–1,000 laser shots.

3. Results and Discussion

A TOF mass spectrum of the clusters formed by laser ablation of CdSe powder is shown in Figure 1 (top). The spectrum consists of a series of main peaks separated by 191.34 amu, which corresponds to the mass of CdSe (191.37 amu). The main peaks are attributed to stoichiometric $(CdSe)_n$ clusters from n = 11–34. The peaks between the main peaks correspond to nonstoichiometric clusters containing one extra Se atom. Clusters of enhanced stability (magic clusters) are identified at n = 13, 19, 33, and 34. Overall, the mass spectrum is similar to the results reported for laser ablation of cadmium selenide bulk powder.[2]

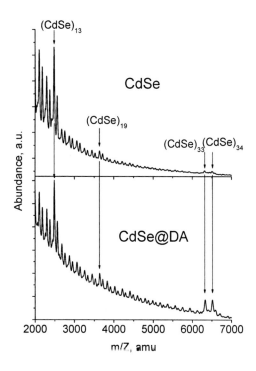

Figure 1. TOF mass spectra of the clusters formed by laser ablation of CdSe bulk powder (top) and CdSe bulk powder mixed with decylamine (DA) (bottom).

Figure 1 shows at the bottom a TOF mass spectrum of the clusters obtained by laser ablation of cadmium selenide bulk powder mixed with decylamine (DA). It consists of the same series of peaks of $(CdSe)_n$ as in the top spectrum. The identical position of the peaks indicates that DA molecules do not bind to the CdSe clusters, while the appearance of the same magic numbers suppose that the structure of the magic clusters formed in the presence of DA is the same. However, the peaks of the magic clusters are much more distinguishable in this series, suggesting a catalyst role of DA in the formation of magic CdSe clusters.

Laser ablation experiments of bulk zinc oxide powder, both reported in Refs. 4,5 and ours (not shown here), revealed a series of $(ZnO)_n$ clusters up to about 20 ZnO monomers only, and a lack of any magic numbers in that mass range. Zinc peroxide has essentially different properties from zinc oxide, namely a much lower thermolysis temperature ($150°C$ for ZnO_2 and $1,975°C$ for ZnO) and density ($1,570$ kg/m^3 and $5,600$ kg/m^3, respectively). The differences indicate much weaker interatomic bonds in ZnO_2 than in ZnO. Practically, that makes zinc peroxide powder a much more efficient precursor for ZnO clusters than the zinc oxide powder used in previous laser ablation experiments. The excess oxygen in zinc peroxide may volatilize during laser ablation. Therefore, using zinc peroxide is a crucial factor for efficient production of ZnO clusters by laser ablation.

Figure 2 (top) shows a TOF mass spectrum of the clusters obtained by laser ablation of zinc peroxide powder. It consist of a series of peaks separated by 81.39 amu, which exactly corresponds to the mass of ZnO, indicating the formation of stoichiometric $(ZnO)_n$ clusters only, and proving that excess oxygen does indeed volatilize during laser ablation. The complete series of $(ZnO)_n$ clusters can be reliably resolved up to about a hundred of ZnO monomers, which is a big step toward larger ZnO cluster production, compared to the results of zinc oxide powder ablation. Some peaks in the series are slightly higher than their neighbors, namely at n = 34 and 60, however the differences are only minor.

Addition of dodecylamine (DDA) to the zinc peroxide powder significantly increases the abundance of the magic clusters, making their peaks easily distinguishable in the series, as shown in Figure 2 (bottom). The position of the peaks remains the same as in the top spectrum (without DDA), indicating that DDA molecules do not bind to ZnO clusters, analogous to the case of CdSe and DA. Therefore, it can be considered that DDA plays a catalyst role in the formation of magic ZnO clusters.

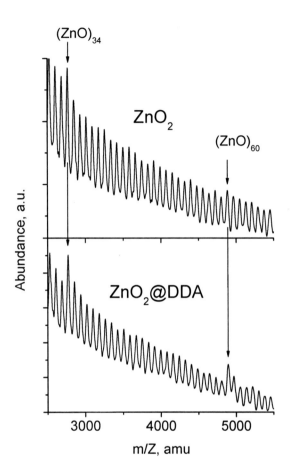

Figure 2. TOF mass spectra of the clusters formed by laser ablation of ZnO_2 bulk powder (top) and ZnO_2 bulk powder mixed with dodecylamine (DDA) (bottom).

The structure of ZnO magic clusters is not clear yet, however, we suppose that $(ZnO)_{34}$ has the same structure as reported for $(CdSe)_{34}$ and found commonly for the other II–VI group compounds.[2,6] As for $(ZnO)_{60}$, the nested cage structure $(ZnO)_{12}@(ZnO)_{48}$ is most probably the most stable, supported by recently reported[7] and our own ongoing first-principles calculation of the isomers. A comprehensive study on the structures of ZnO magic clusters will be reported elsewhere.

4. Conclusion

The laser ablation method was applied for the production of CdSe and ZnO clusters. Bulk powders of cadmium selenide and zinc peroxide have been used as precursors, respectively. In the case of ZnO, using the efficient precursor zinc peroxide allows the acquisition the mass spectra of $(ZnO)_n$ clusters of more than a hundred of monomers, which is a big step towards larger ZnO cluster production as compared to yet reported results. Clusters of enhanced stability (magic clusters) have been observed in both, the CdSe and the ZnO series of clusters. It has been found that addition of alkylamine to the bulk powder precursors significantly increases the abundance of the magic clusters in the series, which allows reliable identification of hitherto unknown ZnO magic clusters, namely $(ZnO)_{34}$ and $(ZnO)_{60}$. It is suggested that the alkylamines play a catalyst role in the formation of magic clusters.

We think that our findings will help to develop methods for cluster production in mass quantities. The variety of the magic clusters observed, and moreover the enhanced stability of the rather large $(ZnO)_{60}$ cluster, open huge possibilities for their applications, especially where atomic precision of about one nanometer scale objects is necessary.

References

1. H. W. Kroto, J. R. Heath, S. C. O'Brien, R. F. Curl, and R. E. Smalley, Nature 318, 162 (1985).
2. A. Kasuya, R. Sivamohan, Yu. A. Barnakov, I. M. Dmitruk, T. Nirasawa, V. R. Romanyuk, V. Kumar, S. V. Mamykin, K. Tohji, B. Jeyadevan, K. Shinoda, T. Kudo, O. Terasaki, Z. Liu, R. V. Belosludov, V. Sundararajan, and Y. Kawazoe, Nature Mater. 3, 99 (2004).
3. T. P. Martin, Phys. Rep. 273, 199 (1996).
4. Burnin, J. J. BelBruno, Chem. Phys. Lett. 362, 341 (2002).
5. M. Kukreja, A. Rohlfing, P. Misra, F. Hillenkamp, K. Dreisewerd, Appl. Phys. A 78, 641 (2004).
6. V. R. Romanyuk, I. M. Dmitruk, Yu. A. Barnakov, A. Kasuya, and R. V. Belosludov, paper to be published in JNN.
7. M. Zhao, Y. Xia, Zh. Tanb, X. Liu, L. Mei, Phys. Lett. A 372, 39 (2007).

SYNTHESIS AND CHARACTERIZATION OF SILVER/SILICA NANOSTRUCTURES

K. KATOK[*], V. TERTYKH, V. YANISHPOLSKII
*O.O. Chuiko Institute of Surface Chemistry,
Ukranian National Academy of Sciences, 17 Naumov St.,
03164 Kyiv-164, Ukraine*

Abstract. The peculiarities of the formation of silver nanoparticles on the surface of porous silica with grafted silicon hydride groups were studied. The conditions of the synthesis of silver nanoparticles with different sizes and their structural and optical properties have been investigated. A reproducible method for the formation of stable colloids of nanosized silver covered with silica shells simultaneously with the reduction of the metal and the formation of silica as a result of hydrolysis and polycondensation of triethoxysilane was elaborated.

Keywords: silica; modified surface; silicon hydride groups; silver; *in situ* reduction; core/shell structure

1. Introduction

Nanoparticles of noble metals possess unique optical properties which are connected with the presence of surface plasmon resonances in the visible or near-by IR region in absorption and scattering spectra.[1] These bands are caused by localized plasmon resonances of metal nanoparticles which are generated by a collective coherent oscillation of free electrons in the electric field of the light wave. This behavior makes metal particles an important object of contemporary nanobiotechnology.[2] In particular, the dependence of localized plasmon resonances on a local dielectric surrounding the particles is used for the design of biosensors of a new generation, which are able to detect optically interactions of biomolecules in the vicinity of the surface of nanoparticles.[3]

---

*smpl@ukr.net

J.P. Reithmaier et al. (eds.), *Nanostructured Materials for Advanced Technological Applications,*   207
© Springer Science + Business Media B.V. 2009

2. Experimental

The synthesis of ordered mesoporous silica of the type MCM-41 has been reported earlier.[4] For the preparation of mesoporous silica with reducing properties we applied a surface modification with an organosilicon polymer with ≡SiH groups, formed as a result of hydrolysis and polycondensation of triethoxysilane. Then the modified silica was impregnated with silver nitrate solutions at room temperature and dried for 24 h in an oven at 150°C.

The synthesis of silver nanoparticles covered with silica shells in colloid solutions were performed by interaction of 0.05 ml of a triethoxysilane solution with a concentration of 0.54 mmol/l and 0.1 ml of a AgNO₃ solution with a concentration of 0.11 mmol/l. The solution obtained gained a yellow color as a result of intensive stirring.

The metal nanoparticles of were characterized using X-ray powder diffraction (DRON-4-07, CuK_α-radiation). Silica/silver structures of the core/shell type were identified by TEM (JEM 100CXII) and UV–VIS spectra recorded with a Carl Zeiss Jena spectrophotometer.

3. Results and Discussion

The metal nanoclusters were immobilized due to a reduction of the metal complexes immediately in a place of attaching of a reducer (surface ≡SiH groups). The reduction process is caused by the properties of the ≡SiH groups, accompanied by their hydrolysis and the formation of highly dispersed metal nanoparticles.

It is possible to regulate the sizes of the metal particles by varying the concentration of the metal salt used for the reduction. The formation of nanoparticles with smaller sizes is accompanied by visible changes of the color of the metal-containing composites in UV–VIS spectra. A blue shift of the surface plasmon band (Figure 1a) is observed as a result of the gradual decrease of the concentration of AgNO₃ from 11.25 to 1.25 mmol; the corresponding sizes of the nanoparticles decrease from 20 to 15 nm (Figure 1b).

Silver nanoparticles of smaller sizes can be synthesized by reducing the time of interaction between the modified mesoporous silica and the AgNO₃ solution. The formation of nanoparticles of larger sizes is accompanied by the appearance of a long-wave wing in the spectrum and a small blue shift of the surface plasmon band in the region of 370–400 nm (Figure 2a). In accordance with XRD data, the formation of silver nanoparticles in the range from 7 to 17 nm (Figure 2b) is observed.

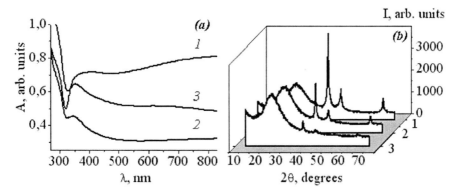

Figure 1. (*a*) UV–VIS spectra of colloid silver obtained after dissolution of the silica matrices modified with triethoxysilane after interaction for 24 h with AgNO₃ solutions with concentrations of 1: 11.25 mmol/l, 2: 3.75 mmol/l, 3: 1.25 mmol/l; (*b*) XRD measurements of the silver-containing silicas before dissolution.

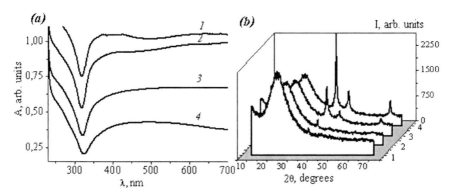

Figure 2. (*a*) UV–VIS spectra of colloid silver obtained after dissolution of silica matrices modified with triethoxysilane after their interaction with AgNO₃ solution (11.25 mmol/l) for 1: 0.5 h, 2: 1.5 h, 3: 5 h, 4: 24 h; *(b)* XRD spectra of the silver-containing silicas before dissolution.

The reduction of metal ions in the moment of hydrolysis and poly-condensation of triethoxysilane, and the formation of silica nanoparticles is one of the possible ways to stabilize silver nanoparticles. The formation of silver nanoparticles covered with silica shells was observed as a result of the interaction of triethoxysilane with a silver nitrate solution. A distinctive feature of these dispersions is the presence of a spectral maximum corresponding to the surface plasmon of isolated, weakly interacting particles whose number is initially the largest in the system.

The intensity of this maximum decreases with increasing degree of sol aggregation due to a decrease of the relative fraction of such particles (Figure 3a, curves 1–4). Increase of the AgNO₃ concentration (0.11 mmol/l) by 1.5–2 times at the same concentration of triethoxysilane (0.54 mmol/l) is accompanied by a drastic broadening and decrease of the intensity of the plasmon band (Figure 3, curves 6, 7). TEM data (Figure 3b) confirm the fact of formation of silver nanoparticles covered with silica shells.

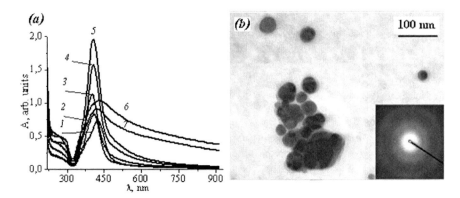

Figure 3. (a) Absorption spectra of colloid silver dispersions formed as a result of an interaction of triethoxysilane and solutions with various $AgNO_3$ concentrations: 1: 0.0033 mmol/l, 2: 0.011 mmol/l, 3: 0.033 mmol/l, 4: 0.078 mmol/l, 5: 0.11 mmol/l, 6: 0.16 mmol/l, 7 : 0.22 mmol/l; *(b)* TEM images of silver nanoparticles with silica shells.

4. Summary

Methods of synthesis of silver-containing silicas were elaborated. It is possible to regulate the sizes of the metal particles by varying the concentration of the metal salt taken for the reduction, and the time of reduction. The process of silver ion reduction in the moment of silica formation as a result of hydrolysis and condensation of triethoxysilane was investigated, and stable colloid solutions of silver covered with silica shells were obtained.

References

1.   U. Kreibig, M. Vollmer, *Optical properties of metal clusters* (Springer, Berlin 1995).
2.   J.L. Gong, J.H. Jiang, Y. Liang, G.L. Shen, R.Q. Yu, J. Colloid Interface. Sci. 298, 752 (2006).
3.   Y.C. Cao, R. Jin, C.S. Thaxton, C.A. Mirkin, Talanta 67, 449 (2005).
4.   A.N. Pavlenko, V.V. Yanishpolskii, V.A. Tertykh, V.G. Il'in, R. Leboda, J. Skubiszewska-Zieba, Chem. Phys. Technol. Surf. 7–8, 19 (2002).

ZnO NANOSTRUCTURES GROWN BY THERMAL
EVAPORATION AND THERMAL DECOMPOSITION METHODS

F. KYRIAZIS[1], S.N. YANNOPOULOS[*1],
A. CHRISSANTHOPOULOS[2], S. BASKOUTAS[2],
N. BOUROPOULOS[2]
[1]*Foundation for Research and Technology Hellas,
Institute of Chemical Engineering and High Temperature
Chemical Processes, FORTH/ICE-HT, P.O. Box 1414,
GR-26 504 Patras, Greece*
[2]*Department of Materials Science, University of Patras,
GR-26 504, Patras, Greece*

Abstract. We report on the growth of novel ZnO nanostructures using thermal evaporation and thermal decomposition methods. Thermal evaporation is used to produce ZnO/carbon nanotubes heterostructures, while ZnO nanocrystals were produced by thermal decomposition of zinc alginate gels. The nanostructures were characterized by a variety of techniques including x-ray diffraction, scanning electron microscopy, Raman scattering, and photoluminescence spectroscopy. Semi-empirical molecular orbital calculations gave information on the nature of the binding between ZnO and carbon nanotubes.

Keywords: ZnO nanostructures; CNT; nanocomposites; thermal decomposition

1. Introduction

Zinc oxide (ZnO) has recently attracted considerable attention due to its unique ability to form a dazzling variety of nanostructures with different morphologies.[1] ZnO is an exceptional material (a direct band gap of 3.37 eV at ambient temperature and a high ecxiton energy of 60 meV) that exhibits semiconducting, piezoelectric, and pyroelectric properties. Under specific, controlled growth conditions, ZnO nanostructures such as nanocombs, nanorings, nanobelts, nanowires, and nanocages have been synthesized.[1]

---

[*]sny@iceht.forth.gr

J.P. Reithmaier et al. (eds.), *Nanostructured Materials for Advanced Technological Applications*,   211
© Springer Science + Business Media B.V. 2009

Nanostructured ZnO finds novel applications in optoelectronics, sensors, transducers, and biomedical sciences because it is bio-compatible.[2]

Thermal evaporation (TE) and hydrothermal methods are among the most frequently employed techniques to grow ZnO nanostructures. Parameters such as temperature, heating time, vapor pressure of Zn or ZnO, the presence of catalysts, the oxygen partial pressure and gas flow rate, and the nature of the substrates, are only few among the variety of parameters, which influence the nucleation, growth and morphologies of nanostructures grown by TE. Si, Si/Au, Al_2O_3, SiO_2 are the most frequently employed substrates. In this contribution we present results for ZnO nanostructures grown on carbon nanotubes (CNTs) which have until now not been used as a substrate for ZnO growth.[3] CNTs have themselves a number of attractive properties and a plethora of applications. However, it is widely believed that their performance would be improved if nano-heterojunctions of CNTs with other nanostructures could be synthesized. In addition, a novel method for the formation of zinc oxide nanocrystals which is based on the thermal decomposition of a precursor zinc compound (zinc alginate) is presented.[4]

2. Experimental

For the TE method an equimolar mixture of ZnO and graphite fine powder was loaded on a quartz boat placed at the closed end of a silica tube. A pellet of multi-walled CNTs was placed into the silica tube a few centimeters away from the reacting mixture near the furnace exit in order to achieve the desired temperature gradient. The mixture was inserted for a few minutes in a preheated (at ~1,000°C) horizontal furnace, while the CNT pellet was placed at various distances from the reacting mixture in the temperature range of 850–950°C. Under these conditions, carbothermal reduction of ZnO occurs and the Zn produced, which has a significant vapor pressure at this temperature, is transferred through the gas phase and re-oxidizes. The zinc could be fully or partially oxidized in the air stream, leading to the formation of stoichiometric ZnO or zinc suboxides.

The thermal decomposition method is based on the thermal treatment of zinc alginate gels. A zinc solution was prepared either from zinc nitrate or from zinc acetate by dissolving the respective salts in ultrapure water. Sodium alginate solutions at a concentration of 1% w/w were prepared in the same way. The zinc alginate beads were produced by dropwise addition of 10 ml of alginate solution into 20 ml of zinc solution. The gel beads formed were maintained in the gelling medium for 30 min under gentle stirring; then they were separated from the solution by a stainless steel grid, placed in a porcelain crucible and heated at 450 or 800°C for 24 h with a heating rate of 10°C/min.

3. Results and Discussion

The morphological characterization of the ZnO nanostructures was performed using scanning electron microscopy (SEM). Figure 1 shows typical SEM images of nanocrystals grown by the two methods described in the previous section. In all cases the XRD patterns were indexed to hexagonal wurtzite structured ZnO. An analysis of the particles size of the ZnO nanocrystals obtained by thermal decomposition revealed a significant increase of the size with the heating temperature (Figure 1a, b). A number of experiments has been undertaken in the TE method where the only parameters changed were the duration of the experiment (4, 8 and 15 min) and the substrate temperature (850–950°C). In this way we were able to obtain a series of nanostructures with different morphologies.[4] For a substrate temperature of ca. 950°C the ZnO nanostructures were found to saturate quickly; their shape is mainly that of polypods. In the case of short-time experiments (4 min) isolated nanorods of ZnO are found dispersed in the CNT matrix (Figure 1c). For substrate temperature <850°C the rate of ZnO nanostructures formation was found to be low.

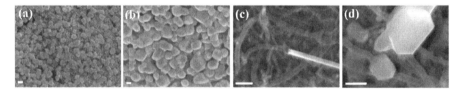

Figure 1. ZnO nanocrystals produced by thermal decomposition of zinc alginate at 450°C (a) and 800°C (b). SEM micrographs of ZnO nanostructures grown on CNTs (c) and (d). Scale bar: 200 nm.

In general, the main building blocks of the nanostructures are rod-like, independent of the growth conditions. These rod-like nano-objects can be found either isolated or self-assembled to mesoscopic structures such as polypods (having a low number of legs, up to 4–5). The legs have in general a hexagonal cross-section and may terminate either with a nearly flat surface, a nanopencil geometry, or a needle-like configuration. In the cases where the growth starts at the end of a CNT, the formation of single nanorods is highly likely. Actually, Figure 1c shows such a case where a nanorod and a bundle of CNTs with similar diameters are joined as the arrow shows.

The ZnO crystal (C_{6v} point group symmetry) has six Raman-active modes. The energies and their symmetries are as follows (Figure 2a): 101 cm^{-1}, E_2 (low); 381 cm^{-1}, A_1(TO); 407 cm^{-1}, E_1(TO); 437 cm^{-1}, E_2 (high); 574 cm^{-1}, A_1(LO) and 583 cm^{-1}, E_1 (LO). The main difference between the spectra of bulk and nanosized ZnO is the intensity of the mode at 583 cm^{-1}.

Based on the significant dependence of this Raman peak on the stoichiometry of oxygen, several works have associated the 583 cm^{-1} peak with oxygen vacancies, zinc interstitials or combination of the two.

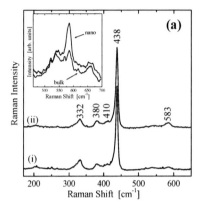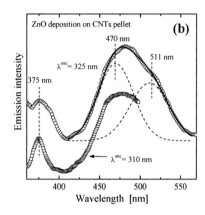

Figure 2. (a) Raman spectra of (i) bulk ZnO and (ii) nanosized ZnO obtained from zinc alginate decomposition at 800°C. (b) PL spectrum of ZnO nanostructures on CNT substrate.

The optical properties of the ZnO nanostructures were studied by photoluminescence (PL) spectroscopy. In Figure 2b PL spectra of ZnO, excited by 310 and 325 nm radiation, are presented. They are composed of two main peaks; a narrow one located at about 375 nm, which corresponds to the near-band edge emission of the wide band gap ZnO, and a broad asymmetric band containing two components at 470 and 511 nm. The latter is usually referred to as green emission and is attributed to the singly ionized oxygen vacancy in ZnO where the emission results from the radiative recombination of a photo-generated hole with an electron occupying the oxygen vacancy.[5] The peak at 470 nm or the blue emission band is not frequently observed in ZnO nanostructures; its origin is still not known. This peak has been assigned to the existence of oxygen-depleted interface traps in ZnO_x.

References

1. Z.L. Wang, J. Phys.: Condens. Mat. 16, R829 (2004).
2. Y.W. Heo, D.P. Norton, L.C. Tien, Y. Kwon, B.S. Kang, F. Ren, S.J. Pearton, and J.R. LaRoche, Mater. Sci. Eng. R. 47, 1 (2004).
3. A. Chrissanthopoulos, S. Baskoutas, N. Bouropoulos, V. Dracopoulos, D. Tasis, and S.N. Yannopoulos, Thin Solid Films 515, 8524 (2007).
4. S. Baskoutas, P. Giabouranis, S.N. Yannopoulos, V. Dracopoulos, L. Toth, A. Chrissanthopoulos, and N. Bouropoulos, Thin Solid Films 515, 8461 (2007).
5. K. Vanheusden, W.L. Warren, C.H. Seager, D.R. Tallant, J.A. Voigt, and B.E. Gnade, J. Appl. Phys. 79, 7983 (1996).

ANTIMICROBIAL EFFECTS OF SILVER NANOPARTICLES SYNTHESIZED BY AN ELECTROCHEMICAL METHOD

R.R. KHAYDAROV[*1], R.A. KHAYDAROV[1],
O. GAPUROVA[1], Y. ESTRIN[2], S. EVGRAFOVA[3],
T. SCHEPER[4], S.Y. CHO[5]
[1]*Institute of Nuclear Physics, Uzbekistan Academy
of Sciences, 702132 Tashkent, Uzbekistan*
[2]*ARC Centre of Excellence for Design in Light Metals,
Department of Materials Engineering, Monash University
and CSIRO Division of Materials Science and Engineering,
Clayton, Victoria, Australia*
[3]*V.N. Sukachev Institute of Forest SB RAS, Krasnoyarsk,
Russia*
[4]*Institute of Technical Chemistry, Leibniz University
Hannover, Germany*
[5]*Yonsei University, Seoul, South Korea*

Abstract. The paper deals with the authors' research in the field of antimicrobial properties of silver nanoparticles obtained by a novel electrochemical technique, which provides low minimum inhibitory concentration (MIC) values as well as a high efficacy of nanosilver as an antimicrobial agent against a range of microbes on the surface of water paints, textile fabrics, and fiber sorbents.

Keywords: silver; nanoparticles; electrochemical synthesis; nanomaterials

1. Introduction

Silver nanoparticles, objects from 1 to 100 nm in size, were extensively investigated over the last decades due to their unique size-dependent optical, electrical and magnetic properties. To date most of the applications of silver nanoparticles are connected with their use as antibacterial/antifungal

---

[*]renat2@gmail.com

J.P. Reithmaier et al. (eds.), *Nanostructured Materials for Advanced Technological Applications*,   215
© Springer Science + Business Media B.V. 2009

agents in biotechnology and bioengineering,[1] medicine,[2] textile engineering,[3] water treatment,[4] and silver-based consumer products.[5] Numerous synthesis approaches were developed to achieve a controlled production of Ag nanoparticles,[6,7] but the quest is still on for easy and inexpensive methods for the production of silver nanoparticles and silver-based materials with highly efficient anti-microbial properties.

2. Materials and Methods

Silver nanoparticles were obtained by our recently reported three-stage process based on the electroreduction of silver ions in water.[8] In order to impregnate a commercially available cotton fabric and water-purifying fiber sorbent[9] samples with silver nanoparticles, the simple padding procedure[3] was used. In a separate experiment, a commercially available water paint was mixed with silver nanoparticles solutions at a ratio of 7:1 in order to impart antimicrobial properties on the paint. The morphology of the silver-based materials obtained was studied using transmission electron microscopy (TEM), scanning electron microscopy (SEM), and atomic force microscopy (AFM) measurements. To evaluate the antibacterial and antifungal properties of Ag nanoparticles, *Escherichia coli* was used as a representative Gram-negative bacterium; *Staphylococcus aureus* and *Bacillus subtilis* were used as Gram-positive bacteria; and *Aspergillus niger* and *Penicillium phoeniceum* were employed to represent cosmopolitan saprotrophic fungi. The antibacterial and antifungal properties of Ag nanoparticles added to cotton fabric/water paint/fiber sorbent samples were evaluated using standard microbiological procedures.[10] The MIC of silver nanoparticle solutions, defined as the lowest concentration of the agent that completely inhibited microbial growth, was determined for various microbes using the standard macrodilution broth susceptibility test.

3. Results and Discussion

The studies revealed that silver nanoparticles suspended in water solution in a three-stage electrochemical technique are nearly spherical, their average diameter being 7 ± 3 nm. Varying the process conditions one can also get particles up to 100 nm in the concentration range of 20–40 mg/l. By AFM measurements it was demonstrated that the applied method had provided silver nanoparticles being sufficiently stable for at least 7 years, even being stored under ambient conditions (Figure 1).

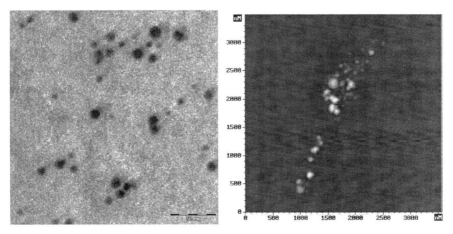

Figure 1. TEM image of silver nanoparticles obtained by the electrochemical method (left) and an AFM image of the same nanoparticles that have been stored for 7 years. (right).

Due to the high purity of the obtained nanoparticles low MIC values for *E. coli* (5 mg/l), *S. aureus* (2 mg/l) and *B. subtilis* (19 mg/l) cultures have been obtained. These values are very low in comparison with those obtained for the same bacteria by Ruparelia and co-authors,[11] although they studied smaller silver nanoparticles (3.32 ± 1.13 nm). It was also observed that an increase of the mean size of Ag nanoparticles in the samples tested causes a rise of the corresponding MIC values, i.e. smaller silver nanoparticles have a greater antimicrobial efficacy.

We have also impregnated cotton fabrics, water paints and water purifying fiber sorbents with nanosized silver colloids synthesized by the electro-chemical technique. In spite of the observed agglomeration of silver nano-particles into larger clusters, the modified materials demonstrate a pronounced antifungal/antibacterial efficacy on *Aspergillus niger, Penicillium phoeniceum* and *S. aureus* cultures (Table 1).

TABLE 1. Antimicrobial effect of samples of cotton fabric, water paint and fiber sorbent impregnated with nanosilver.

Microbe	Non-treated Sample	Modified Cotton Fabric (20 mg/l)	Modified Paint (3 mg/l)	Modified fiber sorbent (20 mg/l)
Aspergillus niger	+	–	–	–
Penicillium phoeniceum	+	–	–	no data
S. aureus	+	–	+	–
Control (samples on beef-extract agar)	+	–	–	–

"+" : growth on beef-extract agar;   "-" : absence of growth on beef-extract agar

4. Summary

The results obtained show that silver nanoparticles produced by our novel electrochemical technique can be used for modifying commercially available water paints, cotton fabrics, and fiber sorbents to prevent bacterial/ fungal growth.

ACKNOWLEDGMENTS

R. R. Khaydarov acknowledges partial support of this work through the INTAS Fellowship Grant No. 5973 for Young Scientists under the "Uzbekistan – INTAS 2006" program.

References

1. C.M. Niemeyer, *Ang. Chem. Int. Ed.* 40 (22), 4128 (2001).
2. P.H. Hoet, I. Brüske-Hohlfeld, and O.V. Salata, *J. Nanobiotechnol* 2, 1 (2004).
3. H.J. Lee and S.H. Jeong, *Textile Res. J.* 75, 551 (2005).
4. A.Y. Solov'ev, T.S. Potekhina, I.A. Chernova et al., *Russ. J. Appl. Chem.* 80, 438 (2007).
5. C. Buzea, I.I. Pacheco, and K. Robbie, *Biointerphases*, 2, 17 (2007).
6. R. Prucek, L.Kvítek, and J. Hrbáč, *Facultas rerum naturalium, Chemica* 43, 59 (2004).
7. K.V. Katok, V. Tertykh, and V. Yanishpolskii, in *Metathesis Chemistry*, edited by Y. İmamoğlu and V. Dragutan (Springer, Netherlands), 471 (2007).
8. R.R. Khaydarov, R.A. Khaydarov, O. Gapurova, Y. Estrin, and T. Scheper, submitted to *J. Nanoparticle Res.* Doi:10.1007/s11051-008-9513-x (2008).
9. R.A. Khaydarov, O. Gapurova, R.R. Khaydarov, S.Y. Cho in: *Modern Tools and Methods of Water Treatment for Improving Living Standards*, edited by A. Omelchenko et al. (Springer, Netherlands), 101 (2005).
10. Y. Estrin, R.R. Khaydarov, R.A. Khaydarov, O. Gapurova, S.Y. Cho, T. Scheper, and C. Endres, in: Proceedings of the 2008 International Conference on Nanoscience and Nanotechnology, 25–29 February 2008, Melbourne, Victoria, Australia, 44 (2008).
11. J.P. Ruparelia, A.K. Chatterjee, S.P. Duttagupta, and S. Mukherji, *Acta Biomaterialia* 4, 707 (2008).

THE EFFECT OF Au NANOCLUSTERS IN TIN OXIDE FILM GAS SENSORS

G.A. MOUSDIS[*1], M. KOMPITSAS[1], I. FASAKI[1],
M. SUCHEA[2,a], G. KIRIAKIDIS[2,b]
[1]NHRF-National Hellenic Research Foundation,
Theoretical and Physical Chemistry Institute-TPCI, 48 Vass.
Constantinou Ave., Athens 11635, Greece
[2]IESL-FORTH and University of Crete [a]Chemistry
Departmant and [b]Physics Departmant, Greece

Abstract: The effect of Au nanoparticles in SnO_2 was investigated for gas sensor applications. The films were prepared by the sol–gel method. $HAuCl_4$ was added to a tin alkoxide solution, the mixture was hydrolyzed and spin coated on glass substrates. The samples were then thermally treated to remove the organics. The films were characterized by thermogravimetric analysis, scanning electron microscopy, and X-ray diffraction. Additionally, the optical absorbance and the reflectivity were measured. The films were tested against hydrogen. The change of the electrical conductivity was used to detect the gas. The response of SnO_2 and SnO_2–Au to H_2 was investigated at different temperatures and concentrations. With the addition of Au nanoparticles the detection limit decreased, the working temperature was reduced (from 180°C to 140°C), and there was a ten times increase of the signal.

Keywords: SnO_2 thin films; Au nanoparticles; hydrogen sensor

1. Introduction

Gas sensors based on metal oxide sensitive layers are playing an important role in the detection of toxic pollutants and inflammable gases. Metal oxides and especially tin dioxide is widely used as a basic material for the preparation of gas sensing devices operating in these applications.[1–3] The effect of

---

[*]gmousdis@eie.gr

J.P. Reithmaier et al. (eds.), *Nanostructured Materials for Advanced Technological Applications*, 219
© Springer Science + Business Media B.V. 2009

the addition of many different metallic particles on the gas sensing properties of metal oxides has been widely studied, but the results depend on the experimental conditions and the method of fabrication.[4]

Hydrogen is an abundant, renewable, efficient, clean energy source. As an industrial gas, it is currently used in a large number of areas, e.g. chemistry (crude oil refining, plastics, as a reducing environment in float glass industry, etc.), food products (hydrogenation of oils and fats), semiconductors (as a processing gas in thin film deposition and annealing atmospheres), and transportation (as fuel in fuel cells and space vehicle rockets).

All these applications require the development of hydrogen sensing devices that allow safe control of the gas usage. Devices capable of detecting the presence of hydrogen above the low explosion limit (LEL) of 40,000 ppm have become indispensable to prevent explosions.[5]

A pure alkoxide method has been used to prepare SnO_2 thin films undoped and doped with Au nanoparticles. The films prepared were tested as H_2 sensors.

2. Experimental

The substrates (glass slides) were cleaned by soaking for 24 h in a sulforochromic bath and kept in isopropanol until used. A $HAuCl_4$ solution in ethanol was prepared by dissolving Au in $HNO_3/3HCl$, then dried under vacuum; the remaining yellow crystals of $HAuCl_4$ were dissolved in ethanol to obtain a 0.3M solution. The tin alkoxide starting solution was obtained from pure $SnCl_4$ using the NH_3 gas method. The working solution was prepared by mixing the tin alkoxide starting solution with the $HAuCl_4$ solution. The solution obtained was aged by stirring at room temperature for 24 h. After spinning, the tin oxide gel films were dried at R.T. for 24 h and then heat-treated for 2 h at 510°C in air. The thickness of the films was approximately 90 nm, as measured by means of an alpha step apparatus.

The films were characterized by thermogravimetric analysis (TGA), scanning electron microscopy (SEM) and X-ray diffraction (XRD). Additionally, the optical absorbance and the reflectivity were measured.

Hydrogen sensing tests were performed in an aluminum vacuum chamber. The chamber was evacuated down to 1 Pa, then filled with dry air at atomspheric pressure. The SnO_2 and Au–SnO_2 thin films were tested in the temperature range of 147–180°C at hydrogen concentrations from 10,000 to 500 ppm. The hydrogen concentration was calculated on the basis on the partial pressures of the sensing gas and air inside the chamber. A bias of 1 V was applied, and the current through the film was measured with a Keithley Mo. 485 Picoammeter. Current modifications helped to monitor the hydrogen sensing.

3. Results and Discussion

Thermogravimetric analysis showed that the isopropanol and the water evaporate at 118°C while the remaining organics are burned up to 400°C. At 400°C the HAuCl$_4$ is reduced to metallic Au. Over 500°C there is no loss of mass (Figure 1a). In the XRD diagram a clear tendency of texturing in the [101] crystalline direction of the tetragonal rutile structure can be seen (Figure 1b). A homogeneous dispersion of irregularly shaped Au nanoparticles with dimensions of few hundreds nanometers is observed in the SEM image (Figure 2a). The surface plasmon resonance (SPR) is the peak at 560 nm in the absorption spectrum (Figure 2b). According to the FWHM, the size of the Au nanoparticles is 3.5 nm.

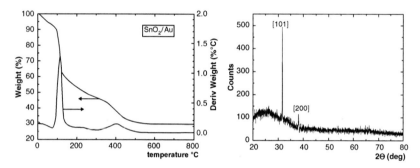

Figure 1. TGA (left) and XRD diagram (right) of a SnO$_2$–Au film.

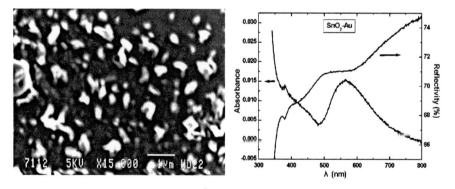

Figure 2. SEM image (left) and UV–VIS spectrum (right) of a SnO^2–Au film.

The response of SnO$_2$ and SnO$_2$–Au to H$_2$ was investigated at different temperatures and concentrations. SnO$_2$ films did not respond at all at temperatures lower than 180°C. With the addition of Au nanoparticles, the detection limit decreased, the working temperature was reduced (from 180°C to 140°C), and there was a ten times increase of the signal (Figure 3).

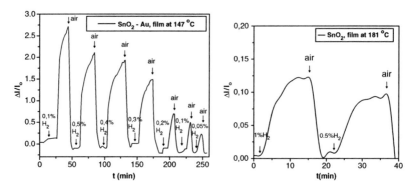

Figure 3. Response of Au–SnO$_2$ and SnO$_2$ thin films at 147°C and 181°C, respectively.

4. Conclusions

The doping of SnO$_2$ with Au nanoparticles decreased the sensor working temperature with respect to the undoped one by 40°C, while the response increased by more than ten times.

ACKNOWLEDGEMENTS

The authors would like to acknowledge the support of the Hellenic General Secretariat for Research and Technology through a bilateral Greek–Romanian cooperation program.

References

1. E. Comini, Anal. Chim. Acta 568, 28 (2006)
2. G.J. Li and S. Kawi, Mater. Lett. 34, 99 (1998)
3. D. Kohl, Sensor. Actuat. 18, 71 (1989)
4. A. Cabot, J. Arbiol, J.R. Morante, U. Weimar, N. Bârsan and W. Göpel, Sensor. Actuat. B: Chem. 70, 87 (2000)
5. Fuel Cell Standards Committee, "*Basic Consideration for safety of Hydrogen Systems*", Technical Report ISO TC 197 N166, International Standards Organization, 2001.

PREPARATION AND SOME CHARACTERISTICS OF POLYEPOXYPROPYLCARBAZOLE THIN FILMS WITH EMBEDDED Au–Ag NANOPARTICLES

A.V. KUKHTA[*1], E.E. KOLESNIK[1], A.I. LESNIKOVICH[2], S.A. VOROBYOVA[2], M. NICHICK[2], A.P. LUGOVSKII[3], V. YEROKHOV[4]

[1]*B.I. Stepanov Institute of Physics, National Academy of Sciences of Belarus, Nezalezhnastsi Ave. 68, 220072 Minsk, Belarus*
[2]*Research Institute for Physical Chemical Problems, Belarusian State University, Nezalezhnastsi Ave. 2, 220080 Minsk, Belarus*
[3]*Research Institute for Applied Physical Problems, Belarusian State University, Kurchatova St. 7, 220064 Minsk, Belarus*
[4]*National University "Lviv Polytechnic", P.O. Box 1050, 79045 Lviv, Ukraine*

Abstract. Colloidal dispersions and thin films containing poly-N-(epoxypropyl)carbazole (PEPK) and Ag–Au nanoparticles (2–3 nm in diameter) were prepared and investigated by TEM, optical and IR spectroscopy, and electrophysical tools. The introduction of nanoparticles results in cooperative phenomena between the polymer and the Ag–Au particles; as a consequence these particles increase the electrical conductivity of the nanocomposite films by at least one order of magnitude, even in very low concentrations.

Keywords: poly-N-(epoxypropyl)carbazole; Ag–Au nanoparticles; thin film; TEM; optical spectroscopy; conductivity

[*] kukhta@imaph.bas-net.by

J.P. Reithmaier et al. (eds.), *Nanostructured Materials for Advanced Technological Applications*, 223
© Springer Science + Business Media B.V. 2009

1. Introduction

Conducting polymers and other organic π-conjugated compounds containing metal nanoparticles have wide prospects for applications in different branches of science and technology. The presence of metal nanoparticles strongly changes both, the optical and the electrophysical properties of the polymers. Additional organic shells on the metal nanoparticle surface results in further changes of the nanocomposite properties. Up to date, the most studies have been performed with aliphatic thiols, though for aromatic thiols the properties changes are more dramatical.

In this paper some spectroscopic and electrophysical properties of solutions and thin films of poly-N-epoxypropylcarbazole (PEPK) containing Ag–Au nanoparticles are presented. This work is aimed at a possible development of non-curable low cost printing conductors. In Au–Ag alloys, Au increases the stability, while Ag results in lower work functions.

2. Experimental

Colloidal dispersions of silver and gold in organic media were synthesized by interphase interactions between silver and gold complexes with quaternary ammonium salts and sodium borohydride, which were dissolved in organic and aqueous phases, accordingly. Surfactants were formed as a result of the interphase interactions and stabilize the colloidal dispersion "in situ". The preparation of organosols and nanoparticles is similar to that described in details earlier.[1,2] The composite films were prepared by the spin coating method. 2-(4-mercaptomethylphenyl)-5-pheniloxazole has been used as aromatic thiol.

3. Results and Discussion

The red-brown silver–gold colloidal solutions were formed in toluene and in PEPK/toluene solutions as a result of interphase interactions. The color of the solutions started to change from yellow to reddish-brown after the reaction had proceeded for 1–2 min. The TEM images of thin polymer films presented in Figure 1 show that in the thin composite films discrete crystalline particles of 1.8 nm diameters with a very narrow size distribution ($\sigma = 0.19$) and a high density are formed. At the same time, thin films with thiol-stabilized nanoparticles include two types of particles. These nanoparticles are crystalline and nearly spherical with mean diameters of 1.4 nm ($\sigma = 0.22$) and 4.9 nm ($\sigma = 0.24$). An important feature of the thin film material prepared is that the individual metal particles do not aggregate but seem to be well-protected and linked to each other by their thiol shells.

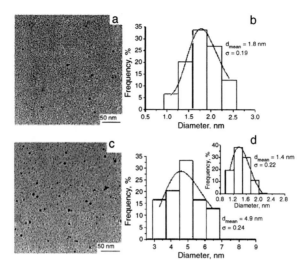

Figure 1. TEM images and histograms of thin polymer films with Ag–Au nanoparticles (a, b) and thiol stabilized Ag–Au nanoparticles (c, d).

In the UV–VIS spectra of these films a broad peak is observed in the 450–500 nm region, owing to a surface plasmon resonance. The peak positions for both films, with nanoparticles and thiol-stabilized particles, are different despite the fact that the particle sizes are comparable as shown by the TEM analysis. Consequently, the exact origin of the differences in the UV–VIS spectra is the presence of two types of nanoparticles.

The details of IR spectral features of PEPK films containing particles and those without them are quite different. Symmetric and asymmetric stretching vibration bands related to the pyrimidine group appear in the 1000–1500 cm^{-1} region. Addition of the freshly prepared Ag–Au colloid to the polymer shifted the bands to the $1,600$–$1,700$ cm^{-1} region, which is characteristic for stretching vibrations of π–system of benzene rings. The region of $1,000$–$1,150$ cm^{-1} is characteristic for stretching vibrations from tert C–N and –CH$_2$–O–CH$_2$– groups present in the polymer chains. Addition of thiol-stabilized nanoparticles to the polymer films results in changes near $1,150$ cm^{-1} and a band duplication at $1,130$ and $1,120$ cm^{-1}, as well as at $1,063$ and $1,070$ cm^{-1}. A possible consequence of the interaction of metal particles with thiol is the formation of an extra peak at 771 cm^{-1}. These changes can be apparently attributed to organic-nanoparticle interactions.

The film conductivity has been measured at different temperatures (290–370 K) in vacuum and air. Figure 2 illustrates that the incorporation of metallic nanoparticles enhances the electrical conductivity of the films. The concentration of nanoparticles is much lower than required for the percolation threshold. The observed temperature dependences are approximately linear in accordance with the hopping conductivity mechanism. The results of this study suggest that the nanoparticles themselves can act as conductive junctions in the composite material. Replacing organic aliphatic shells by aromatic thiols increases the conductivity. The value of the increase depends on the temperature.

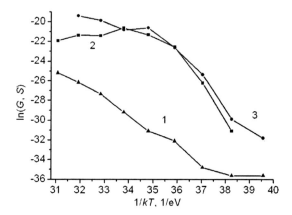

Figure 2. Temperature dependences of conductivity of pure PEPK (1) and Ag-Au+PEPK nanocomposites (2,3) thin films with nanoparticle concentrations of 4.5×10^{-4} M for Ag and 1.9×10^{-4} M for Au; 3: with aromatic thiol based shells.

The temperature dependence of the thin film conductivity presented, measured at a pressure of 10^{-2} Pa during the cooling process, has usual an activation character with $E = 0.66$ eV. The value of the activation energy of composite sample with thiol-stabilized nanoparticles is about 2 eV. This indicates that with the temperature change in the sample not only electronic processes but also structural reorganization occur. The conductivity behaviour in the region of the glass transition (in the middle of Figure 2) is rather unusual. Combining the above mentioned facts allows to assume that a heating process in an electric field at temperatures near the glass the transition (70°C) causes an aggregation of thiol-stabilized metal nanoparticles.

References

1. M.N. Nichick, S.A. Vorobyova, A.I. Lesnikovich, A.V. Kukhta, E.E. Kolesnik. In: V.E. Borisenko, S.V. Gaponenko, V.S. Gurin (Eds.), *Physics, Chemistry and Applications of Nanostructures* (World Scientific, 2007), p. 336.
2. A. Kukhta, E. Kolesnik, D. Ritchik, A. Lesnikovich, M. Nichik, S. Vorobyova. Mat. Sci. Eng. C 26, 1012 (2006).

NANOSTRUCTURAL CARBONACEOUS FILMS WITH METAL (Pd, Ni) NANOPARTICLES

E. CZERWOSZ[1], M. KOZŁOWSKI[1], P. DŁUŻEWSKI[2],
J. KĘCZKOWSKA[3], M. SUCHAŃSKA[3]
[1]Tele & Radio Research Institute, ul. Ratuszowa 11, 03-450
Warszawa, Poland
[2]Institute of Physics PAN, Al. Lotnikow 36/46, 02-668
Warszawa, Poland
[3]Kielce University of Technology, al. 1000 -lecia P. P. 7,
25-314 Kielce, Poland

Abstract. Nanomaterials containing metal particles are interesting from both the scientific and the practical points of view. The aim of our work is to describe our recent results concerning preparation, characterization and possible applications of nanostructural carbonaceous films with metal (Pd, Ni) nanoparticles.

Keywords: nanomaterials; heterostructural carbon films; metal nanocrystallites

1. Introduction

For many years scientists have been interested in developing new nano-materials. Carbon is a particularly interesting material due to its ability to form various chemical bondings. For this reason carbon can create structures with different physical properties, starting from insulators (e.g. diamond) and ending with conductors (e.g. nanotubes). Metallic nanostructures introduced into a carbonaceous matrix can not only modify the electron structure of the system, but also generate quantum size effects and change certain properties, not observed on a macro scale (e.g. light or electron emission).

J.P. Reithmaier et al. (eds.), *Nanostructured Materials for Advanced Technological Applications*, 227
© Springer Science + Business Media B.V. 2009

2. Preparation of Metal-Carbon Nanocrystalline Films

Our nanostructural carbonaceous films with metal (Pd, Ni) nanoparticles were obtained by physical vapour deposition (PVD) from two separate sources containing C_{60} mixtures and metal organic compounds (palladium acetate or nickel acetate). C_{60} and the metal compound were placed in separate Mo boats. We used glassy carbon, Si and silica substrates, depending on the experimental characterization method applied. The substrates were degreased before placing them in the vacuum chamber. The deposition process was carried out under a dynamic pressure of 10^{-3} Pa and at a temperature of 330–340 K (the temperature measured on the surface of the substrate). The films have various structures and morphologies, depending on the film composition and the technological process. Details of the technological process are presented in Refs. 1, 2.

3. Nanostructural Carbonaceous Films with Metal (Pd, Ni) Nanoparticles – Structure and Properties

A variety of different measurement techniques, e.g. transmission electron microscopy (TEM), atomic force microscopy (AFM), chemical analysis methods, and optical absorption spectroscopy have been used to study the physical properties of these carbonaceous composite materials.[3,4]

Heterostructural carbon films with metal nanocrystallites are composed of grains built of carbon and metal nanocrystallites. This means that they contain metal, fullerens, graphite or amorphous carbon nanograines. The structure and size of the crystallites are different for films containing different metal concentrations. The observed metal nanograin sizes vary from 1.5 to 8 nm. Carbon grains, with sizes ranging from 10–100 nm, form a matrix for the metal nanograins. Depending on the type of metal, the observed growth can be columnar (for palladium) or insular (for nickel). In the case of carbon-palladium films, the palladium is dispersed in the carbonaceous lattice and has probably the form of clusters with diameters not exceeding 1 nm. In this case the carbon matrix is of the fullerene fcc type. Changing the technological process parameters (deposition time, substrate temperature) leads to palladium nanograins of various sizes. Films containing nickel have a multi-phase structure and include Ni grains with sizes of 2–6 nm and carbonaceous grains of various carbon allotropes (amorphous carbon, graphite, fullerenes). A detailed description of both structures can be found in our studies.[5-11]

4. Possible Applications

Carbonaceous composite materials may have only recently caught the attention of the world but much advancement has been made since their discovery about a decade ago. Due to these qualities the field of applications is almost endless.

Within the framework of a cooperation of the Kielce University of Technology and the Tele&Radio Research Institute in Warsaw research is carried on the application of heterostructural carbon materials with an admixture of metal nanocrystallites (Ni or Pd) for constructing cold cathode elements which emit electrons due to the field emission properties.[3,4] Heterostructural carbon films possess a high electron emission efficiency depending on their structure, although the electron field emission in these films is less efficient than that of nanotubes or layers of carbon nanofibre emitters.[10-14] Cathodes of nanostructural carbonaceous films with metal nanoparticles can be also used in portable x-ray machines, electroluminescent light sources, and displays.[15-17]

References

1. E. Czerwosz, M. Adydan, P. Dłuzewski, W. Gierałtowski, M. Kozłowski, E. Starnawska, and H. Wronka, Electr. Sci. Works: Vac. Tech. Vac. Technol. 123, 83 (1999).
2. E. Czerwosz, P. Dłuzewski, W. Gierałtowski, J. W. Sobczak, E. Starnawska, and H. Wronka, J. Vac. Sci. Technol. B18, 1064 (2000).
3. E. Czerwosz, R. Diduszko, P. Dłużewski, J. Kęczkowska, M. Kozłowski, J. Rymarczyk, and M. Suchańska, Vacuum 82, 372 (2007).
4. E. Czerwosz, P. Dłużewski, J. Kęczkowska, M. Kozłowski, J. Rymarczyk, and M. Suchańska, Proc. SPIE 6347, 6347–50 (2006).
5. E. Czerwosz, P. Dłuzewski, and R. Nowakowski, Vacuum 54, 57 (1999).
6. E. Czerwosz, B. Surma, and A. Wnuk, J. Phys. Chem. Solid 61, 1973 (2000).
7. E. Czerwosz and P. Dłuzewski, Diamond Relat. Mater. 9, 901 (2000).
8. E. Czerwosz, P. Dłuzewski, W. Gierałtowski, J.W. Sobczak, E. Starnawska, and H. Wronka, J. Vac. Sci. Technol. B18, 1064 (2000).
9. E. Czerwosz, P. Dłuzewski, J.P. Girardeau-Montaut, D. Vouagner, and K. Zawada, Vacuum 63, 355 (2001).
10. Y. Sohda, D.M. Tanenbaum, S.W. Turner, and H.G. Craighead, J. Vac. Sci. Technol. B 15, (1997).
11. G.A.J. Amaratunga and S.R.P. Silva, Appl. Phys. Lett. 68, 2529 (1996).
12. W.L. Geiss, J.C. Twichell, J. Macaulay, and K. Okano, Appl. Phys. Lett. 67, 1328 (1995).
13. L.A. Chernozatonskii, Y.V. Gulayev, Z.J. Kosakovskaja, N.I. Shinitsyn, G.V. Torgashov, Y.F. Zakharchenko, E.A. Fedorov, and V.P. Valchuk, Chem. Phys. Lett. 233, 63 (1995).

14. H.H. Busta and R.W. Pryor, J. Appl. Phys. 82, 5148 (1997).
15. H. Sugie, M. Tanemura, V. Filip, K. Iwata, K. Takahashi, and F. Okuyama, Appl. Phys. Lett. 78, 2578 (2001).
16. W. Knapp, D. Schleussener, A.S. Baturin, I.N. Yeskin, and E.P. Sheshin, Vacuum 69, 339 (2003).
17. E.P. Sheshin, A.S. Baturin, K.N. Nikolskiy, R.G. Tchesov, and V.B. Sharov, Appl. Surf. Sci. 251, 196 (2005).

5.3. NANOCOMPOSITES

NANOSTRUCTURED IONIC CONDUCTORS: INVESTIGATION OF SiO_2–MI (M = Li, Ag) COMPOSITES

A. PRADEL[*], P. YOT, S. ALBERT, N. FROLET, M. RIBES
*Institut Charles Gerhardt Montpellier, UMR 5253
CNRS-UM2-ENSCM-UM1, Equipe Physicochimie des
Matériaux Désordonnés et Poreux (PMDP), CC 1503,
Université Montpellier 2, Place E. Bataillon, 34095
Montpellier Cedex 5, France*

Abstract. Synthesis and characterization of porous Vycor®7930–MI (M = Li, Ag) composites are discussed. Two types of composites were prepared: sintered composites obtained by classical sintering (CS) and composites obtained by electro-crystallisation (EC). In the case of LiI composites an increase of the conductivity of two orders of magnitude as compared to the conductivity of pure LiI was observed. In such a case good coating of insulating particles and filling of the pores was observed. In the case of AgI the main important finding is the presence of an hysteresis phenomenon in the conductivity versus temperature curve at a temperature close to the transition $\alpha \leftrightarrow \beta$ AgI.

Keywords: ion conductors; nanostructured composites; porous oxide glasses

1. Introduction

All solid state batteries are interesting devices as power sources for electric equipments. Compared to batteries including a liquid electrolyte, they show no leakage and a high selectivity of the charge carrier leading to increased safety and lifetime. Moreover they allow miniaturization. However, in order to reach the stage of production, some problems have to be overcome: in particular the electrolyte conductivity should be improved and so the quality of the "electrolyte/electrode" interfaces. About 35 years ago, C.C. Liang

---

[*] apradel@lpmc.univ-montp2.fr

J.P. Reithmaier et al. (eds.), *Nanostructured Materials for Advanced Technological Applications*, 233
© Springer Science + Business Media B.V. 2009

reported that composites made of LiI and Al_2O_3 have a ionic conductivity higher by two orders of magnitude than that of pure LiI.[1] Later on, several groups worked on the ionic conductivity of halide salts (LiI, AgI, AgCl, AgBr) added to inorganic oxides (Al_2O_3, SiO_2) and showed that composites formed by an ionic conductor and an insulating matrix are generally more conductive than the ionic conductor itself.[2–5] More recent work has demonstrated that the enhancement in ionic conductivity could be even larger when the insulating matrix was a porous material, e.g. porous alumina or silica, thus revealing a pore size effect.[6–9] Such composites appear interesting materials as electrolytes for solid state batteries.

While the insulating matrix was a crystalline phase in most of the previous studies (Al_2O_3, porous silica, etc.), a porous glassy matrix (Vycor®7930)[10–12] was chosen as the insulating matrix for composite preparation in the present work. Porous Vycor® glass 7930 (VPG) is a commercial glass from Corning©. It is an almost pure vitreous silica with a composition of 96.3 SiO_2-2.95 B_2O_3-0.04 Na_2O (wt %).[13,14] The glass was prepared by the original process discovered by Hood and Nordberg.[15,16] Due to a porosity of about 28% and an average pore diameter of 4 nm, the porous Vycor® has an internal surface area of approximately 250 m^2/g.[13,14]

Synthesis and characterization of composites obtained by two synthesis processes, classical sintering (CS) and electro-crystallisation (EC), will be discussed in the next part of the paper.

2. Synthesis of Nanocomposites

CS has been used for the preparation of almost all composites described in literature.[1,5–11,12] EC developed by M. Nagai et al.[17,18] during the 1990, was used to synthesize AgI-composites with bulk porous Vycor®7930.[11] Due to its extremely low solubility in water ($\approx 10^{-8}$ moll^{-1}) silver iodide is the only candidate to be inserted inside Vycor®7930 pores.

2.1. CLASSICAL SINTERING

xMI-(1-x)VPG (M = Ag, Li; x = 0.4, 0.5, 0.6) powders were first prepared from silver iodine (Fluka 99.99%) or lithium iodide (Aldrich 99.99%) and nanoporous Vycor®7930 glass. Preliminary, the VPG was ground in an agate mortar and dried at 673 K. The VPG and AgI or LiI powders weighted in stoichiometric proportion were then mixed inside a glove box under argon atmosphere (H_2O < 5 ppm) to avoid water contamination. The mixtures were placed in a silica ampoule, evacuated and sealed under a secondary vacuum ($\approx 10^{-5}$ mbar). They were heated up to a temperature higher by 20° than the melting temperature of AgI (T_f = 830 K) or LiI (T_f = 750 K) with a heating rate of 10 K/h. They were maintained at this temperature for 20 h to insure the soaking of the particles of VPG, and then cooled down to room temperature.

The obtained powders were then ground in an agate mortar and further cold-pressed (5 t) to give pellets of 10 mm in diameter and less than 2 mm in thickness. One pellet was used as such to perform a first series of conductivity measurements. Another one was sealed in an evacuated quartz tube (secondary vacuum $\approx 10^{-5}$ mbar). A second heat treatment at 830 K (AgI) and 750 K (LiI) for 20 h was then carried out in order to increase the density prior to a second series of electrical characterizations. The two kinds of pellets will be called sintered and non-sintered pellets, respectively, further in the text.

2.2. ELECTRO-CRYSTALLISATION

In contrast to the previous process, electro-crystallisation can only be used for the synthesis of AgI based composites. The AgI electro-crystallisation in the pores of the glass was carried according to the process described in Figure 1.

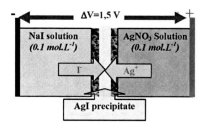

Figure 1. Schematic representation of the electro-crystallisation process.

A Vycor® disk (13 mm in diameter and less than 2 mm in thickness) was placed in between two Pyrex tubes. One tube was filled with a solution of 0.1 moll^{-1} NaI and the other one with a solution of 0.1 moll^{-1} AgNO$_3$. Platinum electrodes were plunged into the two solutions. During the electro-crystallisation process carried out at room temperature for times varying from ¾ to 16 h, a voltage of 1.5 V was applied between the two solutions. Because of the polarization, Ag$^+$ from the AgNO$_3$ solution and I$^-$ from the NaI one penetrated through the nanopores inside the matrix and reacted with each other. It resulted in the formation of a precipitate close to the middle part of the glass as indicated by a change in color of the porous glass from white to yellow. Further to the electro-crystallisation the glassy disk was washed several times with distilled water and dried at 373 K.

3. Physical and Chemical Characterizations

X-ray diffraction (XRD) was used to identify the phases present in the initial powders and in the obtained composites. A PANalytical X'PERT diffractometer was used to perform the XRD measurements using a Cu Kα source (λ = 1.5406 Å). Fresh fractures of the obtained pellets were then analyzed by scanning electron microscopy using a Hitachi S-4500 I or a S360 Cambridge Instrument microscope.

3.1. SINTERED COMPOSITES

XRD was carried out on heat treated powders before and after sintering to identify the crystalline phases and to determine the size of the ion conducting particles. Figure 2 shows the X-rays diffraction patterns of 0.5MI-0.5VPG (M = Ag, Li) powders prepared according to the procedure described in Sect. 2.1.

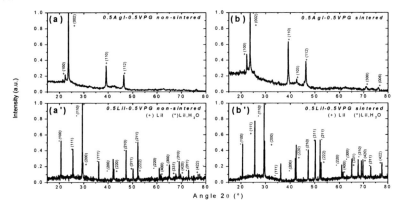

Figure 2. X-ray diffraction patterns of 0.5AgI-0.5VPG composites (non-sintered (a) and sintered at 850 K for 20 h (b)) and of 0.5LiI-0.5VPG composites (non sintered (a') and sintered at 743 K for 20 h (b')).

β-AgI is the only phase identified in the XRD patterns of non-sintered (a) and sintered (b) VPG-AgI composites The diffraction patterns of non-sintered (a') and sintered (b') VPG-LiI composites correspond to that of a mixture of two phases: anhydrous LiI (+) with some traces of mono-hydrated LiI, H_2O (*). LiI was partially hydrated even though all the operations were carried out in a glove box. This result could be explained by the fact that the porous matrix was not completely free of water. Because of the nanometric size of the VPG pores one might suggest that the water adsorbed on the surface and in the pores of the silica could only be removed under more drastic drying conditions than those used in this study.

The apparent crystallite sizes were calculated from the XRD patterns by the Scherrer method using

$$d = \frac{180}{\pi} \frac{K \cdot \lambda}{B \cdot \cos\theta} \tag{1}$$

where B is the peak width at half maximum, θ the angle of diffraction of the highest intensity diffraction peak, K = 0.9 and λ = 1.5406 Å. The crystallite size for 0.5VPG-0.5AgI composites was ≈60 nm before and after the second sintering process. The crystallite size for 0.5VPG-0.5LiI slightly increased during the sintering process, changing from ~90 to ~110 nm after the second heat treatment.

Scanning electron microscopy was used to evaluate the homogeneity of composites as well as the quality of the coating of the insulating particles. Figures 3 and 4 present SEM images of cross sections of MI-VPG composites at various magnifications. For the two types of composites an increase of the pore size (to reach ~100th of nanometers) during heat treatment is observed. Such an increase can be explained by the presence of residual water and an eventual pore coalescence when the heat treatment temperature becomes close to the glass temperature transition of Vycor. A similar observation has been made on a Vycor sample (free from LiI or AgI) when it was heat treated under the same conditions.[10]

Figure 3. SEM images: cross section of a 0.5VPG-0.5AgI sintered pellet (a) and surface of a porous Vycor® particle (b).

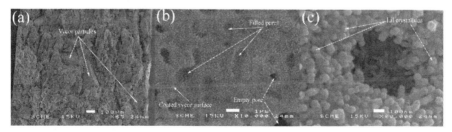

Figure 4. SEM images: cross section of a 0.5VPG-0.5LiI sintered pellet (a) and (b) a coated porous Vycor® particle (c).

Figures (3a) and (4a) show clearly the homogeneity of the 0.5VPG-0.5MI pellets (M = AgI or LiI). Magnified images of Vycor particles show a clear difference between the two types of composites. LiI covers the particle surface, fills the pores and is interconnected throughout the material (Figure 4b, c). At the opposite AgI does not fill the pores and appears to coat only poorly the surface of the Vycor grain (Figure 3b).

3.2. ELECTROCRYSTALLISATION

Figure 5 shows an X-ray diffraction pattern of a composite elaborated by EC (b) along with the diffraction pattern of an AgI precipitate (a). It shows the presence of hexagonal β-AgI and some traces of cubic γ-AgI. The diffraction peaks in the composite pattern are much broader and less resolved than those of the pure precipitate due to the presence of the glassy matrix. The Scherrer method has been used to evaluate the size of the AgI crystallites, i.e. ~11 nm. The SEM images shown in Figure 6 give evidence of the complex structure of the composite. In these images one can see the AgI precipitate which had developed close to the middle of the glassy Vycor. A time dependence of the thickness of the AgI layer has been evidenced by scanning electron microscopy. The layer thickness increases with time to reach a maximum of ≈230 μm after 16.5 h of deposition (Figure 7). Such a behavior can be easily explained by the blocking character of the AgI solid layer to Ag^+ and I^- ions.[11] The composition of the layer containing both conducting and insulating phases has been derived from the SEM images and the thickness of the layer. Whatever the deposition time the composition was found to be 0.21AgI-0.79VPG (in molar fraction).

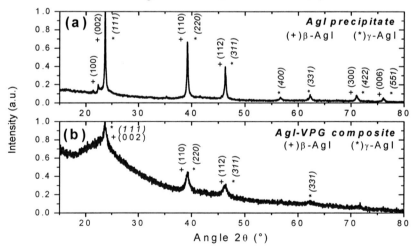

Figure 5. X-ray diffraction patterns of a pure AgI precipitate (a) and of a composite obtained after 16.5 h of electro-crystallisation (b).

Figure 6. SEM images of composites obtained after 6 h of deposition at various magnifications ((a) ×413, (b) ×867 and (c) ×1,520).

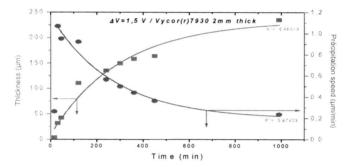

Figure 7. Thickness and growth rate of AgI precipitase inside Vycor®7930 versus deposition time.

Electron probe micro analysis (EPMA) experiments were carried out to obtain the elemental distribution of silver (Figure 8b) and iodine (Figure 8c). Silver and iodine are homogeneously distributed in the samples. This is in agreement with the XRD results which indicate the presence of AgI.

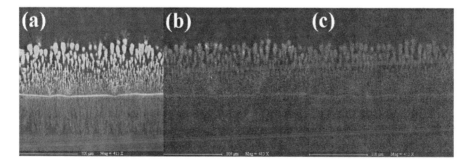

Figure 8. Microstructure (a) and elemental distribution of Ag (b) and I (c) at a cross section of a composite obtained after 6 h of electro-crystallisation (magnification ×413).

4. Electrical Measurements

The ionic conductivity of the composites was analysed using a Solartron SI-1620 by measuring the a.c. impedance at frequencies ranging from 10 MHz to 1 Hz. The measurements were carried out with a homemade sealed conductivity cell with stainless steel electrodes under a primary dynamic vacuum from room temperature to 443 K. The two parallel faces of the pellets were coated by 100 nm thick sputtered platinum electrodes. The d.c. conductivity was calculated from the Cole–Cole plots of the a.c. impedance measurement using a non-linear least square fitting program included in the ZView software.

4.1. CLASSICAL SINTERING

Figure 9 shows the ratio of the conductivities of xVPG-(1-x)MI and pure MI for x = 0.4, 0.5, 0.6 and M = Ag, Li.

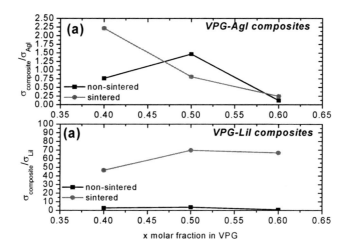

Figure 9. Ratio of the conductivities of xVPG-(1-x)MI and pure MI for x = 0.4, 0.5, 0.6, M = Ag, Li and T = 300 K (classical sintering).

In the case of AgI, the obtained results are rather scattered with no clear evolution versus the AgI content or the preparation procedure. Only a small increase of the conductivity compared to that of pure AgI ($\sigma \times 2.5$) could be obtained. Such a result is in agreement with the poor coating of the insulating particles and the poor filling of the pores. In contrast a clear picture arises from the conductivity measurements of LiI containing composites: the non-sintered composites exhibited conductivities rather similar to that of the pure ionic conductors (1×10^{-8} Scm^{-1} against 2.7×10^{-9} Scm^{-1} for LiI).

Clearly the absence of densification of the pellet is detrimental to an increase of the conductivity. However, after the second heat treatment, in agreement with the better filling of the pores by LiI, the conductivity of the composites increased significantly by two orders of magnitude for the best conductor, i.e. 0.5VPG-0.5LiI.

4.2. ELECTRO-CRYSTALLISATION

The conductivity measurement of samples obtained by electro-crystallisation required a specific preparation of the sample. The method can be understood on the basis of the schema shown in Figure 10a. The classical measurement of conductivity would require deposition of sputtered electrodes on the faces of the pellet. However, the AgI layer being only present in the centre of the pellet we had to use the "transverse method". It required first cutting the sample in a parallelepiped form and then depositing the electrodes on the cross section at the place where AgI was present.

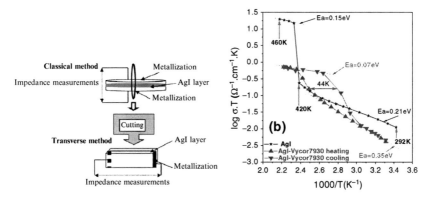

Figure 10. Schema of classical and transverse methods used to measure electrical conductivities (a). Temperature dependence of the electrical conductivity, of AgI-Vycor® composites obtained by electro-crystallisation (b).

The temperature dependence of the conductivity of the AgI composite and pure β-AgI is shown in Figure 10b. The composites do not show improved ionic conductivities compared to pure AgI. Upon heating the composite presents a phase transition at about 405 K while the phase transition $\alpha \leftrightarrow \beta$ AgI occurs at 420 K. Upon cooling the transition is observed at lower temperature ($\approx$360 K). Such an hysteresis has already been observed and was attributed either to the presence of AgI-polytypes[19] or to the compression effect of the silica matrix.[20]

5. Conclusion

Two series of xMI-(1-x)Vycor®7930 (x = 0.4, 0.5 and 0.6; M = Ag, Li) composites were obtained by impregnating molten AgI or LiI into a nanoporous glassy SiO_2 matrix. While a first heat treatment had only minor effects on the conductivity of the LiI composites, a second treatment improved the coating of the insulating particles and filling of the pores. An increase in conductivity of two orders of magnitude has been obtained for 0.5VPG-0.5LiI composite. In the case of AgI, the lack of improvement of the conductivity and the scattered results can be explained by the poor coating of the insulating grains and the poor filling of pores.

Electro-crystallization was then used to insert AgI inside a nanoporous Vycor glassy matrix. An AgI layer precipitates in the center of the insulating matrix. The precipitate thickness is time dependent and shows a leveling off after few hours of deposition. The conductivity of the composite exhibits an anomalous behavior with a hysteresis phenomenon related to the phase transition α ↔ β AgI. The cause of such a behavior, AgI polytypes or effect of the compression of the glassy matrix, is currently under investigation.

References

1. C.C Liang, J. Electrochem. Soc. 120, 1289 (1973).
2. J.B. Phipps, D.L. Johnson, D.H. Whitmore, Solid State Ionics 5, 393 (1981).
3. F.W. Poulsen, N.H. Andersen, B. Kindl, J. Schoonman, Solid State Ionics 9 & 10, 119 (1983).
4. J. Maier, J. Phys. Chem. Solids 46, 309 (1985).
5. G. Ardel, D. Golonitsky, E. Peled, Y. Wang, G. Wang, S. Bajue, S. Greenbaum, Solid State Ionics 113–115, 477 (1998).
6. H. Yamada, A. J. Bhattacharyya, J. Maier, Adv. Funct. Mater. 16, 525 (2006).
7. H. Maekawa, R. Tanaka, T. Sato, Y. Fujimaki, T. Yamamura, Solid State Ionics 175, 281 (2004).
8. H. Maekawa, Y. Fujimaki, H. Shen, J. Kawamura, T. Yamamura, Solid State Ionics 177, 2711 (2006).
9. H. Yamada, I. Moriguchi, T. Kudo, Solid State Ionics 176, 945 (2005).
10. S. Albert, N. Frolet, P. Yot, A. Pradel, M. Ribes, Solid State Ionics 177, 3009 (2006).
11. S. Albert, N. Frolet, P. Yot, A. Pradel, M. Ribes, Micropor. Mesopor. Mater. 99, 56 (2007).
12. S. Albert, N. Frolet, P.G. Yot, A. Pradel, M. Ribes, Mater. Sci. Eng. B 150 (3), 199 (2008).
13. T.H. Elmer (Ed.), *Engineered Materials Handbook, Porous and Reconstructed Glasses, Vol. 4: Ceramics and Glasses* (ASM International, Materials Park, OH, USA, 1992), p. 427.
14. Corning©, http://www.corning.com/lightingmaterials/images/Vycor_7930.pdf.
15. H.P. Hood, M.E. Nordberg, "Treated Borosilicate Glass," U.S. patent 2,106,744, Feb. 1938.

16. H.P. Hood, M.E. Nordberg, "Method of Treating Borosilicate Glasses," U.S. patent 2,286,275, June 1942.
17. M. Nagai, T. Nishino, Solid State Ionics 53–56, 63 (1992).
18. M. Nagai, T. Nishino, J. Mater. Synth. Process. 6 (3), 197 (1998).
19. J.-S. Lee, S. Adams, J. Maier, J. Phys. Chem. Solids, 61, 1607 (2000).
20. M. Tatsumisago, Y. Shinkuma, T. Minami, Nature 354, 217 (1991).

PREPARATION AND PROPERTIES OF FLEXIBLE NANOCOMPOSITES, OBTAINED BY A COMBINATION OF COLLOIDAL CHEMISTRY AND SOL-GEL APPROACH

V.I. BOEV[*1], A. SOLOVIEV[2], C.J.R. SILVA[3], M.J.M. GOMES[3], J. PÉREZ-JUSTE[4], I. PASTORIZA-SANTOS[4], L.M. LIZ-MARZÁN[4]
[1]Institute of Electrochemistry and Energy Systems, Bulgarian Academy of Sciences, Acad. G. Bonchev St., bl. 10, 1113 Sofia, Bulgaria
[2]Center of Microtechnology and Diagnostics, St. Petersburg State Electrotechnical University, 197376 St. Petersburg, Russia
[3]Escola de Ciências, Universidade do Minho, 4710-057 Braga, Portugal
[4]Departamento de Química Física, Universidade de Vigo, 36200, Vigo, Spain

Abstract. Transparent di-ureasil hybrid materials, doped with CdS, Au and Ag nanoparticles (NPs) were obtained by mixing preformed colloidal dispersions with a ureasilicate precursor, followed by gelation and drying. The nanocomposites were characterized by optical spectroscopy in the UV and visible region and by steady-state photoluminescence.

Keywords: di-ureasil; ureasilicate; hybrids; nanocomposites; semiconductors; metals; nanoparticles; sol-gel; colloids; organic-inorganic; flexible material

1. Introduction

An important area for potential application of nanosized materials, particularly semiconductor and metal NPs, is associated with optics. These materials possess specific linear optical properties which strongly depend on the particle

---

[*]v_boev@yahoo.com

size and shape. Moreover, they are also attractive because of their large third order nonlinear susceptibility ($\chi 3$) and their ultra-fast nonlinear optical response.

Usually, semiconductor and metal NPs with different sizes and shapes can be obtained in a form of colloidal dispersions by methods of colloidal chemistry. These colloidal solutions are not convenient for producing optically active devices for optoelectronics. Thus, a very important technological task, which can greatly expand the application range of NPs based materials, is the incorporation of NPs in a solid matrix by transferring them from a liquid colloidal dispersion into a processable matrix.

The matrix itself must satisfy a number of requirements such as a high optical transparency in the visible region, and mechanical and chemical stability. Sol-gel processes are a versatile approach for making matrix materials, as they combine mild synthesis conditions with the ability to incorporate organic and inorganic species into the host medium during the synthesis process. Pure inorganic oxide glasses, doped with CdS, ZnS and PbS NPs[1-3] were prepared by this way. However, such inorganic structures are susceptible to cracking during the drying stage.

A combination of organic and inorganic matrix components results in enhanced resistance against cracking and processability of the final material, while conserving the high optical transparency.[4] Typical representatives of hybrid organic inorganic materials are the so-called ureasilicates or di-ureasils, which consist of a reticulated siliceous backbone covalently bonded to polyether-based segments by urea (-NHCONH-) linkages. These materials were initially used for the immobilization of highly luminescent Eu[+] salts,[5] organic dyes[6] and lithium salts.[7]

Recently, ureasilicate nanocomposites, doped with semiconductor and metal NPs have been successfully fabricated by mixing preformed NPs colloidal dispersions with ureasilicate precursors prior to sol-gel transition.[8,9] It is shown in this article that the final optical quality of the doped ureasilicate xerogels strongly depends on the preparation conditions, such as pH of the medium and the type of catalyst.

2. Experimental

2.1. PREPARATION OF DI-UREASIL XEROGEL, DOPED WITH CADMIUM SULFIDE NPS

CdS-containing ureasil nanocomposites were prepared according to the procedure reported by Boev et al.[8] Briefly, the formation of the ureasilicate precursor was obtained by reacting 0.936 mmol of O,O'- bis(2-amino-propyl)-polyethylene glycol-600 (Jeffamine-ED-600, Fluka) and 1.870 mmol of 3-isocyanatepropyl triethoxysilane (ICPTES, Aldrich). Later 1.12 mmol of tetraethoxysilane (TEOS, Aldrich) was added in order to improve the mechanical properties of the matrix. The pH value of the mixture was measured by indicator paper (Merck). Then 1.0 ml of a colloidal dispersion of fluorothiophenol-capped CdS NPs in tetrahydrofuran (THF, Merck) was added dropwise. The mixture was poured in a Teflon mould with an inner diameter of 42 mm; the mould was kept inside a closed compartment under ammonia atmosphere up to gelling. After subsequent drying at 40°C a flat, uniformly transparent and flexible xerogel in the form of a disk of ca. 0.45 mm thickness and a diameter of 39 mm was obtained. Undoped matrices have been synthesized by the same protocol, but without addition of the CdS colloid before gelation.

Absorption (UV–VIS) and luminescence spectra were recorded at room temperature with Shimadzu UV-2501 PC and Spex Fluorolog spectro-photometers, respectively.

2.2. IMMOBILIZATION OF GOLD AND SILVER NPS IN AN UREASILICATE MATRIX

A detailed description of the preparation procedure has been published elsewhere.[9] Briefly, the procedure includes the formation of the ureasilicate precursor in a first step. Subsequently added NH$_4$OH (500 µl, 4.1 M) served as a catalyst for the sol-gel process. In a second step, a preformed colloidal solution of Au NPs, covered with silica shells (denoted as Me@SiO$_2$, where Me is Au or Ag), was added to the mixture. The particles were dispersed in a pure grade ethanol (Scharlab). The mixture was poured in a polycarbonate cuvette (10 mm path length), covered with Parafilm® and dried within an oven at 35°C for 36 h. Crack-free, transparent monoliths with a base of 7×7 mm^2 and height of 15 mm were obtained. Doping of the ureasilicate matrix with Ag@SiO$_2$ NPs was performed by the procedure described above, but citric acid (0.22 M) was added as a catalyst instead of ammonia.

3.   Results and Discussion

Figure 1 shows optical absorption and PL emission spectra of the doped nanocomposite and, for comparison, the spectra of the original CdS colloidal solution used for the synthesis. As in the original solution, the spectrum of the nanocomposite demonstrates a blue shift of the absorption edge compared to that of bulk CdS (about 500 nm) and a well-resolved absorption peak due to the first excitonic transition. The particle size, determined from the position of the exciton resonance energy state, is about 2.0 nm.

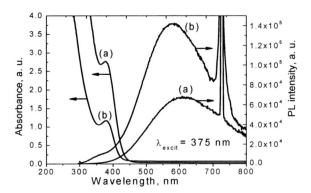

Figure 1. Absorbance and PL spectra for CdS NPs in the original solution (a) and a ureasilicate matrix, doped with CdS NPs (b).

The observed broad PL emission of the NPs both, in the solution and in the matrix is significantly red shifted (272 and 246 nm) with respect to the absorption peaks. Such a PL behavior can be attributed to the recombination of charge carriers immobilized in traps of different energies in the forbidden zone.

The good optical quality of the nanocomposites is ensured through a reliable control of the starting pH of the ureasilicate precursor. It has been established that at pH > 7 flocculation of the NPs occurred. It leads to an increase of scattering in the low energy range of the absorption spectrum. The effect of flocculation may be connected with the presence of residual non-reacted amino groups in the mixture. Under this condition, the stabilizing fluorophenyl ligands attached to the surface of the NPs can be replaced by amine moieties. A pictorial representation of the reaction is shown in Figure 2.

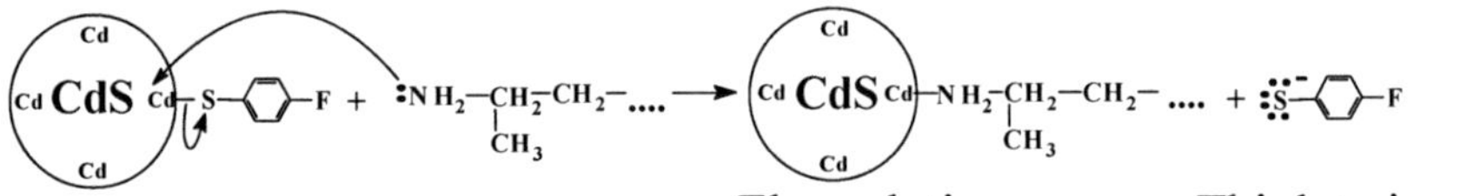

Figure 2. A model of the reaction promoting the flocculation of fluorothiophenol-capped CdS NPs.

The stability of the NPs against flocculation has been achieved by neutralization of the residual amine moieties by adding of some additional amount of ICPTES, which results in a decrease of the pH down to 4.5.

The use of a catalyst is very important for producing optically clear ureasilicate xerogels with homogeneously distributed metal NPs. If only water is used, a change of the optical absorption spectrum after gelation indicates a pronounced decrease of the plasmon band intensity, caused by a gradual sedimentation of the NPs (Figure 3 (left)).

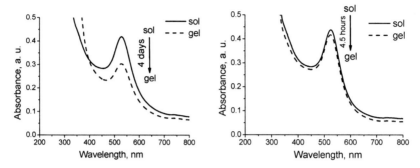

Figure 3. Optical absorption spectrum of the ureasilicate precursor containing $Au@SiO_2$ NPs before and after gelation without ammonia (left) and before and after gellation in the presence of ammonia (right). The particles size is about 15 nm.

In the presence of ammonia, the gelation rate increased up to 20 times. Consequently, the viscosity of the ureasilicate sol increases rapidly within a short period of time, which influences the mobility of the gold particles, decreasing the probability of a collision. This factor prevents, in fact, their aggregation and subsequent sedimentation in contrast to the system without ammonia catalyst. As it is shown in Figure 3 (right), the intensity, position and width of the plasmon band of the NPs remains almost unchanged at the end of gelation.

Figure 4 illustrates the variation of the absorption spectrum with time for silver containing ureasilicate gels, obtained by using ammonia and citric acid as catalysts.

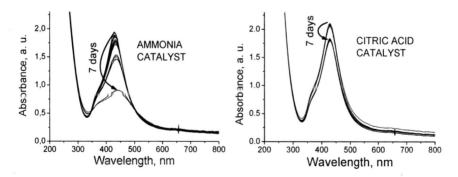

Figure 4. Evolution of the absorption spectra with time for a ureasilicate gel doped with $Ag@SiO_2$ NPs in the presence of ammonia (left) and a ureasilicate gel doped with $Ag@SiO_2$ NPs in the presence of citric acid (right). The particle size is about 20 nm.

It is clear that in the presence of ammonia the silver cores are dissolved due to oxidation into the soluble $Ag(NH_3)_2^+$ complex (Figure 4, left). The reaction can be expressed as follows:

$$Ag + 2NH_3(aq) \rightarrow Ag(NH_3)_2^+$$

If citric acid is used instead of ammonia, the shape and the intensity of the plasmon band remains almost unchanged (Figure 4, right). This is evidence that $Ag@SiO_2$ NPs are not affected by the processes of dissolution or aggregation.

4. Conclusions

The combination of sol-gel methods and colloidal chemistry was successfully applied for the incorporation of CdS, gold and silver NPs in an ureasilicate matrix. Optical absorption spectra demonstrate that the optical properties of starting colloids are fully retained in the solid state material. A precise control of the pH value of the reaction medium guarantees the preservation of quantum size effect of the semiconductor NPs. The choice of the catalyst is very important for achieving optimal conditions for producing gold and silver doped ureasilicate nanocoposites.

ACKNOWLEDGMENT

This work has been supported by the Portuguese Fundação para a Ciência e a Tecnologia (FCT). V. Boev and A. Soloviev thank the FCT for research grants SFRH/BD/3188/2000 and SFRH/BPD/18098/2004, respectively.

References

1. M. Nogami, K. Nagasaka, E. Kato, J. Am. Chem. Soc. 73, 2097 (1990).
2. Y. Zhang, N. Raman, J.K. Bailey, C.J. Brinker, J. Phys. Chem. 96, 9098 (1992).
3. Y. Zhang, C.J. Brinker, J.K. Bailey, N. Raman, R.M. Crooks, C.S. Ashley, Proc. SPIE 1758, 596 (1992).
4. C.-Y. Li, J.Y. Tseng, K. Morita, C. Lechner, Y. Hu, J.D. Mackenzie, Proc. SPIE 1758, 410 (1992).
5. L.D. Carlos, Y. Messaddeq, H.F Brito, R.A. Sá Ferreira, V. de Zea Bermudez, S.J.L. Ribeiro, Adv. Mater. 12, 594 (2000).
6. E. Stathatos, P. Lianos, Langmuir 16, 8672 (2000).
7. V. de Zea Bermudez, L. Alcácer, J.L. Acosta, E. Morales, Solid State Ionics 116, 197 (1990).
8. V.I. Boev, A. Soloviev, C.J.R. Silva, M.J.M. Gomes, Solid State Sci. 8, 50 (2006).
9. V.I. Boev, J. Pérez-Juste, I. Pastoriza-Santos, C.J.R. Silva, M.J.M. Gomes, L.M. Liz-Marzán, Langmuir 20, 10268 (2004).

TEMPLATE SYNTHESIS OF MESOPOROUS SILICAS INSIDE NANOREACTORS BASED ON LARGE PORES OF SILICA GEL

I.S. BEREZOVSKA[*], V.V. YANISHPOLSKII,
V.A. TERTYKH
*O.O. Chuiko Institute of Surface Chemistry, Ukrainian
National Academy of Sciences, General Naumov St. 17,
03164 Kyiv, Ukraine*

Abstract. The peculiarities of template synthesis of mesoporous silicas inside large pores of silica gel are studied. The silicas samples synthesized are characterized by low-temperature adsorption–desorption of nitrogen, X-ray diffraction as well as by thermogravimetric analysis.

Keywords: mesoporous silica; template synthesis; silica gel

1. Introduction

The control of the porous structure and the structural stability of mesoporous silicas are current trends in template synthesis of ordered porous materials[1-5]. However, the range of applications is limited by the structural hydrothermal instability of such materials due to hydrolysis of Si–O–Si bonds in the presence of adsorbed water. The structure of ordered mesoporous silicas is destroyed by mechanical compression due this hydrolysis. The stability of mesoporous silicas can be improved by increasing the hydrophobicity as well as by increasing the degree of condensation due to silanol bonds during the mesostructure formation. An increase of the pore wall thickness also effectively stabilizes the ordered structure against thermal and hydrothermal instabilities. Here we report an approach of improving the mechanical stability of mesoporous silicas by means of a template synthesis inside the pores of inorganic matrices with high structural stability.

---
*berrina2003@rambler.ru

J.P. Reithmaier et al. (eds.), *Nanostructured Materials for Advanced Technological Applications,* 251
© Springer Science + Business Media B.V. 2009

2. Experimental

The synthesis of mesoporous silicas in nanoreactors based on large silica gel pores was carried out by a step-by-step incorporation of micellar solutions of tetraethoxysilane (TEOS) and cetyltrimethylammonium bromide (CTAB) inside pores of silica gels with a measured surface area of 115 m^2/g and an average pore diameter of 24 nm. The procedure of preparing the micellar solution with a molar ratio of 1TEOS:0.18CTAB:5NH$_3$:75H$_2$O is described elsewhere.[6] After template elimination the synthesized materials were characterized by low-temperature adsorption–desorption of nitrogen (ASAP-2000) for an estimation of the surface area (BET equation) as well for a determination of the pore size distribution (DFT method). X-ray diffraction was used for structural analysis of the samples synthesized. Thermogravimetric studies of uncalcined samples provided information about their organic template content.

3. Results and Discussion

Results of the thermal analysis of the template-containing mesoporous silicas synthesized inside the pores of silica gel are presented in Figure 1. Thermogravimetric analysis (TG) and differential thermogravimetric (DTG) curves show a typical decomposition profile with four distinct weight loss steps regarding (1) ammonia and water evaporation (up to 200°C), (2) organic template decomposition (200 up to 360°C) and (3) thermal oxidation of residual organic compounds (360–550°C). For temperatures above 550°C a slight weight reduction (4) corresponding to water losses due to condensation of silanol groups to siloxane bonds is observed. About 19% of the total weight loss of 25.9% correspond to the organic template content. According to the molar ratio of the micellar solution prepared, the theoretical quantity

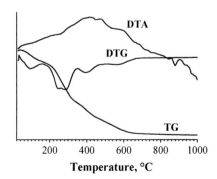

Figure 1. TGA, DTG and DTA curves of mesoporous silicas synthesized inside the pores of silica gel.

of the mesoporous silicas equals 3% of the total weight of the material, which is in good agreement with the results of the by thermal analysis.

The nitrogen adsorption–desorption isotherm (Figure 2a) for silica gel is of type III according to the IUPAC classification.[7] The hysteresis loop at relative high pressures reveals the existence of large pores in the samples. The nitrogen adsorption–desorption isotherms of the silica gel after template synthesis inside the nanopores are characterized by the appearance of capillary condensation regions ($0.3 < p/p_o < 0.4$), a proof for the presence of mesopores. The sharpness of these regions increases after each synthesis; it is accompanied by a simultaneous decrease of the hysteresis loop, which can be explained by filling of the silica gel pores with mesoporous silicas.

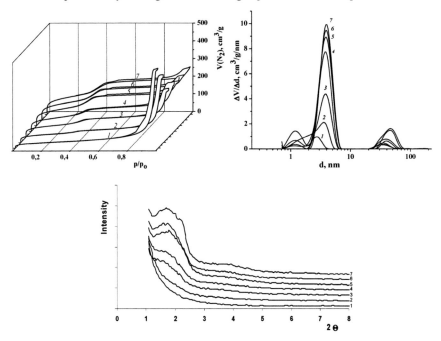

Figure 2. Upper left: nitrogen adsorption–desorption isotherms, upper right: differential curves of the dependence of the pore size on pore volume, and lower panel: diffractograms for silicas synthesized inside large pores of silica gel: initial silica gel (*1*) and silica gels after the first to sixth (*2–7*) introduction of mesoporous materials.

The specific surface area of the silica gel is 115 m^2/g (according to the BET equation); the DFT pore size distribution curve for the silica gel shows a broad peak for pore sizes in the range between 30 and 50 nm (Figure 2b, curve 1). Samples with incorporated mesoporous silicas show an increase of the specific surface area from 115 up to 377 m^2/g (Figure 2a). The pore size distribution curves for silica gels after template syntheses inside their pores reveal an average pore size of 3 nm (Figure 2b, curves 2–7).

The absence of peaks in the diffractogram of the initial silica gel confirms its amorphous nature (Figure 2c, curve 1). The X-ray diffraction patterns of the silica gel after introduction of the micellar solution cause the appearance of diffraction peaks in the lower-angle regime (Figure 2c, curves 2–7). These broad angle diffraction peaks there correspond to the formation of wormlike pore structures, whereas the weak diffraction intensities are caused by the small quantities of incorporated silicas. Structural adsorption characteristics of the synthesized samples are presented in Table 1.

TABLE 1. Structural-adsorption characteristics of synthesized silicas (DFT-method).

Samples		Structural-adsorption characteristics						
		S_{meso} m^2/g	S_{macro} m^2/g	S_{micro} m^2/g	S_{total} m^2/g	V_{meso} cm^3/g	V_{macro} cm^3/g	V_{micro} cm^3/g
Initial silica gel		66	14	23	103	0.46	0.23	0.007
Number of	1	105	9	32	146	0.43	0.15	0.012
template syn-	2	156	3	46	205	0.36	0.05	0.015
theses inside	3	248	1.5	12	262	0.37	0.03	0.004
the pores of	4	292	0.8	0.7	293	0.39	0.03	0
the silica gel	5	295	0.7	0.3	296	0.38	0.02	0
	6	313	1	0.3	314	0.39	0.03	0

4. Summary

In conclusion we presented a successful synthetic procedure that combines the template synthesis with the conventional sol-gel process. The template synthesis of mesoporous silicas with pore diameter of 2.5 nm were carried out inside nanoreactors based on large pores of silica gel. Furthermore the surface area of the initial silica gel after the introduction of micellar solution increased the surface from 115 to about 377 m^2/g.

References

1.   U. Ciesla, F. Schuth, *Microporous Mesoporous Mater.* 27, 131 (1999).
2.   A. Corma, *Chem. Rev.* 97, 2373 (1997).
3.   Z. Zhang, Y. Han, F.-S. Xiao, *J. Am. Chem. Soc.* 123, 5014 (2001).
4.   S.-C. Shen. S. Kawi, *J. Phys. Chem. B* 103, 8870 (1999).
5.   P.T. Tanev. T.J. Pinnavaia, *Chem. Mater.* 8, 2068 (1996).
6.   X. Wang, W. Li, G. Zhu, S. Qiu, D. Zhao, B. Zhong, *Micropor. Mesopor. Mater.* 71, 87 (2004).
7.   S.J. Gregg, K.S.W. Sing, *Adsorption, Surface Area and Porosity* (Academic. London, 1967), p. 303.

5.4. THIN FILMS

NANOCRYSTALLINE METASTABLE HARD COATINGS

S. ULRICH[*], H. HOLLECK, M. STÜBER, H. LEISTE,
J. YE, C. ZIEBERT
*Research Center Karlsruhe, Forschungszentrum Karlsruhe,
Institut für Materialforschung I, Hermann-von-Helmholtz-
Platz 1, 76344 Eggenstein-Leopoldshafen,
Germany*

Abstract. Hard coating materials are usually categorized according to their dominating atomic binding structure into three groups: metallic, covalent and ionic hard materials. Each group is commonly related to a typical lattice structure and, hence, to preferable material properties characteristic for this group. Metallic hard coating materials like TiN, CrN, VC and TiC exhibit the f.c.c. structure and show in contrast to the other two groups a high toughness as well as good adhesion on metallic substrates. The zinc-blende structured materials such as diamond, cubic boron nitride and aluminum nitride are belonging to the covalent hard materials, generally exhibiting a high hardness and in addition high temperature strength. Ionic hard coatings, for example alumina and silicon oxide, typically show a high thermal stability and chemical inertness. Besides this artificial classification, however, novel synthetic film materials have emerged, and have been validated by thermodynamic calculations recently, in the form of metastable, nanocrystalline mixed crystals that combine the advantageous properties of different groups. The formation of these metastable structures requires ultrahigh quenching rates on the order of 10^{13} K/s, and can be usually realized by various physical vapor deposition methods. In this article, four metastable f.c.c.-structured nanocrystalline coating systems are described, including (Ti,Al)N, (Cr,Al)N, (V,Al)(C,N) and (Ti,Si)(C,O), all deposited by magnetron sputtering. The coatings have been characterized with regard to their structure, chemical composition, as well as mechanical properties. The conditions for the formation of metastable phases will be discussed.

Keywords: metastable coatings; nanocrystalline materials; magnetron sputtering

---

[*]Sven.ulrich@imf.fzk.de

J.P. Reithmaier et al. (eds.), *Nanostructured Materials for Advanced Technological Applications*, 257
© Springer Science + Business Media B.V. 2009

1. Introduction

Properties of a solid decisively depend on the type of atoms, of which it is composed, as well as on the structure and arrangement of the atoms. Diamond and silicon both have the same zinc blende crystal structure, and all atoms are sp^3-hybridized, but in the former case, the material comprises of carbon atoms, in the latter, of silicon atoms. *The collective arrangement of the individuals is identical, while the individuals themselves differ, with drastic effects on the properties of the solid.* Diamond, for instance, has a hardness of 100 GPa, while that of silicon equals 10 GPa. Diamond is an insulator, silicon a semiconductor. When comparing diamond with graphite, both solids are made up of the same type of atoms, carbon, but they differ by their crystal structures, either zinc blende or hexagonal. *The individuals are identical, their collective arrangement differs, again with major effects on their properties.* Diamond is the hardest solid whatsoever, graphite is a soft material with a hardness of 0.12 GPa only. *Tailored development of novel materials, in particular of hard materials, requires the consideration of both the atomic composition and the microstructure.*

2. Innovative Concepts for Multifunctional Hard Materials

Three groups of hard materials may be distinguished, depending on which type of bonding prevails: metallic, covalent, and ionic hard materials. *Metallic hard materials* mostly possess a face-centered cubic (f.c.c.) microstructure and are characterized by a high toughness and good adhesion to metal substrates. *Covalent hard materials* mostly have a zinc blende structure, a very high hardness, and high temperature strength. *Ionic hard materials* have the highest chemical stability of all hard materials.[1]

To develop materials of high hardness and toughness, it would therefore be desirable to synthesize covalent hard materials with a crystal structure typical of metallic hard materials, i.e. in a face-centered cubic lattice. Novel chemically stable and, at the same time, tough materials might be ionic hard materials with an f.c.c. lattice. However, no thermodynamic equilibrium conditions can be found in the respective phase diagrams that allow for such materials. Still, two possibilities of implementing the above material concepts exist: *Nanostabilization* and the synthesis of *nanocrystalline, metastable mixed crystals.*

Nanostabilization: If a covalent or ionic hard material grows on an f.c.c. metallic hard material, nanostabilization may occur due to limited deposition kinetics of the substrate. This means that a covalent or ionic hard material grows with a metastable f.c.c. crystal structure up to a certain layer thickness.

Before this crystal structure turns into a thermodynamically stable structure at a layer thickness characteristic of the respective material, another thin layer of the metallic hard material has to be applied. In this way, a thick multi-layer composite is designed, which is either hard and tough or chemically stable and tough.

Nanocrystalline, metastable mixed crystals: Up to certain, very small concentrations, atomic constituents of a covalent or ionic hard material can be dissolved in an f.c.c. lattice of a metallic hard material, such that either hardness or chemical stability of the tough metallic hard material is increased. If a high quenching rate is applied during synthesis, this concentration range can be extended considerably, which positively affects the properties of the metastable mixed crystal. Based on thermodynamic calculations, a favorable combination of a covalent or ionic hard material with a metallic hard material has to be found, for which the concentration range for producing a metastable f.c.c. mixed crystal is as large as possible. In addition, a synthesis method has to be applied, which allows for very high quenching rates. A suitable method is physical vapour deposition, during which quenching rates of up to 10^{13} K/s can be achieved.[2]

3. Thermodynamic Calculations for the Synthesis of Nanocrystalline, Metastable Mixed Crystals

Using a temperature/concentration phase diagram and the concentration dependence of the free enthalpies of all potential phases occurring, the formation of stable and metastable phases shall be explained.[2] Then, a quantitative key parameter, the interaction energy, will be derived on the basis of the regular solution model. This parameter describes the probability of formation of nanocrystalline, metastable mixed crystals.[3,4]

3.1. CONDITIONS OF FORMATION OF STABLE AND METASTABLE PHASES

The diagram in Figure 1 (above) describes schematically the formation of thermodynamically stable phases as a function of temperature and concentration of a quasibinary, eutectic system consisting of a metallic hard material (e.g. TiN, TiC, CrN, VC) and a covalent hard material (e.g. AlN). At low concentrations, the atoms of the covalent hard material are dissolved in the lattice of the metallic hard material. Typically, this is a face-centered cubic lattice. At high concentration, the situation is reversed: Atoms of the metallic hard material are dissolved in the lattice of the covalent hard material. The lattice of the covalent AlN is hexagonal. In the moderate concentration range, two phases are formed. There, the eutectic point is found. The explanations given below will not change in principle, if the covalent hard material is replaced by a heteropolar hard material.

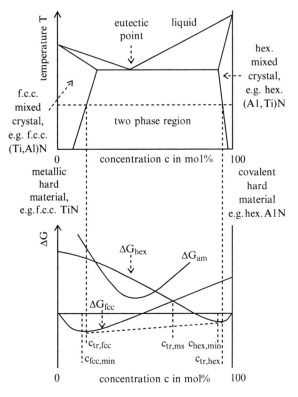

Figure 1. Temperature/concentration phase diagram (above) and the concentration dependence of the free enthalpies (below).

The diagram below in Figure 1 shows the free enthalpies ΔG_i of various phases i as a function of the concentration c at constant temperature. First, the two ΔG functions for the formation of crystalline structures of the metallic hard material ΔG_{fcc} and the covalent hard material ΔG_{hex} shall be considered. If atoms of the covalent material are dissolved in the lattice of the metallic material, ΔG_{fcc} first decreases with increasing concentration. At the concentration $c_{fcc,min}$, a minimum is reached. Then, the value increases again. In the case atoms of the metallic hard material are dissolved in the lattice of the covalent hard material, ΔG_{hex} exhibits the same qualitative behavior with decreasing concentration. The minimum is reached at $c_{hex,min}$. Both functions intersect at $c_{tr,ms}$.[†]

If mobility and time are sufficient, those phases are formed, for which ΔG assumes the lowest values. To determine the stable phases, a secant line is drawn to the curves ΔG_{fcc} and ΔG_{hex}. The boundary points define two concentrations: $c_{tr,fcc}$ and $c_{tr,hex}$. If and only if $\Delta G_{fcc}(c_{fcc,min})=\Delta G_{hex}(c_{hex,min})$, these points agree with both minima of the curves ΔG_{fcc} and ΔG_{hex}. Otherwise: $c_{fcc,min}<c_{tr,fcc}$ and $c_{hex,min} < c_{tr,hex}$. From 0 to $c_{tr,fcc}$, a mixed crystal of the metallic hard material is formed, e.g. a face-centered cubic mixed crystal. From $c_{tr,fcc}$ to $c_{tr,hex}$, a two-phase mixture represents the thermodynamically stable solution, since the mean value of $\Delta G_{fcc}(c_{tr,fcc})$ and $\Delta G_{hex}(c_{tr,hex})$ (these values lie on the secant) weighed according to the concentrations always is smaller than ΔG_{fcc} or ΔG_{hex}. Between $c_{tr,hex}$ and 1, a mixed crystal of a covalent hard material forms again, e.g. a hexagonal lattice. If mobility is limited, also metastable phases (phases with the next higher ΔG value) may form. In our case, mixeds crystals of the metallic hard material forms between $c_{tr,fcc}$ and $c_{tr,ms}$, while mixed crystals of the covalent hard material are formed between $c_{tr,ms}$ and $c_{tr,hex}$. In this way, the range of formation of the mixed crystal of the metallic hard material would reach its maximum extension. If the quasibinary system has a low eutectic point, however, an amorphous network may form. This means that the curve ΔG_{am} of the amorphous phase has to be considered in the case of $\Delta G_{am} < \Delta G_{fcc}$ or $\Delta G_{hex} < \Delta G_{am}$.

3.2. PROBABILITY OF FORMATION OF NANOCRYSTALLINE, METASTABLE MIXED CRYSTALS BASED ON THE REGULAR SOLUTION MODEL

In thermodynamic equilibrium a multi-particle system always assumes the state, in which the Gibbs free energy ΔG is minimum.[3,4] Taking into account the lattice stabilities and the regular solution model, ΔG may be expressed as

$$\Delta G = \sum_i n_i \cdot \Delta G_i + \varepsilon \cdot c \cdot (1 - c) - T \cdot R \cdot (c \cdot \ln(c) + (1 - c) \cdot \ln(1 - c),$$

where ε is the interaction energy, c the concentration of the metallic hard material, $(1-c)$ the concentration of the covalent or ionic hard material, and T the temperature. The lattice stability is considered by ΔG_i. For metallic hard materials in the structure of covalent or ionic hard materials, it typically amounts to +70 kJ/mol. For covalent or ionic materials in the structure

---

[†]Here and in the following: *tr* = transition, *ms* = metastable, *am* = amorphous.

of the metallic material, it is about +30 kJ/mol. For amorphous networks, it ranges between +20 kJ/mol and +50 kJ/mol. The interaction energy may be estimated using the following empirical formula based on experimental results[5,6]:

$$\varepsilon = 160\,\frac{kJ}{mol} \cdot \Delta d - 30\ kJ/mol$$

Here, Δd is the difference between the lattice parameter of the metallic material and the theoretical lattice parameter of the covalent material, if the latter would assume the structure of the metallic hard material. Figure 2 reveals the calculated interaction energies of 12 quasi-binary systems, consisting of a metallic hard material and AlN.[5,6] The smaller the interaction energy, the lower is the probability of formation of an amorphous phase, and the higher is the concentration $c_{tr,ms}$, at which a metastable mixed crystal is formed in the structure of the metallic material. For stoichiometric fcc-(Ti,Al)N, this concentration $c_{tr,ms,TiAlN}$ amounts to 0.65. Compared to (Ti,Al)N, only (V,Al)N, (Cr,Al)N, and (V,Al)(C,N) are more suitable.

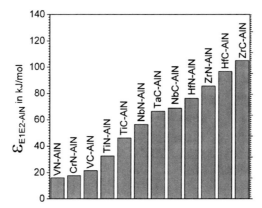

Figure 2. Calculated interaction energy for different AlN-based material system.

4. Synthesis of Nanocrystalline Metastable Mixed Crystals

Three AlN-based and one O-based metastable materials will be discussed in the following: (Ti,Al)N, (Cr,Al)N, (V,Al)(C,N), and (Ti,Si)(C,O).

4.1. METASTABLE (Ti,Al)N

The first successfully synthesized,[7,8] modeled,[2–6] and industrially applied metastable material in this context was (Ti,Al)N. Compared to standard PVD TiN, (Ti,Al)N has a higher hardness (depending on composition and residual stress state, between 2200 HV0.05 and 3000 HV0.05) and a better oxidation resistance. Typically, magnetron sputtering (targets: Ti–Al–N or Ti–Al, process gas: Ar, reactive gas N_2) is used for the synthesis of (Ti,Al)N. Recently, extensive experimental activities using this material system were reported.[9–11] In particular, the maximum solubility of the covalent AlN in the TiN lattice was determined: $c_{tr,ms}$ amounts to 0.66 ± 0.01.[11] New ab-initio calculations[12] confirm this experimental result: $c_{tr,ms}$ is 0.69 ± 0.05. Further development of (Ti,Al)N is aimed at reducing the friction coefficient. If a low carbon concentration is supplied during the coating process, for instance, by using the reactive gas CH_4, this carbon is incorporated at nitrogen lattice sites and the metastable, face-centered cubic (Ti,Al)(C,N) is formed. When the CH_4 fraction is further increased, the following microstructures develop: f.c.c.-(Ti,Al)(C,N) nanocrystallites with separated amorphous carbon clusters, f.c.c.-(Ti,Al)(C,N) nanocrystallites completely enclosed by a thin carbon skin, and an amorphous carbon matrix with homogeneously dispersed f.c.c.-(Ti,Al)(C,N) nanocrystallites. Thus, the friction coefficient may be adjusted specifically between 0.1 and 0.7.[13]

4.2. METASTABLE (Cr,Al)N

As compared to (Ti,Al)N, development of metastable face-centered cubic (Cr,Al)N is expected to result in an increased corrosion resistance[8] due to the chromium content and the higher solubility of AlN in the CrN lattice according to thermodynamic estimations (see Sect. 3.2 and Refs. 5, 6). By reactive r.f. magnetron sputtering of a composite target (CrN : AlN = 70:30 mol %) in a pure nitrogen plasma at a pressure of 0.9 Pa and a substrate temperature of 260°C, the metastable structure was synthesized successfully.[14,15] The elemental composition of the layers was studied by electron microprobe analysis. The metal ratio was Cr:Al = 70.3:29.7. The nitrogen fraction slightly exceeded 50 at. %. The metastable structure was detected by means of X-ray diffraction, transmission electron microscopy, and high resolution TEM.[15] The coatings exhibited a Vickers hardness between 2250 HV0.05 and 3050 HV0.05 and residual stresses below 2 GPa.[15] This high hardness is attributed to solid solution strengthening (AlN: 1200 HV and CrN: 1100 HV0.05 – 2300 HV0.05). Current research activities concentrate on optimizing the process parameters,[15] modeling,[16,17] modification of (Cr,Al)N by additional elements like Si[18,19] or V,[20] as well as on corrosion studies.

4.3. METASTABLE (V,Al)(C,N)

According to the thermodynamic modeling (see Sect. 3.2), (V,Al)(C,N) is an AlN-based four-component system in which AlN reaches the highest solubility in the structure of the metallic hard material, here VC. In addition, this four-component system allows to tailor the friction coefficient similar to the Ti–Al–C–N system (see Sect. 4.1). Moreover, the friction behavior may be influenced positively by the formation of vanadium oxide when the material is exposed an oxygen-containing atmosphere at higher temperatures.

For production, a ceramic composite target (VC:AlN = 60:40 mol %) was subjected to r.f. magnetron sputtering in a pure argon plasma at a pressure of 1.1 Pa. The target power was 250 W at a frequency of 13.56 MHz, the substrate temperature amounted to 150°C. The energy of the ions hitting the substrate during film growth was varied systematically between 25 and 175 eV.

The composition of the coatings reflected that of the target, with a slight enrichment of carbon of about 5 at. %. Argon incorporation and oxygen impurities together amounted to less than 2 at. %. The SEM picture in Figure 3 (left) shows the morphology of the film surface (top) and the fracture surface of the finely columnar (V,Al)(C,N) layer (middle) as well as of the hard metal substrate (bottom).

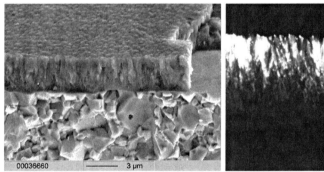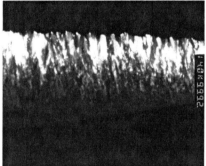

Figure 3. SEM micrograph of the morphology of a (V,Al)(C,N) surface and cross-sections of coating and hard metal substrate (left); TEM dark field image of the coating (right).

Both XRD and TEM studies reveal a face-centered cubic phase with a lattice parameter of 0.4159 ± 0.0020 nm. 15 TEM diffraction rings were evaluated, 15 values were determined for the lattice parameter. Then, the mean value and the standard deviation were calculated. The right image of Figure 3 shows a TEM dark field image of the (V,Al)(C,N) coating.

Depending on the process parameters selected, Vickers hardnesses and valies of the reduced Young's modulus may by adjusted between 1850 and 3260 HV0.05 and between 450 and 520 GPa, respectively.[21]

4.4. METASTABLE (Ti,Si)(C,O)

Metastable thin films combining a metallic hard material with an ionic hard material were synthesized successfully. Their face-centered cubic microstructure was characterized and, hence, this class of new materials class was presented for the first time in Ref. 6. In thermodynamic equilibrium, the coexistence of three phases is expected for a TiC–"SiO" material: TiC, SiO_2, and Si.

The material was grown by r.f. magnetron sputtering (300 W, 13.56 MHz) of a ceramic composite target (TiC : "SiO" = 80: 20 mol%) in a pure argon plasma at 0.15 Pa, a substrate bias of −50 V, and a substrate temperature of 500°C.

By means of XRD and TEM investigations, only one phase was detected, a face-centered cubic lattice with a lattice parameter of 0.4245 ± 0.0065 nm. The crystallite size calculated from the width of the X-ray diffraction peaks according to the Debye-Scherrer formula was 3.9 nm. A TEM diffraction image of this layer in shown in Figure 4.

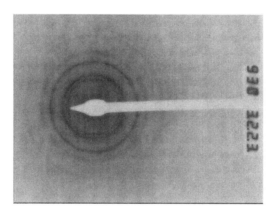

Figure 4. TEM diffraction image of a metastable f.c.c. (Ti,Si)(C,O) coating.

Double solid solution strengthening resulted in a Vickers hardness of 4400 HV0.05. Hence, the range of superhard materials was reached. The reduced Young's modulus was 608 GPa, the extremely high residual stress amounted to 23 GPa, and the critical load of failure in scratch tests was found to be 23 N.

5. Conclusions

Metastable materials as mixed crystals being synthesized from a metallic hard and a covalent or ionic hard material can be produced successfully with PVD methods, utilizing the possibility to realize a limitation of kinetics.

As a basis, thermodynamic modeling was applied, which allows to predict the solubility of the covalent or ionic material in the lattice of the metallic material. These metastable materials are multi-functional and may reach a good adhesion to metallic substrates, a high toughness and hardness as well as an exvellent high-temperature and chemical stability.

ACKNOWLEDGMENTS

The authors gratefully acknowledge the financial support of the Helmholtz-program NANOMICRO.

References

1. H. Holleck, J. Vac. Sci. Technol. A 4, 2661 (1986).
2. H. Holleck and H. Schulz, Thin Solid Films 153, 11 (1987).
3. H. Holleck, Surf. Coat. Technol. 36, 151 (1988).
4. H. Holleck, High Temp. Sci. 27, 295 (1998).
5. H. Holleck, H. Leiste, M. Stüber, and S. Ulrich, Invited talk: Constitution- and Nano-structured Moddeling in Thin Films, *International Conference on Metallurgical Coatings and Thin Films 2003*, San Diego, CA, USA (24.04.2003).
6. H. Holleck, M. Stüber, S. Ulrich and C. Ziebert, Invited talk: Materials Aspects of the Nanoscale Thin Film Design, *International Conference on Metallurgical Coatings and Thin Films 2006*, San Diego, CA, USA (04.05.2006).
7. G. Hakansson, J.E. Sundgren, D. McIntyre, J.E. Greene, and W.D. Münz, Thin Solid Films 153, 55 (1987).
8. O. Knotek, M. Atzor, A. Barimani and F. Jungblut, Surf. Coat. Technol. 42, 21 (1990).
9. L. Hultman, Non-Equilibrium Mater. 103, 181(1995).
10. R. Cremer, M. Witthaut, A. von Richthofen, and D. Neuschutz, Fresenius J. Analyt. Chem. 361, 642 (1998).
11. G. Erkens, R. Cremera, T. Hamoudia, K.-D. Bouzakis, I. Mirisidis, S. Hadjiyiannis, G. Skordaris, A. Asimakopoulos, S. Kombogiannis, J. Anastopoulos, and K. Efstathiou, Surf. Coat. Technol. 177, 727 (2004).
12. P.H. Mayrhofer, D. Music, and J.M. Schneider, Appl. Phys. Lett. 88, 071922 (2006).
13. M. Stüber, U. Albers, H. Leiste, S. Ulrich, H. Holleck, P.B. Barna, A. Kovacs, P. Hovsepian, and I. Gee, Surf. Coat. Technol. 200, 6162 (2006).
14. C. Kunisch, R. Loos, M. Stüber, and S. Ulrich, Z. Metallkd. 90, 847 (1999).
15. S. Ulrich, H. Holleck, J. Ye, H. Leiste, R. Loos, M. Stüber, P. Pesch, S. Sattel, Thin Solid Films 437, 164 (2003)
16. R.F. Zhang and S. Veprek, Acta Mater. 55, 4615 (2000).
17. P.H. Mayrhofer, D. Music, T. Reeswinkel, H.G. Fuss, and J.M. Schneider, Acta Mater. 56, 2469 (2008).
18. J.L. Endrino, S. Palacin, M.H. Aguirre, A. Gutierrez, and F. Schafers, Acta Mater. 55, 2129 (2007).
19. J. Soldan, J. Neidhardt, B. Sartory, R. Kaindl, R. Cerstvy, P.H. Mayrhofer, R. Tessadri, P. Polcik, M. Lechthaler, and C. Mitterer, Surf. Coat. Technol. 202, 3555 (2008).

20. R. Franz, J. Neidhardt, B. Sartory, R. Kaindl, R. Tessadri, P. Polcik, V.H Derflinger, and C. Mitterer, Tribo. Lett. 23, 101 (2006).

21. S. Ulrich, M. Stüber, C. Ziebert, and H. Holleck, Magnetron sputtered nanocrystalline, metastable (V,Al)(C,N) hard coatings, *International Conference on Metallurgical Coatings and Thin Films 2008*, San Diego, CA, USA (29.04.2008).

INFLUENCE OF THE PRECURSOR MATERIALS ON THE STRUCTURAL PROPERTIES OF POLY-Si THIN FILMS OBTAINED BY ALUMINIUM-INDUCED CRYSTALLIZATION

D. DIMOVA-MALINOVSKA[*]

Central Laboratory for Solar Energy and New Energy Sources, Bulgarian Academy of Sciences, 72 Tzarigradsko Chauss Blvd., 1784 Sofia, Bulgaria

Abstract. The process of aluminium-induced crystallization (AIC) has attracted increasing scientific interest in the last decade, because of the possible application of the resulting poly-Si films in different devices, e.g. solar cells, thin film transistors (TFTs), image sensors, bio-application, etc. In the present paper, a study of the influence of the precursor layers – the interfacial oxide layer formed by different methods, and a-Si:H – on the structural properties of poly-Si films prepared by AIC is presented. The influence of the temperature of deposition and the hydrogen content in the a-Si:H precursor layers and the optical and structural properties of the obtained poly-Si films are presented as well.

Keywords: aluminium induced crystallization; polycrystalline Si films; Raman spectra; optical band gap; reflectance

1. Introduction

The method of aluminium-induced crystallization (AIC) is being investigated for the preparation of poly-Si thin films for thin film transistors, solar cells and bio-applications.[1–4]

Poly-Si thin films can be prepared by different methods, for example by solid state crystallization (SPC), where amorphous Si is transformed to poly-Si by annealing at about 600^0C and higher, or by laser crystallization

---

[*]doriana@phys.bas.bg

J.P. Reithmaier et al. (eds.), *Nanostructured Materials for Advanced Technological Applications*,   269
© Springer Science + Business Media B.V. 2009

(LC). These methods have disadvantages – high temperature processing (SPC) or the use of expensive equipment (LC). The AIC is characterized by a lower thermal budget than the other methods of preparation of poly-Si films and by the possibility to use low-cost substrates such as glass. The grains generated in the obtained films are larger than the film thickness, which is important for device application. Since the poly-Si films prepared by this method are used as seeding layers for epitaxial growth of thicker Si films, one of the most important parameter of the films is the quality of their surface and the crystalline structure.

The preparation of poly-Si films by MIC (M = metal) is based on an annealing step of amorphous Si films, deposited by different methods (e-beam evaporation, sputtering, PECVD). When Si is in contact with certain metals, the temperature for crystallization of a-Si is found to be drastically reduced. During the annealing pronounced reactions take place between the semiconductor and the metal film. According to reported results[5] the reaction between metal and a-Si occurs at the interlayer by diffusion. During the isothermal heat treatment, the semiconductor film is dissolved into the metal film where it diffuses and precipitates. The metal film is saturated with Si, the dissolved semiconductor becomes supersaturated and supersaturation can be relieved by crystallites growth. Crystallite formation is possible at constant temperature (which is the case in MIC) due to the higher free energy of amorphous semiconductors as compared to that of single crystals. The concentration of Si dissolving in the metal solvent from amorphous material will be higher than if dissolution were occurring from crystalline material. In dilute solutions the concentration of the solute is proportional to $\exp(G/kT)$, here G is the free energy per solute atom. Therefore, the excess free energy ΔG, associated with the amorphous material in contact with the solvent should enhance the solute concentration. These processes are solid-solid reactions, since this behaviour is observed at temperatures lower that the eutectic one of the system. During the annealing, crystalline silicon grains are formed in the bottom layer, while Al atoms move to the top surface, resulting in a layer exchange. After annealing, the aluminium is removed from the surface of the poly-Si films obtained by etching in a chemical solution based on phosphoric acid.

It has been reported that the parameters of the AIC process and the preparation of the precursor layers such as Al and a-Si:H films are of great importance to obtain good properties of the resulting poly-Si layers[6-13]. In this paper the influence of the interfacial oxide layer, formed by different methods, on the structural properties of poly-Si films prepared by AIC is described. The dependence of the deposition temperature (T_s^{a-Si}) of unhydrogenated (a-Si) and hydrogenated (a-Si:H) amorphous silicon layers prepared by magnetron sputtering on the structural properties of poly-Si films obtained by AIC of glass/Al/a-Si(a-Si:H) structures is studied. The influence of the hydrogen partial pressure during the magnetron sputtering of the a-Si:H precursor films[12,13] is also investigated.

The structure of the poly-Si films has been studied by X-ray diffraction (XRD), scanning electron microscopy (SEM), energy dispersive spectroscopy (EDS), and Raman and/or microprobe Raman spectroscopy. The Raman spectra were excited by the 488 nm line of an Ar^+ laser. The peak positions and the full width at half maximum (FWHM) of the Raman bands and of the 2θ reflections (in the case of XRD) were determined, respectively. The surface topography of the Al films has been studied by optical microscopy and atomic force microscopy (AFM). Finally, hemispherical UV reflectance spectra in the range 250–400 nm were measured.

2. Influence of the Al Precursor and the Interfacial Al_2O_3 Layer on the Properties of Poly-Si

One of the factors influencing the AIC process is the interfacial aluminium oxide layer, which has an effect on the nucleation rate of the Si crystallites and on the Si or Al diffusion through the oxide membrane.[7] It has been observed that a thicker oxide leads to longer annealing times.[8]

2.1. PREPARATION OF THE SAMPLES

In order to investigate the effect of the properties of the Al precursor two deposition methods were applied: thermal evaporation and sputtering. The Al layers of some of the samples were exposed to air for 2 or 24 h prior to the deposition of the hydrogenated amorphous silicon (a-Si:H) films; the results were compared with that of samples without exposure to air. a-Si:H films were deposited on Al coated glass substrates by RF magnetron sputtering of a CZ c-Si target (p-Si, Wacker, 9–12 Ωcm) at a substrate temperature of 250°C in a mixture of Ar (0.5 Pa) and H_2 (0.05 Pa). The base pressure in the chamber was 10^{-4} Pa. These conditions resulted in hydrogen concentrations (measured by elastic recoil detection analysis (ERDA)) in the a-Si:H of about 6 at. %. The glass/Al/a-Si:H structures were then annealed in air at 530°C for 7 h. To investigate the influence of the interfacial oxide layer on the sputtered Al, three sets of samples were prepared: with Al layers deposited at room temperature or at 300°C and then exposed for 24 h to air, and with a sputtered Al_2O_3 (about 4 nm) on the Al layers deposited at 300°C. The samples were annealed in H_2 at 500°C for 7 h. The Al layer remaining on top of the poly-Si formed due the aluminium-induced layer exchange process was removed using a standard Al etch solution. The technological parameters of this set of samples are presented in Table 1. In the case of evaporated Al precursors deposited at room temperature different exposure times of the Al layers to air were applied: 2 or 24 h. The influence of the annealing atmosphere (air or hydrogen) was also studied.

TABLE 1. Peak position (ω_{TO}) and FWHM of the Si–Si TO-like peak of AIC poly-Si films (annealed in H_2) prepared using different sputtered precursors. The microprobe Raman spectra were measured at grain or inter-grain positions.[13]

Samples	Precursor structure with sputtered Al	T_{Al} (°C)	Exposure of Al to air (h)	ω_{TO} (cm^{-1})		FWHM (cm^{-1})	
				Grain	Inter-grain	Grain	Inter-grain
SPA	Glass/Al/ a-Si:H	RT	24	519.8 ± 0.5	517.2 ± 0.5	9.0 ± 0.5	9.5 ± 0.5
SPB	Glass/Al/ a-Si:H	300	24	521.1 ± 0.5	521.4 ± 0.5	7.2 ± 0.5	7.4 ± 0.5
SPC	Glass/Al/sput. Al_2O_3/a-Si:H	300	–	521.2 ± 0.5	521.0 ± 0.5	8.1 ± 0.5	8.2 ± 0.5
c-Si reference		–	–	520.7 ± 0.5		5.9 ± 0.5	

2.2. RESULTS AND DISCUSSION

The position ω_{TO} of the Si–Si TO-like peak and its FWHM taken from microprobe Raman spectra of poly-Si prepared by AIC of glass/sputtered Al/Al_2O_3/a-Si:H structures with different interfacial aluminium oxide layers are presented in Table 1. The TO-like peak of a c-Si wafer, used as a reference, is observed at 520.7 cm^{-1}, while its FWHM is 5.9 cm^{-1}. It can be seen that the poly-Si films prepared from Al deposited at 300°C and exposed to air for 24 h (sample SPB) or capped by magnetron sputtered Al_2O_3 (sample SPC) exhibit ω_{TO} values very close to that of the c-Si reference value, for both conditions under which Raman signals have been taken: from the grains and from the inter-grain material. The results demonstrate a good homogeneous crystalline structure of these poly-Si films. The spectra of the poly-Si prepared from Al deposited at room temperature and exposed to air for 24 h (sample SPA) exhibit ω_{TO} values of 517.2 and 519.8 cm^{-1}, measured on the grains and inter-grain, respectively.

An estimate of the grain size and the stress in the films can be made from the downshift and the FWHM of the Si–Si TO-like Raman peak.[14,15] It should be noted that accurate values for the grain size cannot be determined from the relationships described previously in the literature, as they depend upon the structure of the grains and their boundaries.[15,16] Despite of this, comparisons between spectra obtained from similar materials are valid. The grain size is inversely proportional to the FWHM of the Si–Si TO-like peak.[15] The FWHM values of the SPB sample are the smallest. It can be deduced that deposition of Al at 300°C results in poly-Si with larger grains compared to the case of Al deposited at room temperature (sample SPA). The slightly higher FWHM of the Si–Si TO-like peak of the sample prepared from a structure containing an about 4 nm thick sputtered interfacial Al_2O_3 layer (sample SPC) indicates the formation of slightly smaller grains compared to that obtained from an Al precursor kept for 24 h in air (SPB). It is known that diffusion of Si through the aluminum oxide membrane during the AIC process strongly depends on its thickness; it suppresses the nucleation rate and the crystallization velocity of the poly-Si films.[7] The thickness of the Al_2O_3 in sample SPB could be estimated to about 2 nm, according to Ref. 8, which is only half of that of sample SPC. The smaller grain size in the sample with thicker, sputtered Al_2O_3 could thus be explained by a suppressed crystallization rate during the AIC process. Thus, the thickness of the sputtered Al_2O_3 layer should be further optimized.

The shift of the Si–Si TO-like peak towards higher or lower wave numbers can be related to the amount of compressive or tensile stress in the poly-crystalline films, respectively.[14] It can be suggested that poly-Si films prepared by AIC using a sputtered Al precursor deposited at room temperature consist of grains with tensile stress, while those prepared from Al deposited at 300°C exhibit a compressive one. Although this result needs further investigation, it is possible to suppose that the presence of Al and more disordered Si in the inter-grain region could be a reason for the presence of the tensile stress.[17]

Figure 1 shows optical images of the same poly-Si films as in Table 1. The films prepared from Al deposited at room temperature and exposed to air, or from Al capped by sputtered Al_2O_3, exhibit a dendritic growth, probably due to the lower crystallization rate of the grains. In contrast, those prepared from Al deposited at 300°C and exposed to air have more dense structures and smoother surfaces.

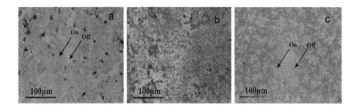

Figure 1. Optical images of AIC poly-Si films prepared using different sputtered Al precursors: (a) room temperature deposition and 24 h exposure to air; (b) 300°C deposition and 24 h exposure to air; (c) 300°C deposition and capping by sputtered Al_2O_3. Arrows "On" point to grains and arrows "Off" to inter-grain material.[12]

Topographical AFM images of the different sputtered Al precursors used for the preparation of the poly-Si films presented in Table 1 and in Figure 1 are shown in Figure 2. The average surface heights of the Al precursor films used in samples SPA, SPB and SPC are 21.3, 120.6 and 23.6 nm, respectively.

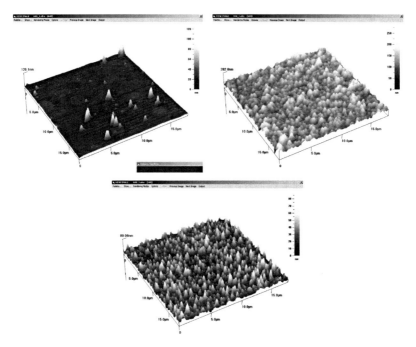

Figure 2. AFM images of different sputtered Al precursors: (a) room temperature deposition and exposure 24 h to air; (b) 300°C deposition and 24 h exposure to air; (c) 300°C deposition and capping by sputtered Al₂O₃.[12]

The respective root mean square (rms) values of these heights are 5.2, 23.5, and 9.6 nm. The greater roughness of the Al deposited at 300°C and exposed to air can be related to the presence of larger Al grains. Due to the fact that grain boundaries in the Al precursor layer are preferential nucleation sites and that films with larger grains contain less grain boundaries, it can be suggested that Al deposited at 300°C is more suitable as a precursor for obtaining poly-Si films with larger grains by AIC.

The other method of Al deposition applied in this work was vacuum evaporation. The FWHMs and ω_{TO} values of the Si–Si TO-like peak of poly-Si films prepared from glass/Al/Al₂O₃/a-Si:H structures with an evaporated Al precursor (exposed to air for 2 or 24 h) are shown in Table 2, together with the annealing conditions. The samples annealed in air exhibit a lower FWHM, indicating a larger grain size. This, however, cannot be a result of the influence of the annealing atmosphere (it has been shown that annealing in the presence of hydrogen results in poly-Si films of higher quality than in the case of annealing in other atmospheres[17]). It can be related to the higher annealing temperature in air (530°C) compared to that in hydrogen (500°C). There is a trend of lower FWHMs and ω_{TO} values closer to that of the c-Si reference (for both annealing atmospheres) for samples prepared from an Al precursor exposed to air for 2 h (samples EVA and EVC). This is an indication of the presence of larger grains in the poly-Si films. The thicker aluminium oxide layer formed during the longer exposure time to air[18] leads to stronger suppression of the diffusion of Si through the oxide membrane, thus reducing the nucleation rate and the crystallization velocity during the AIC process.

TABLE 2. Peak position (ω_{TO}) and FWHM of the Si–Si TO-like peak of AIC poly-Si films prepared from different evaporated (at room temperature) Al precursors, in different atmospheres (air or H_2). The microprobe Raman spectra were measured at grain or inter-grain positions.[13]

Samples	Precursor structure with evaporated Al	Exposure of Al to air (h)	Annealing atomsphere	ω_{TO} (cm^{-1})		FWHM (cm^{-1})	
				Grain	Inter-grain	Grain	Inter-grain
EVA		2	Air	520.8 ± 0.5	520.5 ± 0.5	7.0 ± 0.5	7.5 ± 0.5
EVB	Glass/Al/a-Si:H	24	Air	521.9 ± 0.5	–	6.5 ± 0.5	–
EVC		2	H_2	521.1 ± 0.5	521.1 ± 0.5	7.8 ± 0.5	7.8 ± 0.5
EVD		24	H_2	522.9 ± 0.5	520.5 ± 0.5	8.0 ± 0.5	8.4 ± 0.5
c-Si reference		–	–	520.7 ± 0.5		5.9 ± 0.5	

It should be noted that there were no Raman bands in the spectra taken from the inter-grain material of the poly-Si prepared by in air from an Al precursor exposed for 24 h. This fact could be explained by the presence of highly disordered amorphous Si between the grains, or by a high Al concentration in the inter-grain material. The latter can be excluded taking into account the EDS concentration profiles of Si and Al shown in Figure 3b, which demonstrate that the Al concentration is randomly distributed.

SEM images and the corresponding EDS concentration profiles of the poly-Si films summarized in Table 2 are shown in Figure 3. The use of a precursor exposed to air for 2 h (Figure 3a) leads to poly-Si with large grains (20–30 µm) and a dendritic structure. However, the use of a 24 h exposed Al precursor (Figure 3b) results in poly-Si with more dense but smaller grains (15–20 µm). Polycrystalline Si films with large inter-grain distances (in the range of 10 µm) are produced when the AIC process is performed in air (Figure 3a, b) for both investigated Al precursors (2 or 24 h exposure to air, samples EVA and EVB, respectively). Annealing in hydrogen (Figure 3c, d) yields poly Si films (samples EVC and EVD) with more homogeneous structures and smaller inter-grain distances.

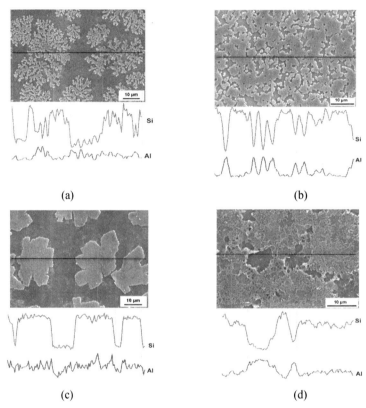

(a)   (b)

(c)   (d)

Figure 3. SEM images and the corresponding EDS concentration profiles of poly-Si films prepared by AIC of different Al precursors evaporated at room temperature, exposed to air for 2 (a, c) or 24 h (b, d). The annealing was performed in air (a, b) or hydrogen (c, d).[13]

The crystal structure of two poly-Si films obtained using Al precursor films deposited at room temperature by the two methods (sputtering and evaporation) has been studied by XRD analysis. Both Al precursor films were kept in air for 24 h before the a-Si:H deposition. The annealing was performed in air for 7 h at 530°C. From the XRD spectra of both films shown in Figure 4 the sizes of the crystallites were estimated by the Debye–Scherer equation to be 46 and 167 nm for the sputtered and evaporated Al precursors, respectively. An AFM image of the precursor layer deposited by evaporation is shown in Figure 5. Its surface average height and rms values are 30.8 and 14.2 nm, respectively. Comparison of the AFM images of both

Al precursor layers (sputtered and evaporated, Figures 2a and 5, respectively) reveals that the evaporated Al has a greater roughness and larger grains. This leads to larger crystallites in the poly-Si films obtained by AIC.

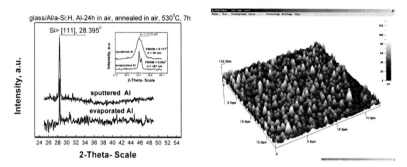

Figure 4 (left). X-ray diffraction spectra of AIC poly-Si films prepared using (a) evaporated or (b) sputtered Al precursors (both deposited at room temperature and exposed to air for 24 h). The inset shows the respective diffraction spectra for the determination of the crystallite size by the Debye–Scherer equation.[12,13]

Figure 5 (right). AFM image of an evaporated Al precursor deposited at room temperature and exposed to air for 24 h.[12,13]

3.   Influence of the Hydrogen Content in a-Si:H Layers

The influence of a-Si:H films with different hydrogen contents on the structural properties of poly-Si films prepared by AIC of glass/a-Si:H/Al samples is presented in this section. For comparison, un-hydrogenated amorphous Si (a-Si) films were also used as precursors.

3.1.   PREPARATION OF THE FILMS

Glass/Al/a-Si and glass/Al/a-Si:H structures were prepared to obtain poly-Si thin films by AIC. The Al precursor films were deposited on glass substrates by rf magnetron sputtering at a substrate temperature of $T_s = 300°C$. They were kept for 24 h in air before the deposition of a-Si films for native aluminum oxide formation. As it has been reported in Sect. 2 the Al precursor films prepared under these conditions result in the best structural properties of the poly-Si films. Unhydrogenated (a-Si) and hydrogenated (a-Si:H) films were deposited on the top of the Al layers by rf magnetron sputtering of a crystalline Si (c-Si) target with 130 W RF power at various substrate temperatures $T_s^{a\text{-}Si:H}$, without and with heating up to 400°C in Ar (0.5 Pa) or (0.5 Pa) + H_2 (0.05 Pa, 0.1 Pa and 0.2 Pa) atmospheres, respectively. The thicknesses of both layers, Al and a-Si:H, were equally 100 nm.

The prepared glass/Al/a-Si samples were isothermally annealed at 530°C for 6 h in forming gas (N_2 + 5% H_2).

3.2. RESULTS AND DISCUSSION

Raman spectroscopy was used to study the influence of the deposition temperature and the hydrogen partial pressure during the sputtering of the a-Si and a-Si:H precursor layers on the structure of the resulting poly-Si films.

The Raman spectra of films obtained from a-Si or a-Si:H precursors deposited at different T_s^{a-Si} and three different hydrogen partial pressures are shown in Figure 6. All of the samples display spectra typical for the crystalline Si structure: a Si–Si TO-like band, centered between 518.5 and 520.5 cm^{-1}. For crystalline silicon measured under the same conditions, the peak is at 521 cm^{-1} and has a FWHM of 4.5 cm^{-1}. The dependence of the peak position and the FWHM (estimated from a Lorenzian fit) on the substrate temperature and the H_2 partial pressure are presented in Figure 7. The accuracy of the values is indicated by the error bars.

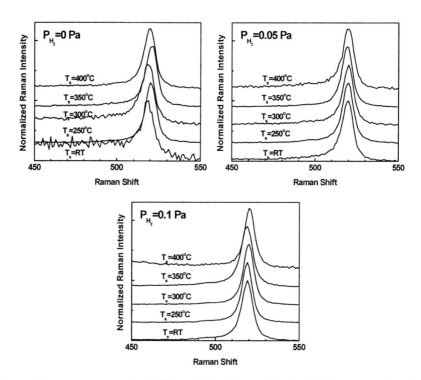

Figure 6. Raman spectra of poly-Si films obtained from a-Si and a-Si:H precursors deposited at different T_s^{a-Si} and different H_2 pressure: 0, 0.05 and 0.1 Pa.

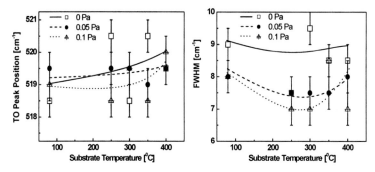

Figure 7. Position (left) and FWHM (right) of the Si–Si TO-like peak from Raman spectra of the samples presented in Figure 6. The lines are just guides to the eye.

The following tendencies can be seen. The position of the Si–Si TO-like peak shifts slightly to higher wave numbers with increasing $T_s^{a\text{-}Si}$. This is an indication of a reduction of the tensile stress in the poly-Si samples with increasing $T_s^{a\text{-}Si}$. A weak tendency for increased stress in the poly-Si films grown with a-Si:H precursors can be noticed and explained by the effusion of H during the AIC. This could leave pinholes and microvoids in the poly-Si films, resulting in higher tensile stresses. The FWHM of the Si-Si TO-like peak passes through a minimum for poly-Si films obtained from hydrogenated precursor layers deposited between 250°C and 350°C; in this case larger grains were obtained in the poly-Si films. It should be noted that we observed similar results for the AIC of glass/a-Si(a-Si:H)/Al structures.[17] It is possible to suppose that the presence of H in the precursor layer enhances the disorder in the a-Si:H precursor during the annealing. This stimulates the dissolution of Si atoms into the Al and the consecutive re-arrangement to a crystalline Si structure, resulting in a higher growth rate and larger grains in the poly-Si films. This suggestion is based on the fact that a higher degree of disorder is energetically more favourable for the transformation of the a-Si:H film into a poly-Si one. It is known that the H concentration in a-Si:H films decreases with increasing substrate temperature, so that the precursors deposited at RT should have a higher H content.[19] In this case, the higher quantity of effusing hydrogen probably creates conditions for an increased diffusion rate of Al and Si. Thus, intermixing of Al and Si would take place within the bulk of the resulting poly-Si film, which inhibits the exchange between Al and Si and leads to a lower rate of crystallite growth and a smaller grain size. The deposition of the precursor layers at temperatures >350°C would result in a reduced H content and in a better short range order of the a-Si:H films, which slows down the re-arrangement into the crystalline structure; the poly-Si films would again have smaller grains.

XRD spectra of the poly-Si films obtained from a-Si:H precursors deposited at $T_s^{a\text{-}Si:H} = 300°C$ with different H_2 partial pressures are shown in

Figure 8. Only one reflection peak at about $2\theta = 28.50°$ is observed for the poly-Si films which correspond to the (111) orientation. The position of this peak in c-Si is at $2\theta = 28.48°$. The intensity of the peak is highest for the poly-Si films obtained from a precursor deposited with $P_{H2} = 0.1$ Pa. This observation is an indication of better crystalline quality. The average grain size D in the films was calculated by the Debye–Scherer equation from the FWHM of the (111) peak $\Delta 2\Theta$.[8] The data obtained with an accuracy of about 10% are presented in Table 3. The larger grains (D = 630Å) are obtained for the poly-Si from a precursor deposited with $P_{H2}= 0.1$ Pa.

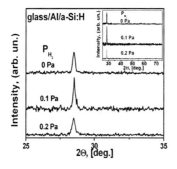 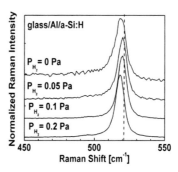

Figure 8 (left). XRD spectra of poly-Si films obtained by AIC from a-Si:H precursor deposited at $T_S = 300°C$ with different H_2 partial pressure.

Figure 9 (right). Raman spectra of poly-Si films obtained by AIC from a-Si:H precursor deposited at $T_S = 300°C$ with different H_2 partial pressure.

Raman spectra of poly-Si films deposited at $T_s^{a\text{-Si:H}} = 300°C$ are presented in Figure 9. All films display Raman spectra typical for the crystalline Si structure: a Si–Si TO band is centered between 518.5 and 519.5 cm^{-1}. The Si–Si TO-like peak for c-Si is at 521 cm^{-1} with a FWHM of 4.5 cm^{-1} measured under the same conditions. The shift of this peak in the poly-Si films to lower wave numbers could be due to tensile stress, as reported earlier.[14] The peak position is slightly shifted to that of c-Si with increasing p_{H2} up to 0.1 Pa. Thus, the incorporation of H with moderate concentrations in the a-Si:H precursor results in lower tensile stress in poly-Si. However, its further increase to 0.2 Pa leads to higher stress values, probably as a result of the creation of pinholes due to the higher effusion of H. The FWHM is the lowest in the films obtained from a-Si precursors deposited with $p_{H2} = 0.1$ and 0.2 Pa (Table 3). It is known that the grain size is inversely proportional to the FWHM of the peak. A tendency of increasing the grain size with increasing p_{H2} to 0.2 Pa could be noticed and explained by the influence of the hydrogen on the process of layer exchange. Higher hydrogen concentrations might accelerate the process of diffusion of Si in the Al layer; as a result grains with larger size are obtained under identical AIC conditions.[20]

TABLE 3. Deposition conditions of the precursor layers. The H concentration φ is calculated from FTIR absorption spectra, ξ – from ERDA, 2θ and Δ2θ from the (111) XRD peak and its FWHM, and average grain size of the poly-Si films, D, from the XRD data. The values of the position of the Si–Si TO-like peak ω_{TO}, and its FWHM are given as well.

T_s^{a-Si} (°C)	T_s^{Al} (°C)	P_{H2} (Pa)	φ (at %)	ξ (at %)	2θ (°)	Δ2θ (°)	D (Å)	σ (MPa)	ω_{TO} (cm^{-1})	FWHM (cm^{-1})
No heat.	300	0.10	19	25	28.44	0.13	630	−680	519	8
250	300	0.10	–	–	–	–	–	–	518.5	7
300	300	0.10	15	19	28.49	0.13	630	−700	519.5	7
400	300	0.10	11	16	28.40	0.15	546	−393	520	7
250	300	0.00	0	4	28.50	0.21	431	−629	520	7.7
250	300	0.05	9	14	28.48	0.23	372	−661	519.5	7.5
300	300	0.2	20	26	28.50	0.24	341	−780	518.5	7

Optical microscopy images of the surfaces of the poly-Si films, obtained from a-Si:H precursor layers deposited at different H_2 pressure, are shown in Figure 10. The sample obtained in pure Ar atmosphere exhibits a high density of Si islands on the poly-Si surface. The surface of poly-Si prepared with P_{H2} = 0.1 Pa has a smooth surface. In the case of the film obtained from a a-Si:H precursor deposited with P_{H2} = 0.2 Pa, the surface becomes rough. It is well known that hydrogen starts to effuse from a-Si:H films during annealing at T > 350–400°C and out-diffuses completely at 700°C in less than 1 h, resulting in the creation of micro-holes in the precursor films. The higher the hydrogen concentration in the precursor films, the higher is the concentration of hydrogen in the microvoids and the higher will be the number of the resulting micro-holes in the obtained poly-Si films. This effect and the presence of Al in the films could be reasons for the increasing stress, the deterioration of the short range order and the increased roughness of the poly-Si.

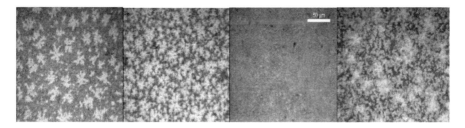

Figure 10. Optical micrographs of poly-Si films obtained by AIC from a-Si:H precursors deposited at T_S = 300°C with different H_2 partial pressure. From left to right: 0, 0.05, 0.1, 0.2 Pa.

We studied hemispherical UV reflectance spectra of the poly-Si films to observe changes of their structure quality, as described in Ref. 21. They are presented in Figure 11. For comparison the corresponding spectrum of a c-polished Si wafer measured under the same conditions is shown, too. The two maxima in the spectrum of single-crystalline Si at 280 and 365 nm are caused by optical interband transitions at the X-point (band E_2) and along the Γ–L axis (band E_1) of the Brillouin zone, respectively.[22] Deviations from the UV-R spectrum of crystalline bulk Si are related to deteriorations of the long-range order or amorphization of the material.[22] The intensity of the E_2 band decreases and that of the E_1 increases with long range disorder according to the theory of long range order relaxation effects. Additionally, at short wavelengths, in particular at 280 nm, the reflectance is largely determined by the high value of the absorption coefficient ($\alpha > 10^6$ cm^{-1}) corresponding to a penetration depth of less than 10 nm. Imperfect crystallinity in the near-surface region will cause a broadening and height reduction of this maximum.[21] It can bee seen from Figure 11 that the UV-R is very low in the whole wavelength range for the poly-Si films obtained from a-Si deposited without hydrogen, and smeared out maxima are observed. An increase of P_{H2} results in an increase of the reflectance and the appearance of the bands at 280 and 365 nm at $P_{H2} = 0.05$ Pa, which become well pronounced at $P_{H2} = 0.1$ Pa. Further increase of P_{H2} leads to a decrease of the reflectance, while the bands in the spectrum disappeared. The poly-Si films have been further characterized using the ratios of the integral reflectances of the E_1 and E_2 peaks and the total integral reflectivity in the UV-R spectra. This method has been applied because it involves both the width and the intensity of the main features related to the optical interband transitions. The dependence of the integral intensities of both bands on the H_2 partial pressure during the deposition of the a-Si:H precursors are shown in Figure 12. The maximum intensity of the E_2 band and the minimum of the E_1 band are observed at $P_{H2} = 0.05$–0.1 Pa. It is possible to conclude that poly-Si films prepared using a-Si precursor with medium H concentrations ($P_{H2} = 0.1$ Pa) have a better long range order and a better structure of the near surface region. This conclusion correlates with the observations from XRD, Raman spectra and optical images. The poly-Si film obtained from a-Si:H precursor deposited with $P_{H2} = 0.1$ Pa has larger size of the crystalline grains and better short range order, according the XRD and Raman spectroscopy analyses (Table 3). Moreover, from the optical microscopy images of the poly-Si films displayed in Figure 10 it can be seen that smoother surface morphology is observed again for the a-Si:H precursor with hydrogen partial pressure $P_{H2} = 0.1$ Pa.

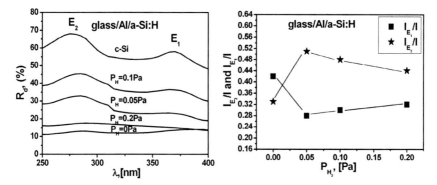

Figure 11 (left). UV reflectance spectra of poly-Si films obtained by AIC from a-Si:H precursors deposited at $T_S = 300°C$ with different H_2 partial pressure.

Figure 12 (right). Relative integral reflectance of the main peaks of optical transitions I_{E1}/I and I_{E2}/I of the poly-Si film vs. H_2 partial pressure during the deposition of a-Si:H precursors.

4. Conclusion

The complex study of the structural properties such as short and long range ordering and the quality of the near-surface region structure demonstrates that the properties of poly-Si films obtained by AIC depend on the conditions of the deposition of the precursors layers Al and a-Si:H. It was shown that the structural quality of poly-Si films prepared by the AIC of glass/Al/a-Si:H stacks depends on the method and the temperature of deposition of the Al precursor film, on the thickness and formation method of the interfacial oxide, and on the annealing atmosphere. The XRD results indicate that poly-Si films prepared using an evaporated Al precursor exhibit three to four times larger crystallites than those prepared using sputtered Al (for the same annealing conditions). In the case of sputtered Al, a deposition temperature of 300°C is found to be more favorable for large grain formation. The preparation of poly-Si films with good structural quality by AIC of glass/Al/Al$_2$O$_3$/a-Si:H stacks with a sputtered interfacial oxide has been demonstrated. In this case, the grains are slightly smaller than those of samples with a native interfacial oxide. Since the thickness of the sputtered oxide is about twice that of the native one, it can be concluded that the thickness of the sputtered Al_2O_3 should be further optimized. It was demonstrated that there is a good correlation between the data obtained from XRD, Raman spectroscopy, optical microscopy and UV-reflectance spectra concerning the structural properties, such as short and long range ordering and the quality of the structure of the near surface region of poly-Si films prepared by AIC from a-Si:H precursors deposited with different hydrogen partial pressures. Poly-Si films with better structural properties and surface morphology are obtained using a-Si:H precursor layers deposited in the temperature interval of 250–350°C and moderate hydrogen partial pressure.

ACKNOWLEDGMENT

This work was funded by the Bulgarian National Scientific Fund (project X-1503).

References

1. A. Straub, P. Widenborg, A. Sproul, Y. Huang, N. Harder, and A. Aberle, J. Cryst. Growth, 168, 265 (2004).
2. T. Voutsas, Thin Solid Films 515, 7406 (2007).
3. L. Pramatarova, E. Pecheva, P. Montgomery, D. Dimova-Malinovska, T. Petrov, A. Toth, and M. Dimitrova, J. Nanosci. Nanotechnol. 8, 924 (2008).
4. E. Pecheva, L. Pramatarova, M. Dimitrova, P. Laquerriere, D. Dimova-Malinovska, in: I. Mihailescu and A. Vaseashta (Eds.), (NATO Series, Springer, Berlin, 2008, in press).
5. J. O. McCaldin and H. Sankur, Appl. Phys. Lett. 19, 524 (1971).
6. O. Nast and A. J. Hartmann, J. Appl. Phys. 88, 716 (2000).
7. J. Schneider, J. Klein, M. Muske, S. Gall, W. Fuhs, J. Non-Cryst. Solids, 338–340, 127 (2004).
8. J. H. Kim and J. Y. Lee, Jpn. J. Appl. Phys. 35, 2052 (1996).
9. J. Rau, J. Klein, J. Scneider, E. Conrad, I. Sieber, M. Stöger-Pollach, P. Schattschneider, S. Gall, and W. Fuhs, Proceedings of the 20th European Solar Energy Conference, Barcelona, 1067 (6–10 June 2005).
10. G. Ornaghi, J. Beaucarne, J. Poortmans, Nijs, and R. Mertens, Thin Solid Films 451–452, 476 (2004).
11. A. Pihan, P. Slaoui, Roca I Cabarrocas, and A. Fosca, Thin Solid Films 451–452, 328 (2004).
12. D. Dimova-Malinovska, H. Nichev, O. Angelov, M. Sendova-Vassileva, M. Sendova, and V. Mikli, J. Optoelectron. Adv. Mater. 9, 359 (2007).
13. O. Angelov, M. Sendova-Vassileva, D. Dimova-Malinovska, A. Bouzekova, and M. Sendova, in: J. Dragieva and E. Balabanova (Eds.), *Nanoscience and Nanotechnology* (Heron Press, Sofia, 2006), p. 81.
14. N. Nickel, P. Lengsfeld, and I. Sieber, Phys. Rev. B61, 15558 (2001).
15. Z. Igbal and S. Veprek, J. Phys. C 15, 377 (1982).
16. H. Campbell and F. M. Fauchet, Solid State Commun. 58, 739 (1988).
17. D. Dimova-Malinovska, O. Angelov, M. Sendova-Vassileva, V. Grigorov, and M. Ka-menova, Proceedings of the 19th European Solar Energy Conference, Paris, 371 (7–11 June 2004).
18. A. Adegboyega, J. Phys. III France 2, 1749 (1992).
19. N. Tzenov, M. Tzolov, and D. Dimova-Malinovska, *Renewable Energy V.5 part III*, (Elsevier Science, Pergamon, UK, 1994), p. 1685.
20. P. Langsfeld and N. Nickel, J. Non-Cryst. Solids 299–302, 778 (2002).
21. Harbeke and L. Jastrebski, J. Electrochem. Soc. 137, 696 (1990).
22. D. L. Greenaway and G. Harbeke, *Optical Properties and Band Structure of Semiconductors* (Pergamon, Oxford, 1968).

THE RENASCENCE OF ZnO – PROPERTIES AND APPLICATIONS

D. DIMOVA-MALINOVSKA[*]
Central Laboratory for Solar Energy and New Energy Sources, Bulgarian Academy of Sciences, 72 Tzarigradsko Chaussee Blvd., 1784 Sofia, Bulgaria

Abstract. ZnO is a material widely used in industry for decades, which experiences a scientific and technical renascence in the last few years. ZnO is a wide band gap semiconductor (3.37 eV at room temperature) with direct electronic transitions, a large exciton energy (60 meV) and hexagonal close-packed structure (wurtzite). Its unique chemical, surface and nano-scale properties give opportunities to find different new applications of ZnO in optoelectronics, planar optical waveguides, bio-application etc. It also defines application of ZnO as a transparent conducting electrode for photo-voltaics (solar cells) and light emitting diodes, and also as phosphorescent and luminescent material in the visible (blue–green) and ultraviolet regions. Recently it attracted considerable attention for gas sensors and antibacterial applications, as it withstands harsh process conditions and is a material safe to human beings and animals. In this contribution the structural and optical properties of nanocrystalline ZnO films deposited by magnetron sputtering are presented and analysed in dependence of the deposition conditions and doping. The sensitivity of undoped and doped ZnO films to ammonia at room temperature is compared and discussed.

Keywords: ZnO; XRD; optical band gap; Urbach energy; gas sensing; nanowires

1. Introduction

Zinc oxide has been recognized as one of the most important semiconductor materials for optoelectronics, solar cells, piezoelectric applications, gas sensing, bio-applications, etc.[1-4] In the last few years, ZnO has been object

---

[*]doriana@phys.bas.bg

J.P. Reithmaier et al. (eds.), *Nanostructured Materials for Advanced Technological Applications*, 285
© Springer Science + Business Media B.V. 2009

of intensive investigation, particularly its low dimensional structures: thin layers with nanometer crystallites, nano-wires, nano-rods or nano-tubes, nano-bells and others. Because of its chemical sensitivity to different adsorbed gases, high chemical stability, amenability to doping, nontoxicity and low costs it attracted much attention for gas sensors. It is one of the earliest discovered gas sensing oxide materials. The sensitivity of nano-grain ZnO gas elements is comparatively high due to the grain-size effect.[5] An increase of the optical transparency and the conductivity by doping with Al,[6] and of the gas sensitivity and selectivity for particular gases by doping with Au, Ag, Pd, Pt, Ru, Rh, Er, Ti, V, etc.[7,8] has been reported. The optical, electrical and structural properties of the films have a direct impact on the parameters of devices involving ZnO. It is known that the introduction of a high concentration of impurities into perfect semiconducting crystals causes a perturbation of the band structure with the result that the parabolic distribution of the states will be broadened and prolonged by a tail extending into the energy gap. The accommodation of impurities in the lattice results also in localized strains. The resulting deformation potential increases (in the case of compressions) or decreases (in the case of dilation) the band gap locally.[9] In the case of non-homogeneous impurity distributions, the local interaction will be more or less strong and will also smear both band edges.

For further development of high sensitivity ZnO gas sensors, the correlation between the sensitivity and the microstructure of nano-grain ZnO gas sensing elements is necessary to understand. In this contribution the dependence of the structural and optical properties of thin nanocrystalline ZnO films deposited by magnetron sputtering on deposition conditions and doping is presented. The relationship between the stress in undoped films and those doped with different elements (Al, Er, H, Ta and Co), calculated from X-ray diffraction (XRD) spectra, and the structural disorder determined from the band tails width is analysed. The sensitivity of the samples prepared under different conditions to ammonia at room temperature is compared and discussed.

2. Deposition of Undoped and Doped ZnO Thin Films

Thin films of ZnO were deposited by r.f. magnetron sputtering of ceramic ZnO or ZnO + 1%Al_2O_3 targets in atmospheres of Ar (0.5 Pa) and Ar (0.5 Pa) + H_2(0.1 Pa) (to study the influence of hydrogen on the properties of ZnO:H films) at substrate temperatures T_s between 150°C and 500°C. Discs of sintered ZnO and (ZnO (96 at. %) + Al_2O_3 (1 at. %)), 100 mm in diameter, were used to deposit undoped and Al-doped ZnO films (ZnO:Al), respectively. The films were deposited on glass substrates at an r.f. power of 180 W. The thickness of the films was in the range 350–400 nm. In the case of Er (ZnO:Er), Ta (ZnO:Ta) and Co (ZnO:Co) doping, chips of these metals were placed symmetrically on the ZnO target in the zone of the maximum erosion. The Al concentration in the films was about 2 at. %. The Er content was 0.20 and 0.40 at. %, that of Ta 1.0–3.0 at. % and finally that of Co 4–5 at. % in two different sets of samples for each dopant, respectively. The H content in the ZnO:H films deposited in an Ar + H_2 atmosphere, measured by elastic recoil detection analysis, was about 2 at. % in the case of T_s = 150°C but decreased to 1 at. % at T_s = 400°C. XRD spectra were obtained using a DRON 3 spectrometer with Cu Kα radiation: λ (CuK$_{\alpha1}$) = 1.540560 Å and λ (CuK$_{\alpha2}$) = 1.544426 Å (intensity half that of CuK$_{\alpha1}$). The instrumental profile broadening was 0.080 in 2Θ geometry. Optical transmission and reflection spectra were obtained in the range 300–1,500 nm, using a CARY UV-VIS-NIR spectrophotometer. The resistivity was measured between co-planar Al electrodes. The sensitivity to exposure to NH_3 has been measured by two methods: by the time-frequency dependence of a quartz crystal microbalance (QCM), and by the ratio of the film resistivity in air to that in the presence of the target gas.

3. Structural and Optical Properties of Undoped and Doped ZnO Thin Films

Figure 1 shows XRD patterns of all sets of films prepared at different T_s. Analysis of the 2Θ scans between 25° and 75° indicates that the deposited films were polycrystalline.[10–13] Reflections corresponding to the (002) (110), (102), (110), (103), (112) and (004) planes of wurtzite ZnO are observed. The reflection corresponding to the (002) plane becomes more pronounced with increasing T_s, while the other almost disappear, which demonstrates a preferential (002) orientation of the grains in the ZnO films deposited at higher T_s. An improvement of the structural properties with T_s is observed for all sets of films. Only in the case of ZnO:Al a deterioration of the crystalline structure is observed with T_s increasing from 150°C to 275°C. However, further increase of T_s result in an improvement of the structure (Figure 1b). The shift of the XRD peaks to lower values with increasing Ts in the cases of ZnO:H and of ZnO:Er films deposited at T_s = 500°C could be due to a possible reaction between ZnO and H_2 and the oxidation of Er, respectively.[11,12]

Table 1 contains the data obtained from XRD and the analyses of optical spectra for all of sets of ZnO samples. The (002) peak shows deviations from 34.44°, which is the value for ZnO powder. Larger deviations were observed for ZnO films deposited at lower T_s and those doped with Al, Er and Ta. They indicate that the interplanar spacing changed relative to that of ZnO powder, which is probably due to factors such as lattice strain and interstitial defects.[7] The values of the strain Δc and the stress σ, derived from the XRD data, are given in Table 1. They were calculated using to the following equation for σ[14]:

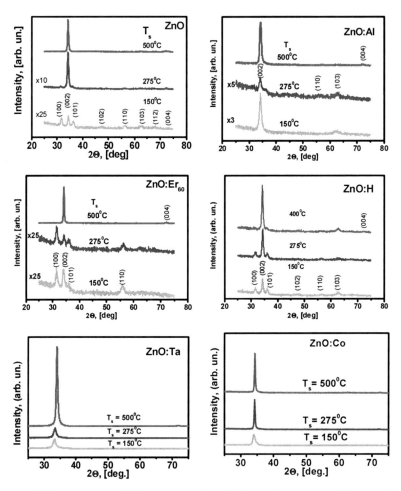

Figure 1. XRD patterns of different series of ZnO films, deposited at different substrate temperatures T_s: (a) ZnO, (b) ZnO:Al, (c) ZnO:Er, (d) ZnO:H, (e) ZnO:Ta, (3–4.5 at. %), and (f) ZnO:Co (4–5 at %).[10–12]

$$\sigma\,[\text{GPa}] = -233 \times (c_{film} - c_{bulk})/c_{bulk}, \qquad (1)$$

where c_{film} is the lattice parameter of the c axis for ZnO films and c_{bulk} that of bulk ZnO, namely the unstrained lattice parameter (c_{bulk} is 0.5206 nm[14]).

TABLE 1. 2θ values of the (002) XRD peak and its FWHM , the average grain sizes D, the strain Δc, the stress σ, the optical band gap E_g, the Urbach energy E_0 and the coefficient B for different series of ZnO films. The concentration of the dopants is given in brackets.

Samples description	T_s (°C)	2θ (°)	FWHM (°)	D (nm)	Δc_{002} (%)	σ (GPa)	E_g (eV)	E_0 (meV)	B^2 (cm² eV)
ZnO	150	34.30	0.52	16.0	0.34	−0.81	3.33	63	3.90E12
ZnO	275	34.40	0.47	17.7	0.08	−0.18	3.30	59	4.70E12
ZnO	500	34.40	0.33	25.2	0.08	−0.18	3.27	60	6.30E12
ZnO:Al (1 at. %)	150	34.20	0.62	13.6	0.65	−1.52	3.36	100	2.60E12
ZnO:Al (2 at. %)	275	34.15	0.61	13.4	0.80	−1.79	3.41	120	7.60E11
ZnO:Al (1 at. %)	500	34.27	0.62	13.4	0.40	−0.98	3.33	91	2.26E12
ZnO:Er (0.2 at. %)	150	34.10	0.62	13.4	0.92	−2.15	3.33	72	2.47E12
ZnO:Er (0.2 at %)	275	34.30	0.62	13.4	0.34	−0.81	3.34	70	3.29E12
ZnO:Er (0.2 at. %)	500	34.20	0.57	14.5	0.65	−1.52	3.28	63	3.15E12
ZnO:Er (0.4 at %)	150	34.16	0.59	14.0	0.80	−1.79	3.34	65	3.14E12
ZnO:Er (0.4 at %)	500	34.11	0.53	15.7	0.88	−2.06	3.30	52	3.15E12
ZnO:H (2 at. %)	150	34.42	0.43	19.3	0	0	3.31	70	5.30E12
ZnO:H (2 at. %)	275	34.37	0.48	17.3	0.15	−0.36	3.31	56	4.25E12
ZnO:H (1 at. %)	400	34.26	0.47	17.7	0.46	−1.07	3.27	56	7.10E12
ZnO:Ta (1 at. %)	150	33.81	0.79	10.5	1.77	−4.12	3.32	78	3.56E12
ZnO:Ta (1 at %)	275	33.86	0.40	20.8	1.61	−3.76	3.31	80	3.01E12
ZnO:Ta (1 at. %)	500	34.02	0.29	28.6	1.15	−2.68	3.31	73	4.50E12
ZnO:Ta (2 at. %)	150	33.09	1.21	6.8	3.92	−9.13	3.44	100	2.00E12
ZnO:Ta (2 at. %)	275	33.47	0.88	9.4	2.77	−6.44	3.44	95	3.77E11
ZnO:Ta (3 at. %)	500	34.07	0.45	18.5	0.52	−2.33	3.37	91	1.57E12
ZnO:Co (5 at. %)	150	34.09	0.81	10.2	0.5	−2.24	3.28	118	9.7E11
ZnO:Co (4.8 at. %)	275	34.27	0.28	30.0	0.22	−0.98	3.31	114	5.54E11
ZnO:Co (4 at. %)	500	34.33	0.25	33.0	0.14	−0.63	3.33	104	1.1E11

The values of σ are negative, demonstrating that the stresses in the deposited films are tensile in the direction of the c-axis. It is well known that defects in the films are formed during growth. These cause lattice disorder, which generates the intrinsic stress in the films. As can be seen, in general, the values of the strain and the stress decrease with increasing T_s. This is due to the fact that the atoms will have higher energies to adjust their position in the lattice, and thus the stress tends to be relaxed in films deposited at higher T_s. Stress and stain are higher in the doped films. Obviously the presence of the doping atoms Al, Er and Ta cause an increase in the lattice disorder of the ZnO films. Films deposited by sputtering in Ar + H_2 have low stress values. Possibly the presence of hydrogen creates better conditions for ZnO film growth. The average grain size D in the films was calculated with an accuracy of about 10%, by applying the Scherrer equation, using the full width at half maximum (FWHM) of the (002) peak.[15] The FWHM decreased with increasing T_s for undoped and Er- and Ta-doped ZnO films, which is an indication of an increased average grain size (Table 1). The incorporation of Al, Er and Ta resulted in smaller average grain sizes.

The transmission of all samples was about 92–93% for wavelengths higher than 600 nm, assuming a value of 100% for the transmission of the glass substrate. The absorption coefficient α was calculated as[16]:

$$\alpha\,(\lambda) = 1/d\,\ln\,[\,(1\text{-}R(\lambda))^2\,/\,T(\lambda)\,], \qquad (2)$$

where d is the film thickness, T the transmittance and R the reflectance.

The spectral dependence of α exhibits two regions: a power law one at high photon energies and an exponential one at lower energies. The formula for allowed direct transitions can be used to obtain the optical gap E_g[16,17]:

$$\alpha\,(h\nu) = B\,[(h\nu - E_g)^{1/2}\,/\,h\nu\,]\,. \qquad (3)$$

The spectral dependence of the absorption coefficient for undoped ZnO and doped ZnO films as well as for ZnO:H films are shown in Figure 2. The calculated energy gaps (3.27–3.41 eV) are typical for ZnO (Table 1). E_g decreases with increasing T_s for all sets of samples (except in the case of Al doping). The optical energy gap of the ZnO:Al thin films increases with T_s up to 275°C, above which it decreases. The observed widening of the band gap in these films could be a result of the Burstein–Moss effect. The blue shift of E_g in the Al-doped ZnO films, as compared to undoped films, could be due to an increase in the donor concentration, related to shallow Al donors. Incorporation of Er into the ZnO films, and sputtering in Ar + H_2, do not influence E_g significantly, compared to the value for undoped ZnO films. The coefficient B in Eq. (3) was calculated; values of B^2 are also given in Table 1. B was assumed to be a parameter of the band tail states.[18,19] According to Refs. 19–21, a decrease of B implies an increase of the disorder of the structure and a consequent increase of the conduction band tail width. In our case, a decrease in B is observed with doping and decreasing T_s, which is accompanied by a deterioration of the crystalline structure of the ZnO films.

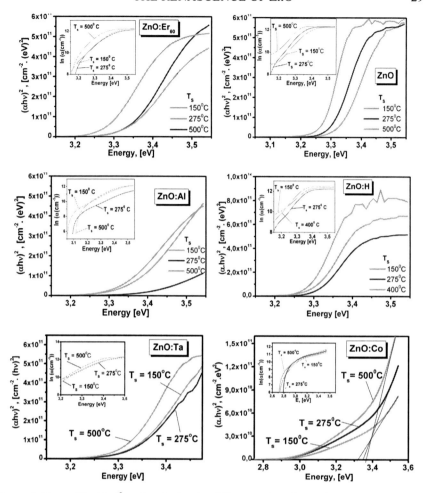

Figure 2. Plots of $(\alpha * h\nu)^2$ against $h\nu$ for the different series of ZnO films, deposited at different substrate temperatures T_s. The insets show plots of $\ln \alpha$ vs $h\nu$: (a) ZnO, (b) ZnO:Al, (c) ZnO:Er (0.2 at %), (d) ZnO:H, (e) ZnO:Ta (3–4.5 at. %) and (f) ZnO:Co (4–5 at. %).[10–12]

The optical transmittance spectra of ZnO and ZnO:Co films in the ultra-violet–visible region are shown in Figure 3. The spectra of ZnO:Co films show characteristic absorption bands at about 567, 613 and 662 nm, resulting from 3d–3d electron transitions ($^4A_2(F) \rightarrow {}^2A_1(G)$, $^4A_2(F) \rightarrow {}^4T_1(P)$ and $^4A_2(F) \rightarrow {}^2E(G)$, respectively) of the tetrahedrally coordinated Co^{2+} ions. It was shown in Ref. 21 that the valence state of Co atoms in the ZnO lattice is Co^{2+}, and that Co^{2+} ions substitute for Zn^{2+}. The optical band of ZnO:Co films increases with increasing T_s probably due to a decrease of the Co concentration, as shown by Rutherford backscattering spectrometry and energy

dispersive X-ray spectroscopy analyses. A decrease of the optical gap of
ZnO–CoO alloys with increasing Co concentration has been reported by
other authors as well.[22,23] The Co doped films exhibit significant band
tailing in the lower energy region, which is not present in the spectral
dependence of α of undoped ZnO films (Figure 2). It can be attributed to
point defects such as Zn vacancies, substitution of Zn^{2+} by Co^{2+} (CoZn) or
interstitial Co^{3+}, as supposed in Ref. 22.

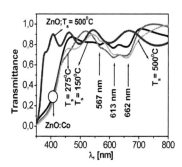

Figure 3. Transmittance spectra of ZnO:Co films deposited at different T_s. The spectrum
of undoped ZnO is given for comparison. The Co^{2+} d–d electron transitions are marked by
arrows.

In the lower energy range where α varies exponentially with photon
energy, it is possible to assume that the spectral dependence of the absorption edge follows the Urbach formula[15]

$$\alpha \,(h\nu) = \alpha_0 \exp \,[(h\nu - E_1)/E_0] \tag{4}$$

where α_0 is the Urbach absorption at the edge E_1, and E_0 is the Urbach
energy width, which is believed to be a function of the structural disorder.

The exponential dependence of the absorption on $h\nu$ in the Urbach
region ($h\nu < E_g$) is due to a perturbation of the parabolic density of the
states at the band edge: increasing structural disorder results in an increase
of E_0.[15] The energy dependences of α at lower photon energies are displayed
in the insets of Figure 2. The calculated Urbach energies are also given in
Table 1. It can be seen that doping with Al produces a significant increase
of the Urbach band tail width, compared to un-doped ZnO films, an indication of increasing structural disorder, which is confirmed by the XRD
results. The incorporation of Er results in a slight increase of E_0. The
ZnO:H films exhibit the lowest values of the Urbach energy. This is related
to the improvement of the crystalline structure, as indicated by the XRD
pattern. Possibly, the hydrogen compensates some structural defects during
the deposition of ZnO:H films. It should be noted that the deposition rate for

this set of samples is the lowest, which should result in better crystallinity. This is more pronounced for T_s values up to 275°C, as indicated by the XRD patterns. The width of the Urbach tail decreases with increasing T_s for all sets of samples, which could be related to an improvement of the structure of the ZnO films, as the XRD data show. The size of the grains with (002) orientation is the largest in ZnO:H films (about 17 nm) and does not change significantly with T_s. The Urbach band gap decreases with increasing T_s and increases with the incorporation of dopants, due to the corresponding changes of the structural disorder.

Figure 4 displays the dependence of the Urbach energy (a) and the coefficient B^2 (b), respectively, on the stress for all ZnO films under investigation. It can be seen that there is a correlation between the stress and the structural disorder in the films: an increase of the stress results in an increase of the Urbach band tail width and a decrease of the B^2. For the samples doped with Al, the values of the Urbach tail width are the highest, those for the coefficient B lowest. However, they follow the same tendency with the stress. Note that we have not observed a correlation between the rate of deposition of the films and the stress. However, the correlation between the average grain size D and the stress shown in Figure 5 demonstrates that the values of the stress decreases with increasing D. A similar result has been reported in Ref. 13. The decrease of the grain size will influence the built-in electrical field at the grain boundaries, and this will result in a perturbation of the density of the states. The increase of the structural disorder due to the doping or due to the lower deposition temperature is accompanied by a decrease of the average grain size, and an increase of the stress and the strain in the ZnO films, which cause a perturbation of the parabolic density of states at the band edges. The changes of the structural properties of magnetron sputtered ZnO films (undoped, doped with Al, Er and Ta, or deposited in presence of H_2) are compared with those of the optical properties. The optical band gap E_g of the films was obtained using the equation for direct transitions; it ranged from 3.27 to 3.44 eV. The Urbach band tail width was also calculated from the exponential dependence of the absorption edge. The lattice parameter of the c axis obtained from XRD data shows the presence of tensile stress in the films. A correlation between the values of the stress and of the band tails was observed: increasing stress is accompanied by increasing values of the band tails width. The tensile stress and the band tail width decrease with increasing substrate temperature. The results were discussed having in mind that accommodation of the impurities in the lattice results in localized strains, reflected in a perturbation of the parabolic density of the states at the band edges.

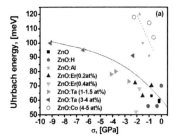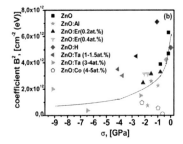

Figure 4. Dependence of the Urbach energy (a) and the coefficient B^2 (b) on the stress in ZnO, ZnO:Al, ZnO:Er, ZnO:H, ZnO:Ta and ZnO:Co films deposited under different conditions. The lines serve to guide the eye.

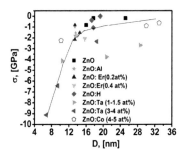

Figure 5. Dependence of the stress on the grain size in ZnO, ZnO:Al, ZnO:Er, ZnO:H, ZnO:Ta and ZnO:Co films deposited under different conditions. The curve serves to guide the eye.

4. Gas Sensing

In this section the sensitivities of undoped films and those doped with H, Er, Ta, and Co to NH$_3$ exposure to were measured by two methods, the quartz crystal microbalance method and the resistivity changes under NH$_3$ exposure.

4.1. QMC MEASUREMENTS

The ability of undoped and doped ZnO thin films to sorb ammonia vapour was evaluated by the QCM method. For this purpose, the resonance frequency shift response was measured. The sorption process was investigated over an aqueous solution of ammonia with a concentration of 100 ppm. The change in the frequency f of the QCM as a function of its mass loading is detected. Figure 6 shows the frequency change Δf as a function of the exposure time to a 100 ppm ammonia solution. A drastic decrease was observed for all samples investigated. After that, the rate slowly decreases and the frequency reaches to a constant value. The rate of change for the first 2 min

S_p was calculated in order to estimate the sensitivity of each kind of ZnO layer. The S_p value and the total sorbed mass Δm was evaluated on the basis of the measured time-frequency characteristics from the relationship between Δf and Δm, given by the Sauerbrey equation[24]

$$\Delta f = - (2.26 \times 10^{-6} f^2 \Delta m)/A, \qquad (5)$$

where f (MHz) is the frequency of the QMB before exposure to ammonia, Δm (g) the mass of the sorbed gas on the surface of the ZnO films and A (cm^2) the surface area of the electrodes.

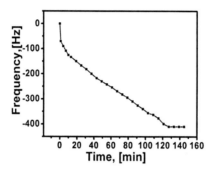

Figure 6. Time–frequency characteristic of a quartz resonator with a ZnO:H film over a 100 ppm NH_3 solution.

The total mass sorbed on the samples was calculated for the whole period until saturation was achieved.[25] The data are presented in Table 2. The lowest rate of sorption and the smallest total sorbed mass (4.5 Hz/min and 4.86 ng, respectively) are registered for ZnO. In the case of films doped with Ta and Co, values of 8.5 Hz/min and 8.76 ng, and 5.65 Hz/min and

TABLE 2. Values of the optical band gap E_g, the average grain size D, the rate of sorption S_p, and the sorbed mass Δm, over 100 ppm NH_3, for different kinds of sample. The values of D were calculated using the 2θ values (given in the brackets) and the corresponding FWHM.

Samples	E_g (eV)	D (Å); 2θ (°)	S_p (Hz/min)	Δm (ng)
ZnO:Er	3.30	102; (31.50°) 98; (34.30°) 90; (35.80°)	13.20	32.20
ZnO:Co	3.352.95	105; (34.04°)	5.65	12.78
ZnO:Ta	3.46	131; (33.10°)	8.50	8.76
ZnO	3.30	160; (31.60°) 177; (34.40°) 110; (36.21°)	4.50	4.86
ZnO:H	3.31	173; (34.37°)	80	37.7

12.78 ng for the sorption rate and mass respectively were obtained. It has to be noted that the largest rate of frequency change of 13.2 Hz/min was measured for the ZnO:Er layers, which corresponds to a total sorbed mass of 32.20 ng, and on ZnO:H layers (80 Hz/min and about 37.7 ng, Figure 6). However, in the case of ZnO:H the rate of adsorption decreased to 2.47 Hz/min after the first 2–3 min. A faster response time of the sensitivity of ZnO:Er films to NO_2 as compared to ZnO and ZnO:Pd has been reported in Ref. 26, as well. However, in this case the sensitivity was defined as the ratio of the resistivity in the target gas to that in air.

4.2. ELECTRICAL RESISTIVITY BASED SENSITIVITY

The resistivity of the films was calculated from I–V characteristics measured in the dark using a Keithley 6517 electrometer. The co-planar evaporated Al electrodes were verified to yield ohmic behaviour. The sensitivity of the ZnO films to NH_3 was determined as the ratio between the resistivities measured in air and in the presence of NH_3. The variation of the sensitivity of undoped and H-doped ZnO films with the NH_3 concentration is shown in Figure 7. The sensitivity increased with the concentration up to about 20,000 ppm. An initial trend of saturation at higher NH_3 concentrations is also observed. It can be seen that the sensitivity of the ZnO:H films is higher than that of the ZnO films. The ZnO:H film deposited at T_s = 400°C exhibited a higher sensitivity than that deposited at 150°C for NH_3 concentrations less than about 18,000 ppm. For the higher sensitivity films, it was found that injection of 120 ppm of NH_3 at room temperature induced a decrease in the resistivity, thus leading to a response magnitude of 2.5, while for an injection of 630 ppm, the response magnitude was about 27. Figure 8 gives an example of response transients of ZnO:H films. It has to be noted that the signal could returned to its initial value after several cycles. This indicates that the adsorption of ŅH on the film surface was reversible.

It is well accepted that the sensitivity of semiconductor gas sensors is attributed to chemisorption of oxygen and OH^- on the oxide surface and subsequent reaction between adsorbed species and the gas tested, which causes the change of the resistance.[27] It is also known that atmospheric oxygen molecules and water vapor are adsorbed on the surface of the ZnO in the form of O_2^- and OH^-, respectively, thereby reducing the electrical conductivity. When a ZnO sensor is exposed to a reductive atmosphere, the test gas like NH_3 reacts with the surface species, the trapped electrons are returned to the conduction band of the ZnO, causing an increase of the conductivity change.

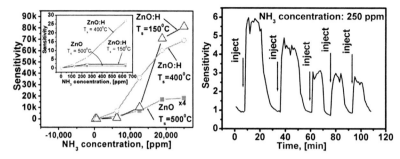

Figure 7 (left). Sensitivity of ZnO:H and ZnO films toward NH_3 at room temperature. The insert shows the sensitivity at lower NH_3 concentration.

Figure 8 (right). Transient response of a ZnO:H film ($T_s = 400°C$) for a NH_3 concentration of 250 ppm.

The different effective surface areas of the films could be one of the reasons for their different sensitivities. However, the effective surface area estimated from the grain size (calculated from XRD spectra) is almost the same for all set of samples studied here. Thus, the results presented here require another explanation. Hydrogen and H_2O (OH^-) species adsorbed in the ZnO:H films during the deposition could explain the higher response of ZnO:H as compared to that of pure ZnO films and of the ZnO:H sample deposited at higher T_s. One of the possible reactions on the substrate during the sputtering process in an $Ar + H_2$ atmosphere is:

$$ZnO + H_{2gas} \leftrightarrow Zn + H_2O_{gas} \qquad (6)$$

Thermodynamic analyses[28] show that this reaction will be shifted to the right side with increasing temperature. According to Eq. (6) deposition at 150°C would results in higher H concentrations, but at 400°C in higher amounts of water vapor in ZnO:H films. Possibly the presence of OH^- accelerates surface reactions with the reducing gas, with NH_4OH as a product, additionally to the NH_3 reaction with adsorbed oxygen species. Thus, the samples with higher OH^- concentrations (deposited at $T_s = 400°C$ in $Ar + H_2$) would be more sensitive to lower NH_3 concentrations.

A comparison of the changes of the sensitivity of ZnO:Co, deposited at different T_s, and ZnO, deposited at $T_s = 500°C$, with NH_3 concentrations up to 18,750 ppm is shown in Figure 10. The undoped ZnO films deposited at low T_s (150°C and 275°C) have a very low sensitivity to the NH_3 concentrations used; their sensitivities are not presented here. The sensitivity of ZnO:Co films increases with increasing T_s. They demonstrates a sensitivity two orders of magnitude higher than that of undoped ZnO films deposited at the same $T_s = 500°C$. A typical transient response curve of a ZnO:Co thin film is shown in Figure 10. Usually the gas sensitivity increases with decreasing the grain size in the films.[29] However, this is not the case here (Table 2 and Figure 9). The higher sensitivity of Co-doped ZnO films reported in this work could be due to the presence of higher concentrations of point defects at the surface of the grains, such as interstitial Co^{3+} and CoZn, where higher concentrations of O^{2-} and OH^- could be adsorbed. It is well accepted that NH_3 reacts with the surface species; the trapped electrons are returned to the conduction band, causing an increase of the conductivity of the ZnO:Co films and, in turn, the sensitivity of the sensor.[27]

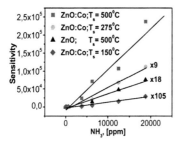 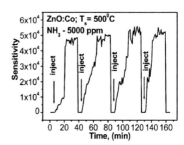

Figure 9 (left). Variation of the sensitivity of ZnO and ZnO:Co thin films versus NH_3 concentration.

Figure 10 (right). Transient response curve of a ZnO:Co thin film to ammonia at room temperature.

5. Conclusion

The sensing characteristics of ZnO based NH_3 sensors were presented. Undoped films and those doped with H, Er, Ta or Co of about 100 nm thickness on the surface of a QCM showed a sensitivity to ammonia. The highest rate of sorbtion was registered in the first 2 min for all investigated samples. Higher sensitivities were observed for H- and Er-doped ZnO films, compared to undoped, ZnO:Co and ZnO:Ta films. The differences in the grain size in the deposited films could be one of the reasons for the different behaviour of the QCM sensors. This study of the sensitivity of ZnO thin films shows that they have a potential for applications as room temperature sensors down to NH_3 concentration of 100 ppm.

6. Perspectives for New Applications of ZnO

Among the known one-dimensional nanomaterials, ZnO has some key advantages: (i) it exhibits both semiconducting and piezoelectric properties that can form the basis for electromechanically coupled sensors and transducers; (ii) ZnO is biosafe and biocompatible and can be used for bio-medical applications; (iii) ZnO exhibits the most diverse and abundant con-figurations of nanostructures know so far, such as nanowires, nanobelts, nanorings, nanobows and nanohelices. It has been shown that ZnO thin film strongly interact with bacteria thus causing destruction of the cell wall and even death of the biological entity.[30] The underlying principle is the electro-static interaction between the charged cell membrane and the ZnO film interface. This interaction is a sort of oxidation-reduction process, during which the bacteria exchange charges with the ZnO grains; these charges can be recorded via the impedance characteristics of the device. The smaller the ZnO grains (nanoparticles), the more pronounced the antibacterial action, i.e. the larger the interaction with the bacteria. The kinetics of this interac-tion should be unique for a given sort of bacteria, because of the unique cell membrane manifested by the bacterial effect of ZnO. Recording this kinet-ics with the time can provide special maps to distinguish between different types of bacteria as well as to measure their minute concentrations from the variation of the signal intensity. Fine-tuning of the maps can be achieved by controlling the morphology and size of the ZnO nanospecies comprising the thin films: nanoparticles or nanowires.

Various synthesis methods have been used to grow ZnO nanorods or nanowires. They include chemical vapor transport and condensation, ther-mal evaporation, metalorganic chemical vapor deposition, hydrothermal methods, sol–gel, and electrochemical deposition. Figure 11 shows SEM images of ZnO nanowires deposited on a glass substrate coated by SnO_2 by an electrochemical method.[31] Figure 12 shows SEM images of *E.coli* before and after treatment with ZnO nanofluids.[30] The considerable damage of some *E.coli* which has caused the breakdown of the membrane of the bacte-ria can clearly be seen. It is supposed that direct nanoparticle-cell mem-brane interactions and generation of active oxygen species by ZnO particles could be mechanisms of the observed antibacterial effect.

Figure 11. SEM images of ZnO nanowires deposited on SnO_2 coated glass by an electro-chemical method.[31]

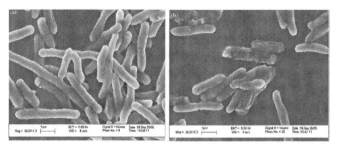

Figure 12. SEM images of E. coli: (a) before antibacterial tests; (b) after treatment with 0.2% ZnO nanofluids for 5 h.[30]

Wireless devices may allow in situ real time biomedical monitoring and detection, but they require a power source. Such devices should be self-powered rather than use a battery. The body provides numerous potential power sources: mechanical energy, vibration energy, chemical energy (glucose), and hydraulic energy, but the challenge is the efficient conversion into electrical energy. If accomplished on the nanoscale, such power sources could greatly reduce the size of integrated nanosystems for optoelectronics, biosensors, resonators, and other applications. The physical principle for creating, separating and preserving piezoelectric charges in nanowires is the coupling of piezoelectric and semiconducting properties. An approach for converting mechanical energy into electrical power using aligned ZnO nanowires[32] has been demonstrated. Thus, ZnO nanowires could be used in new self-powering devices that harvests electricity from the environment for applications such as implantable biomedical devices, wireless sensors and portable electronics.

ACKNOWLEDGEMENTS

This work is being performed with financial support from the Bulgarian National Scientific Fund (projects UF 1505).

References

1. H. Cao, J. Y. Xu, D. Z. Zhang, S. H. Cheng, S. T. Ho, E. W. Seelig, X. Liu, and R. Chang, Phys. Rev. Lett. 84, 5584 (2000).
2. J. Wang and K. Lakin, Appl. Phys. Lett. 42, 352 (1983).
3. D. Dimova-Malinovska, J. Luninescence 80, 207 (1999).
4. C. Eisele, S. Klein, C. Nebel, and M. Stutzmann, Proc. 17th EC PVSEC, Munich, Germany, 3044 (2002).
5. J. Q. Hu, Y. Q. Pan, Y. A. Shun, and Z.-Z. Tian, Sensor. Actuator. B 66, 277 (2000).
6. D. Dimova-Malinovska, N. Tzenov, N. Tzolov, and L. Vassilev, J. Mater. Sci. Technol. B 52, 59 (1998).

7. Y. Cao, L. Miao, S. Tanemura, M. Tanemura, Y. Kuno, Y. Hayashi, and Y. Mori, Jpn. J. Appl. Phys. 45, 1623 (2006).
8. M. Aslam, V. Chaudhary, I. Mulla, S. Sainkar, B. Mandale, A. Belhekar, and V. Vijayamohanan, Sensor. Actuator. 75, 162 (1999).
9. R. H. Parmenter, Phys. Rev. 97, 587 (1955).
10. D. Dimova-Malinovska, O. Angelov, H. Nichev, M. Kamenova, and J.-C. Pivin, JOAM 9, 248 (2007).
11. D. Dimova-Malinovska, H. Nichev, O. Angelov, V. Grigorov, and M. Kamenova, Superlattice Microst. 42, 123 (2007).
12. D. Dimova-Malinovska, O. Angelov, H. Nichev, and J.-C. Pivin, JOAM 1, 248 (2007).
13. J. Mass, P. Brattacharya, and R. Katiyar, Mater. Sci. Eng. B 103, 9 (2003).
14. B. Zhu, X..Sun, S. Guo, X. Zhao, J. Wu, R. Wu, and J. Liu, Jpn. J. Appl. Phys. 45, 7860 (2006).
15. B. E. Warren, *X-Ray Diffraction* (Dover, New York, 1990).
16. J. Pankove, *Optical Processes in Semiconductors* (Prentice-Hall, Englewood Cliffs, NJ, 1971).
17. D. Dragoman and M. Dragoman, *Optical Characterization of Solids* (Springer, Heidelberg, 2002).
18. N. F. Mott and E. A. Davis, *Electronic Processes in Non-Crystalline Materials* (Clarendon, Oxford, 1979), p. 289.
19. N. Saito, J. Appl. Phys. 58, 3504 (1985).
20. L. Nedialkova, D. Dimova-Malinovska, and V. Kudojarova, Sol. Energy Mater. Sol. Cells 27, 37 (1992).
21. D. Dimova-Malinovska, L. Nedialkova,, M. Tzolov, and N. Tzenov, Sol. Energy Mater. Sol. Cells 53, 333 (1998).
22. C. Song, S. Pan, X. J. Liu, X. W. Li, F. Zeng, W. S. Yan, B. He, and F. Pan, J. Phys.: Condens. Mat. 19, 176229 (2007).
23. T. Fukumura, Z. Jin, K. Hasegawa, M. Kawasaki, P. Ahmet, T. Chikyow, and H. Koinuma, J. Appl. Phys. 90 4246 (2001).
24. V. Malov, *Piezoresonansnie datchizi* (Enegyatomizdat, Moscow, 1989), p. 68 (in Russian).
25. H. Nichev, D. Dimova-Malinovska, V. Georgieva, O. Angelov, and J.-C. Pivin, This volume, p. 285.
26. N. Koshizaki and T. Oyama, Sensor. Actuator. B 66, 119 (2000).
27. G. S. T. Rao and D. T. Rao, Sensor. Actuator. B 55, 166 (1999).
28. D. Dimova-Malinovska, Ph.D. thesis (University, Moscow, 1974).
29. S. Kohiki, M. Nishitani, T. Wada, and T. Hirao, Appl. Phys. Lett. 64, 2876 (1994).
30. L. Zhang, Y. Jiang, Y. Ding, M. Povey, and D. York, J. Nanopart. Res. 9, 479 (2007).
31. D. Dimova-Malinovska, P. Andreev, and K. Shturbova, Unpublished data.
32. Z. L. Wang and J. Song, Science 312, 22 (2006).

SENSITIVITY OF ZnO FILMS DOPED WITH Er, Ta AND Co TO NH₃ AT ROOM TEMPERATURE

H. NICHEV[*1], D. DIMOVA-MALINOVSKA[1],
V. GEORGIEVA[2], O. ANGELOV[1], J.-C. PIVIN[3], V. MIKLI[4]
[1]*Central Laboratory for Solar Energy and New Energy Sources, Bulgarian Academy of Sciences, 72 Tzarigradsko Chaussee Blvd., 1784 Sofia, Bulgaria*
[2]*Georgi Nadgakov Institute of Solid State Physics, Bulgarian Academy of Sciences, Bulv. Tzarigradsko chaussee, 72, 1784 Sofia, Bulgaria*
[3]*CSNSM, CNRS-IN2P3, Bat.108, 91405 Orsay Campus, France*
[4]*Centre for Materials Research, Tallinn Technical University, Ehitajate tee 5, 19086 Tallinn, Estonia*

Abstract. The sensitivity of ZnO films, undoped and doped with Er, Ta and Co, to exposure to NH₃ has been studied. The films were deposited by magnetron co-sputtering of a sintered ZnO target with chips of the different doping elements on its surface. The sensitivities of the ZnO films to NH₃ exposure were measured by the quartz crystal microbalance method using the resonance frequency shift. The sorption process was investigated over an aqueous solution of ammonia with a concentration of 100 ppm. The speed and the mass of sorption were calculated on the basis of the time – frequency characteristic.

Keywords: ZnO; magnetron sputtering; gas sensors

1. Introduction

Due to its high chemical stability, low dielectric constant and high optical transmittance, ZnO has numerous applications as a dielectric ceramic, pigment, catalyst, transparent conductive electrode and sensing material in

---

[*] nitschew@yahoo.de

J.P. Reithmaier et al. (eds.), *Nanostructured Materials for Advanced Technological Applications,* 303

electronic and optoelectronic devices, solar cells and gas sensors.[1-4] As a sensor material, ZnO is one of the earliest discovered and most widely applied oxides in gas sensors.[3,4]

In this work, the sensitivities of undoped and doped ZnO films (ZnO:Er, ZnO:Ta and ZnO:Co) at room temperature to exposure to NH_3 were measured by the quartz crystal microbalance (QMB) method.

2. Experimental Details

Thin films of undoped and doped ZnO were deposited by r.f. magnetron sputtering in an atmosphere of Ar as described elsewhere.[5] A disc of sintered ZnO was used to deposit undoped ZnO films. In the case of Er, Ta and Co doping, chips of the doping elements were placed symmetrically on the ZnO target. The concentration of the doping elements was about 0.4–4.5 at %, as determined by Rutherford Back Scattering (RBS) analysis. The sputtering pressure was 0.5 Pa.

The sensors with undoped and doped ZnO films were prepared using rotated Y-cut (the so-called AT-cut) polished quartz crystal plates of 8 mm diameter, upon which silver electrodes of 5 mm in diameter were deposited on both sides by vacuum evaporation. In this way, resonators with a frequency of 14 MHz were obtained. Subsequently, thin undoped and doped ZnO films of 100 nm thickness were deposited on both sides of the resonator at $T_s = 220°C$ only on the metal electrode areas. The resulting structures were kept in air until saturation of the frequency value was reached, and then placed over an aqueous NH_3 solution inside an isolated and thermally stabilized measurement chamber. The NH_3 concentration in the aqueous solution was 100 ppm. The experimental set-up for measuring the sensing properties of the QMB is described in detail in Ref. 6.

3. Results and Discussion

The ability of the undoped ZnO thin films and those doped with Er, Ta and Co to sorb ammonia vapour was evaluated by the QMB method. For this purpose, the resonance frequency shift response was measured. The frequency accuracy was ~0.1 Hz in these experiments.

Figure 1 shows the frequency change Δf as function of the exposure time over a 100 ppm ammonia solution for different samples. The highest rate of the change in frequency S_p was registered in the first 2 min for all samples investigated. Thereafter, the rate slowly decreased and the frequency approached a constant value.

The rate of change for the first 2 min was calculated in order to estimate the sensitivity of each kind of ZnO layer. The S_p value and the total sorbed mass, Δm, were evaluated using the relationship between Δf and Δm for a AT-cut quartz crystal, given by the Sauerbrey equation[7]:

$$\Delta f = - (2.26 \times 10^{-6} f^2 \Delta m)/A, \qquad (1)$$

where f (MHz) is the frequency of the QMB before exposure to ammonia, Δm (g) is the mass of the sorbed gas on the surface of the ZnO films and A (cm^2) is the surface area of the electrodes.

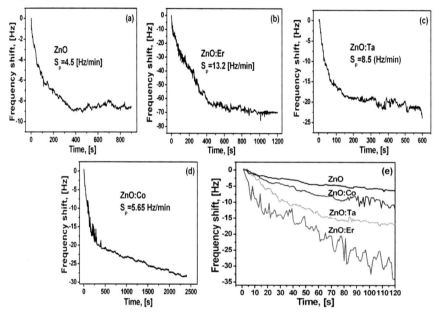

Figure 1. Time-frequency shift characteristics of a quartz crystal microbalance with: (a) ZnO, (b) ZnO:Er, (c) ZnO:Ta, (d) ZnO:Co films; (e) presents the frequency shifts for the all samples for 120 s over a 100 ppm NH₃ solution.

The total sorbed mass on the samples was calculated for the whole period until saturation was achieved. The lowest rate of sorption and the smallest total sorbed mass (4.5 Hz/min, 4.86 ng, respectively) were registered for undoped ZnO films (with a grain size of 25.3 nm). In the case of ZnO films doped with Ta (28.6 nm grain size) and Co (33 nm grain size), values of 8.5 Hz/min and 8.76 ng and 5.65 Hz/min and 12.78 ng for the sorption rate and mass, respectively, were obtained. It has to be noted that the largest rate of change of the frequency (13.2 Hz/min) was measured in the case of sorption on ZnO:Er layers (15.7 nm grain size), which corresponds to a total sorbed mass of 32.20 ng.

Differences in the grain sizes of the different samples could be one of the reasons for the different sorption rates and masses measured by the QMB.

It is well known that the sensitivity of ZnO gas sensors depends upon the grain size: the smaller the grain size, the higher the gas sensitivity.[3] The ZnO:Er sample has the smallest grain size and highest values of the effective surface which, could absorb a larger mass of ammonia, and its sensitivity is highest. Our results show that the value of the sorbed mass increases with decreasing average grain size in the films. Additionally, an influence of catalysts such as Er and Ru[8] on the gas absorption has to be considered.

4. Conclusion

The sensing characteristics of ZnO based NH_3 sensors were studied. Undoped films and those doped with Er, Ta or Co, of about 100 nm thickness, on the surface of a QMB showed a high sensitivity to ammonia. The differences in the grain size in the deposited films could be one of the reasons for the different behaviours of the QMB sensors. The study of the sensitivity of undoped and doped ZnO thin films deposited by magnetron sputtering on the surface of a QMB shows that they have a potential application as room temperature ammonia sensors.

ACKNOWLEDGEMENTS

This work is being performed with financial support from the Bulgarian National Scientific Fund by project UF 05/2005 and partially by project NT-3-03/2006.

References

1.   D. Dimova-Malinovska, N. Tzenov, N. Tzolov, and L. Vassilev, J. Mater. Sci. Technol. B 52, 59 (1998).
2.   Y. Cao, L. Miao, S. Tanemura, M. Tanemura, Y. Kuno, Y. Hayashi, and Y. Mori, Jap. J. Appl. Phys. 45, 1623 (2006).
3.   G. S. T. Rao and D. T. Rao, Sensor. Actuat. B 55, 166 (1999).
4.   M. Aslam, V. Chaudhary, I. Mulla, S. Sainkar, B. Mandale, A. Belhekar, and V. Vijayamohanan, Sensor. Actuat. A 75, 162 (1999).
5.   D. Dimova-Malinovska, H. Nichev, O. Angelov, V. Grigorov, and M. Kamenova, J. Superlattice Microst. 42, 123 (2007).
6.   V. Lazarova, L. Spassov, V. Georgiev, S. Andreev, E. Manolov, and L. Popova, Vacuum 47, 1423 (1996).
7.   V. Malov, Piezoresonanie datchizi (Enegyatomizdat, Moscow, 1989), p. 68 (in Russian).
8.   M. S. Wang, G. H. Jain, D. R. Patil, S. A. Patil, and L. A. Patil, Sensor. Actuat. B 115, 128 (2006).

PULSED LASER DEPOSITION OF LAYERS AND NANOSTRUCTURES BASED ON CdTe AND Bi

A.S. YEREMYAN[*], H.N. AVETISYAN, L.G. ARSHAKYAN
*Institute of Radiophysics and Electronics of NAS of Armenia,
1 Brs. Alikhanian St., 378410, Ashtarak, Armenia*

Abstract. In this work the PLD method was used for the growth of CdTe layers and layered [CdTe/Bi]$_m$ nanostructures. Technological regimes were found for the growth of CdTe layers with the hexagonal wurtzite structure. The observed dependence of the structure of the layers (hexagonal or cubic) on the laser intensity is related to the energy state of the ablated material and its influence on the orientational properties of the substrate surface. It is shown that the temperature for the monocrystalline growth of this meta-stable phase can be decreased significantly (from 300°C to 170°C) by initial deposition of a seeding submonolayer of bismuth at moderate intensities of the evaporating laser. Such a decrease of the growth temperature allows the fabrication of multilayer [CdTe/Bi]$_m$ structures with abrupt interfaces.

Keywords: PLD; CdTe; hexagonal; bismuth nanolayers

1. Introduction

An important peculiarity of pulsed laser deposition (PLD) as a film-growth technique is that as a result of ablation of the target material, the flux of material deposited onto the substrate is characterized by an energy dispersion of constituent particles: ions, atoms, and electrons.[1–3] Variation of the energy of the deposited particles in a wide range (from 25 to 150 eV in most cases) by changing the laser intensity,[1–5] introducing inert gas into the evaporation chamber,[6] etc., provides an additional control parameter and allows to synthesize composite materials and metastable phases, including combinations which are difficult to obtain with equilibrium growth methods. This is the case, in particular, for the CdTe layers and layered structures

[*] arsham@irphe.am

J.P. Reithmaier et al. (eds.), *Nanostructured Materials for Advanced Technological Applications*,   307
© Springer Science + Business Media B.V. 2009

[CdTe/Bi]$_m$ with abrupt interfaces considered in this work. A series of studies have shown that under special non-equilibrium conditions, namely, using the PLD technique, it is possible to obtain CdTe layers with the crystalline structure of hexagonal wurtzite,[5–10] whereas the cubic zinc-blende structure is the usual form of crystallization of this material.[8–10] Along with other parameters such as substrate type and temperature, a key role in determining the growth conditions for the metastable phase is played by the energy state of the plasma of the ablated material. This parameter, which is governed by the intensity of the evaporating laser pulses, is critical for determining the optimal film-growth regime especially in the case of substrates with strongly oriented surfaces (e.g. the mica substrate considered here), which are initially characterized by with a large density of active centers of nucleation.

    Technological regimes where the growth of hexagonal wurtzite-type CdTe can be realized are of special interest in view of the fabrication of multilayer CdTe–Bi–CdTe nanostructures in which the alternation of Bi and CdTe layers, aligned with the close-packing planes, would allow to obtain quantum-well structures with abrupt barriers, and to control the growth direction of the bismuth layers, which is essential due to the large anisotropy of the effective masses of bismuth.[5]

2. Experimental Details

The PLD equipment used consists of a Q-switched Nd^{3+}:YAG laser (emission wavelength 1.06 μm, pulse duration 30 ns, energy per pulse 1.54 J) and a vacuum chamber with a residual gas pressure of $2.6 \cdot 10^{-4}$ Pa. The CdTe layers were obtained on mica substrates by deposition from a CdTe compound target. In order to carry out electron diffraction analyses in the transmission mode, the CdTe and CdTe/Bi layers were detached from the mica substrates. For this purpose, the layers were deposited on mica substrates initially covered by a laser-deposited KBr layer with a thickness of about 10 nm, which is sufficiently thin to maintain the orientational properties of the mica substrate.[11] The subsequently deposited layers are detached from the substrate by dissolving the KBr in water.

    The investigations of the growth regimes of CdTe layers were carried out by variation of the substrate temperature and the laser pulse intensity. The explored range of laser intensities in the irradiation zone of different targets is shown in Figure 1, where the measured thickness of the layers (i.e. the amount of material) deposited per laser pulse is given as a function of the pulse intensity. The thicknesses were measured using an interferometer and the thickness measurement system Filmetrics F20.

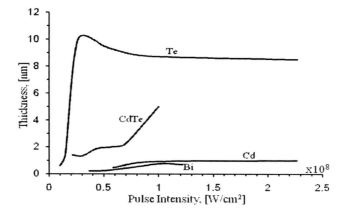

Figure 1. Thickness of different layers deposited per pulse of laser radiation.

3. Results and Discussion

3.1. DEPOSITION FROM COMPOUND CdTe TARGET

Electron diffraction analyses have shown that in the substrate temperature range 100–200°C and at relatively high intensities (> ~$1.5 \cdot 10^8$ W/cm² or a fluence of ~4.5 J/cm²) the layers have a polycrystalline structure. In contrast to the more usual cubic sphalerite structure of CdTe, the observed diffraction rings correspond to reflections from layers with a hexagonal structure. In the temperature range of 200–320°C at the same laser intensities, a dotted (mosaic) pattern is observed (Figure 2, left), corresponding to a hexagonal closepacked structure with packing faults.

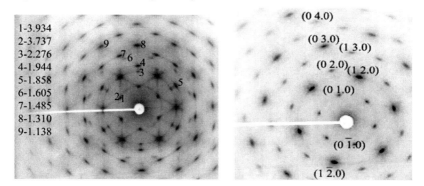

Figure 2. Diffraction patterns of a hexagonal wurtzite CdTe layer deposited on a mica substrate. Left: Sample with twinning (the numbers show the measured interplanar distances); right: indexed pattern of a single-phase oriented layer.

From analyses of patterns recorded at various angles of the incident electron beam it was established that a gliding plane perpendicular to the basis axis of the hexagonal cell exists, i.e. the obtained structure has the space group symmetry P63mc, the same as wurtzite ZnS.

Single-crystalline CdTe layers oriented along the [00.2] axis were obtained at lower laser intensities. An indexed diffraction pattern corresponding to an intensity of $1 \cdot 10^8$ W/cm^2 and a temperature 300°C is shown in Figure 2, right.

The observed dependence of the film structure on the laser intensity can be explained by the influence of the particle flux on the substrate surface; a decrease of the laser intensity corresponds to a decrease of the average kinetic energy of the ions (about 25 eV for $I_{CdTe} \sim 1 \cdot 10^8$ W/cm^2), i.e. a decrease of the amount of particles which induce defects in the near-surface region of the substrate. The high density flux of material results in a lower temperature for crystalline growth, whereas the optimal regimes for the growth of defect-free layers are achieved using moderate intensities, at which the initially highly-oriented properties of the mica substrates are preserved.

3.2. DEPOSITION OF NANOLAYER STRUCTURES Bi-CdTe

As was reported in an earlier publication,[5] at laser intensities of $I_{Bi} = (1.65-3.3) \cdot 10^8$ W/cm^2 (laser fluences 5–10 J/cm^2), single-crystalline layers of bismuth are obtained on (001) splitted KBr substrates, starting at temperatures as low as 60°C. The structure and quality of the layers do practically not depend on the laser intensity in this range and are preserved up to a growth temperature of 180°C, above which an intense re-evaporation from the KBr surface is taking place.

Single-crystalline Bi layers were obtained on strongly oriented mica substrates in the same temperature range but at comparatively low laser evaporating intensities ($\sim 1.0 \cdot 10^8$ W/cm^2). The bismuth nanolayers with thicknesses of ~5 nm are strictly oriented along the trigonal [111] axis of the rhombohedric crystal, which corresponds to the [00.3] direction in hexagonal notations.

However, from the results presented in Sect. 3.1 the growth temperature of ~300°C for monocrystalline wurtzite CdTe is much higher than the maximum temperature (~180°C), above which the quality of the Bi layers deteriorates due to strong re-evaporation from the substrate surface. This significantly complicates the choice of deposition parameters for Bi-CdTe multilayers.

Our investigations have shown that a significant decrease of the growth temperature of CdTe layers can be achieved, when a bismuth submonolayer

(~0.7 nm) is initially deposited on the mica substrate covered with a 10 nm KBr layer (see the experimental details above). Using the same laser intensity as in the case studied in Sect. 3.1 ($1 \cdot 10^8$ W/cm²), the growth of single-crystalline CdTe layers on such substrates occurs at temperatures as low as 170°C. The growth direction of these layers coincides with that observed in the previous case (Figure 2, right).

The diffraction pattern of a layered structure CdTe(35 nm)-Bi(5 nm) obtained with the method described above at a growth temperature of 170°C and at a laser intensity on the Bi target of $I_{Bi} = 5.4 \cdot 10^7$ W/cm² is shown in Figure 3.

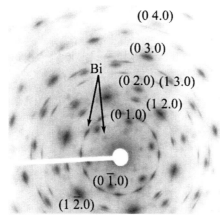

Figure 3. Diffraction pattern of a layered structure CdTe/Bi. The indexed spots are the same as in Figure 2, right and correspond to CdTe layer.

It can be seen from Figure 3, that along with the reflections corresponding to the CdTe layer (cf. Figure 2, right), there are stretched spots which correspond to the hexagonal bismuth layer oriented with the close-packed plane parallel to the substrate surface. The stretching of the spots and the presence of excess reflections in the diffraction pattern (Figure 3) indicates that the crystalline quality of the bismuth layers obtained on CdTe deteriorates compared to the layers deposited directly on mica substrates. This is caused by a deformation of the bismuth lattice along the trigonal axis which results in a transformation of the lattice from the rhombohedric to the hexagonal closed-packed structure. As a result, there are no systematic extinctions on diffraction pattern. As we believe, the quality of bismuth layers on CdTe at such low growth temperatures can be further improved by a variation of the laser intensity on the Bi target.

4. Conclusions

In summary, CdTe layers and $[CdTe/Bi]_m$ multilayer structures were obtained on mica substrates using the pulsed laser deposition technique from a compound target. The technological regime of the growth of CdTe layers with hexagonal wurtzite structure was established. It was shown that a significant decrease (from 300°C to 170°C) of the temperature for monocrystalline growth can be achieved for this metastable phase by the initial deposition of a seeding submonolayer of bismuth at a moderate intensity of evaporating laser. The observed dependence of the layer structure (hexagonal or cubic) on the laser intensity is related to the energy state of the plasma plume and its influence on the orientational properties of the substrate surface. The established technological regimes allow to obtain abrupt multilayer heterostructures with alternating lattice-matched nanolayers of CdTe and Bi.

ACKNOWLEDGEMENT

This work is supported by the Armenian State Programme "Semiconductor Nanoelectronics".

References

1. N. Basov, V. Boykov, O. Krokhin, G. Sklizkov, JETP 51, 989 (1966).
2. S. Fahler, H. Krebs, Appl. Surf. Sci. 96–98, 61 (1996).
3. J. Lunney, Appl. Surf. Sci., 86, 79 (1995).
4. A. Alexanian, N. Aramyan, K. Avjyan, A. Khachatryan, R. Grigoryan, A. Yeremyan, in: R. Potyrailo and W.F. Maier (Eds.), *Combinatorial and High-Throughput Discovery and Optimization of Catalysts and Materials* (CRC Press, Boca Raton, FL, 2006), p. 427.
5. A.G. Alexanian, H.N. Avetisyan, K.E. Avjyan, N.S. Aramyan, G.A. Aleksanyan, R.P. Grigoryan, A.M. Khachatryan, A.S. Yeremyan, Proc. MRS 799, 2003, paper Z2.9.1.
6. I. Spinulescu-Curnaru, Phys. Status Solidi 15 , 761 (1966).
7. F. Jackson, L. Berlousis, P. Rocabois, B. Cavenett, J. Cryst. Growth 159, 200 (1996).
8. Y. Yezhovsky, I. Kalinkin, Thin Solid Films, 18, 12 (1973).
9. D.B. Holt, Thin Solid Films, 37, 91 (1976).
10. Y. Kawai, Y. Ema, T. Hayashi, Thin Solid Films, 147, 75 (1987).
11. G. Distler, S. Kobzareva, *Electron Microscopy* (Maruzen, Tokyo, 1966), p. 493.

5.5. CHALCOGENIDES AND OTHER GLASS SYSTEMS

TERNARY Ge–Se–(B, Ga, Tl) CHALCOGENIDES – PREPARATION, PROPERTIES AND APPLICATIONS

P. PETKOV[*]
*Institute of Electrochemistry and Energy Systems,
University of Chemical Technology and Metallurgy,
Department of Physics, University of Chemical Technology
and Metallurgy, 8 Kl. Ohridski Blvd., 1756 Sofia, Bulgaria*

Abstract. Ternary amorphous chalcogenide Ge–Se–(B, Ga, Tl) glasses and thin films, their preparation, properties, and applications, are reviewed, expanding on recently obtained results. The progress obtained is not only connected with a better understanding of their structure, chemical bonding and properties, but also with applications of these glasses and films in opto-electronics as waveguides and optical memories.

Keywords: chalcogenide thin films; vacuum thermal evaporation; mechanical behaviors, optical properties

1. Introduction

Chalcogenide glasses contain one or more chalcogen elements (sulphur, selenium or tellurium) in combination with elements from the IVth, Vth or VIth groups of the periodic table of elements. They are solids with covalent bonds (the maximum ionic conductivity is 9%), and their properties are significantly different from those of oxide glasses on the basis of SiO_2. The intensive development of modern technologies has led to the synthesis of new glassy chalcogenide materials with quite attractive properties which are already successfully used as photoresists in microelectronics[1] or as media for optical information recording in infrared optics.[2,3] The energy gap of the amorphous chalcogenides, strongly dependent on their composition,[4–6] plays an important role for their electrical and optical properties, as in tetra-hedral semiconductors.

---

[*]p.petkov@uctm.edu

J.P. Reithmaier et al. (eds.), *Nanostructured Materials for Advanced Technological Applications,*   315
© Springer Science + Business Media B.V. 2009

Ge$_x$Se$_{100-x}$ is a good glass former in the chalcogenide family with a glass-forming region for x < 43%. The chalcogenide glasses in the Ge–Se system are prospective materials for infrared optics owing to the broad range of transparency and good mechanical properties, such as hardness, adhesion, low internal stress, and water resistance.[7] Accordingly, the incorporation of a third element into the tetrahedral structure of Ge–Se often causes considerable changes of the structure and subsequently new, promising properties of the material. For instance, chalcogenide glasses from the Ge–Se–Tl system show pressure-induced crystallization, which has found applications in optoelectronics and switching phenomena.[8–10] Ge–Se–Ga(In) glasses are studied as prospective materials for integrated planar optical circuits and their components for routing, amplifying or generating optical signals (mainly in the infrared spectral region), as well as in diffractive optics, in holography, as photo-resists, etc.[11–13]

This paper aims to summarize the results of studies on thin Ge–Se–Y films, where the additive Y is either B, Ga or Tl, including their deposition, characterization and applications.

2. Experimental Part

2.1. PREPARATION OF BULK GLASSES AND THIN FILMS

Glasses from the ternary Ge–Se–Y systems, where Y represents B, Ga or Tl in amounts of 5, 10, 15, or 20 mol %, were prepared by the melt quenching technique. The alloys were synthesized in evacuated (~10^{-3} Pa) and sealed quartz ampoules from elemental Ge, Se and B (or Ga or Tl) with 4N purity by conventional direct monotemperature synthesis in a rotating furnace. The following step-wise heating regimes were employed: (i) heating to 493 K with a rate of 4 K/min and maintenance for 1,800 s, (ii) heating to the melting temperature of B (Ga, Tl) with a rate of 3 K/min, (iii) increasing the temperature up to 973–1,123 K, the melting temperature of germanium, with a rate of 2 K/min (applying vibration stirring), (iv) heating up to 1,273 K with a rate of 1 K/min, (v) quenching in a mixture of ice and water. These temperature regimes and heating rates have been chosen according to the melting temperatures of the starting components. For comparison with the stoichiometric Ge$_{20}$Se$_{80}$, non-stoichiometric Ge$_{17}$Se$_{83}$ and Ge$_{14}$Se$_{86}$ binary glasses have also been synthesized following the above procedure.

Thin films were deposited by vacuum thermal evaporation (VTE) from the respective bulk materials in a standard vacuum set-up with a residual gas pressure of 1.33 × 10^{-4} Pa. The thermal evaporation process was performed from a quasi-closed, inductively heated tantalum evaporator at a distance of 0.12 m to the substrate. The evaporation temperatures were maintained in the range of 850–1,000 K depending on the bulk glass composition, and monitored with a Ni–Ni/Cr thermocouple. Different substrates were selected according to the purposes of the study; they were rotated during the deposition to avoid non-uniformities in the film thickness.

2.2. KINETICS OF EVAPORATION AND CONDENSATION

For thin film deposition from complex bulk glasses composed of elements with very different atomic masses and properties it is of great importance to study the kinetics of the process. Problems with the evaporation, caused by the differences of the atomic masses and evaporation temperatures of the elements, can be avoided utilizing a special evaporator constructed by Vodenicharov.[14]

The deposition from alloys composed of elements with different vapour pressures is another important problem of practical meaning. The deposition of films by vacuum sublimation at substrate temperatures $T_s \ll T_m/3$ (where T_m is the melting temperature) is generally assumed to proceed in the sequence vapor phase → liquid phase → solid (amorphous) phase. The process of film preparation is described thermodynamically by the kinetic gas theory; the specific evaporation rate v_e is given by

$$v = 0.584\alpha_e p_0 \sqrt{\frac{M}{T_e}} \exp\left(-\frac{Q_e}{RT_e}\right) \tag{1}$$

where α_e is the evaporation coefficient, p_o the residual gas pressure, M the molar mass, R the universal gas constant, T_e the evaporation temperature, and Q_e the evaporation energy. This equation refers to free evaporation in vacuum from a small-area evaporator. The dependence of the evaporation rate on the Y content in the investigated glasses is presented in Figure 1.

The condensation process can be considered as a superposition of three mass flows: a flow of condensed particles, a flow of particles arriving to the substrate, and a flow of re-evaporated particles. The specific condensation rate v_c is determined by an equation similar to that for evaporation. The evaporation and condensation energies of the processes are determined from the slopes of the dependencies $(v_e T_e^{1/2}) = f(1/T_e)$ and $(v_c T_s^{1/2}) = f(T_s^{1/2})$, respectively. The energy values increase fast for multicomponent glasses. The relation of the condensation energy on the composition of the films is presented in Figure 2 together with the results for the binary Ge–Se system investigated previously[15] for comparison (in the inset). Our experiments with GeSe vapor deposition illustrate the possibility to obtain polycrystalline films at substrate temperatures lower than $2T_m/3$ (Ref. 16). Thin films of multicomponent glasses prepared under the same conditions (pressure and substrate temperature) are found to be amorphous in the whole concentration region of the investigated glasses.[17,18] The addition of any of the three elements into the Ge–chalcogenide host material causes exactly the same effects on the Q_e values (Figure 2).

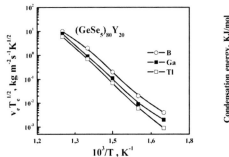 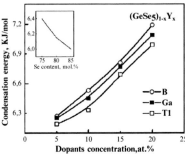

Figure 1. (Left). Dependence of the evaporation rate on the evaporation temperature.

Figure 2. (Right). Condensation energy variation with the B, Ga and Tl content.

According to the band model for amorphous semiconductors proposed by Ovshinsky,[19] in a covalently bonded alloy all atoms have their valences locally saturated. The amorphous materials can be considered as being composed of an interactive matrix whose electronic configurations are generated by free energy forces and can be defined by the chemical nature and coordination of the constituent atoms. The valence band states must be considered neutral when they are occupied, whereas the conduction band states are neutral when they are vacant. Some overlap of the valence band tail and the conduction band is believed to take place in the center of the mobility gap. The resultant charge transfer creates positively charged states above and negatively charged states below the Fermi level. B, Ga or Tl additives of incorporated in the Ge–Se matrix influence the electronic structure of the material. These III group elements are most often positively charged; their bands are located in the upper half of the mobility gap. The introduction of a charged additive disturbs the concentration of the native charged defects, but keeps the charge neutrality. As a consequence, the Fermi level shifts towards the conduction band, and the electronic properties of the multicomponent glasses alter.

2.3. CHARACTERIZATION OF THE FILMS

After the study of film deposition kinetics the thin films were characterized with respect to their morphology, structure, topography, and composition.

The structure was studied by X-ray diffraction (XRD) with a Philips diffractometer using CuK_α excitation radiation in the range of $2\theta = 20$–$65°$. All films exhibited an amorphous structure represented by a broad halo and the lack of diffraction peaks.

The morphology of the films under study was examined by scanning electron microscopy (SEM, Hitachi S4000) and transmission electron microscopy (TEM, Philips). Figure 3 (top) shows a uniform structure of the Ge–Se–Ga chalcogenide films which was verified by a TEM study. They do not possess any crystalline inclusions as visualized by the hallo in the electron diffractograms (Figure 3, down).

Top-view SEM micrographs of films from the Ge–Se–B system revealed a uniform and homogeneous character of the surfaces. Neither defects (liquation separation) nor traces of initial nucleation were visible in the SEM picture of a $Ge_{15}Se_{75}B_{10}$ film (Figure 4).

The topography as examined by atomic force microscopy (AFM) revealed smooth surfaces of the coatings as seen from the typical AFM images shown in Figure 5. The distribution of the roughness across the film surface was extrapolated from the AFM histograms. The average value of the roughness is about 2.4 nm derived from a sample area of 2×2 μm^2, which is less than 3% for films with thicknesses of ca. 500 nm.

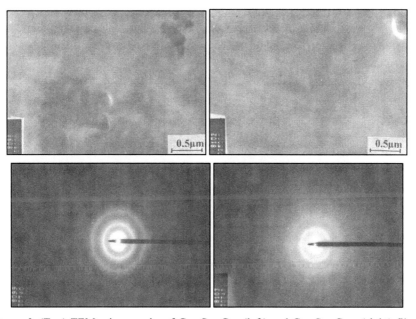

Figure 3. (Top) TEM micrographs of $Ge_{19}Se_{76}Ga_5$ (left) and $Ge_{16}Se_{64}Ga_{20}$ (right) films; (below) electron diffraction pattern of the same samples.

Figure 4. SEM micrograph of a film with a composition of $Ge_{15}Se_{75}B_{10}$.

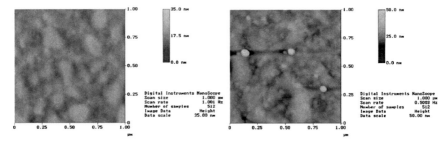

Figure 5. AFM images of $Ge_{20}Se_{80}$ (left) and $Ge_{16}Se_{64}Ga_{20}$ (right) films.

The composition of the films was derived by Auger electron spectroscopy (AES). The concentration profiles were obtained by layer-by-layer etching using an Ar^+ gun with an ion beam energy of 3 keV and an incidence angle of 15° toward the normal. The results obtained show film compositions very close to that of the respective bulk glass, as demonstrated in Figure 6.

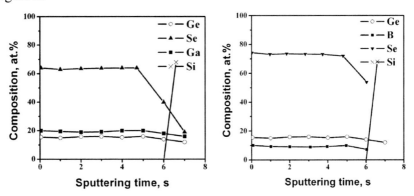

Figure 6. AES analysis of $Ge_{16}Se_{64}Ga_{20}$ and $Ge_{15}Se_{75}B_{10}$ films.

The analysis of the films prepared from the bulk glasses under investigation verified that the preparation conditions selected by the study of the condensation kinetics are favorable for the preparation of films with a homogenous structure, and reproducible composition and properties.

3. Application Relevant Properties

3.1. MECHANICAL PROPERTIES

The stress σ of the deposited amorphous films has been studied ex situ by the deflection of silicon cantilever beams and calculated from Stoney's equation[20]

$$\sigma = \frac{E}{6(1-\nu)}\frac{D^2}{Rd} \tag{2}$$

where d is the film thickness, R the radius of the curvature of the substrate, and E, ν and D are the Young's modulus, Poisson's ratio and thickness of the substrate, respectively. In our case the thickness of the films (up to 0.6 μm) was much smaller than that of the substrate (45 μm), which allowed us to apply the approximated Stoney equation for the stress determination. The results for σ in binary $GeSe_x$ and ternary $(GeSe_5)_{100-y}Ga(Tl, B)_y$ thin films as a function of their composition are presented in Figure 7.

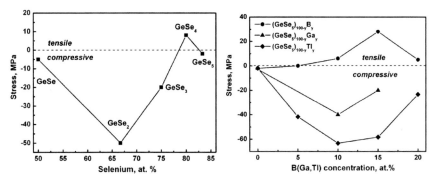

Figure 7. Stress in amorphous $GeSe_x$ (left) and $(GeSe_5)_{100-y}Ga(Tl, B)_y$ (right) thin films.

The stress in the system thin film/substrate is the sum of the thermal stress, originating from the different thermal expansion coefficients of film and substrate, and the intrinsic stress, determined by the structure and the properties of the film. During the deposition the substrates were kept at room temperature, which avoided the occurrence of thermal stress; therefore we assume that the results obtained are due to the appearance of intrinsic stress in the films. In such a case the film density, compactness, elasticity and structural rigidity will be of significant importance for the sign and magnitude of the stress.

The stress relaxation was controlled within a period of 6 months by multiple measurements; the results are shown in Figure 8. Considerable stress relaxation was observed after 3 months for the $GeSe_2$ films (Figure 8a) which were under the highest compressive stress from all samples in the

Ge–Se system after the deposition, while there were almost no changes in the films with negligible stresses (e.g. GeSe$_4$ and GeSe$_5$). A similar trend in the stress relaxation was revealed for the (GeSe$_5$)$_{100-y}$Ga$_y$ films (Figure 8b); the residual internal stresses after 3 months were in most cases on the order of 10 MPa.

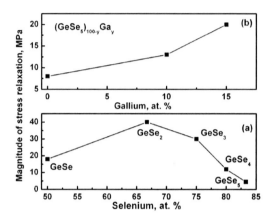

Figure 8. Stress relaxation of (a) GeSe$_x$ and (b) (GeSe$_5$)$_{100-y}$Ga$_y$ thin films.

3.2. OPTICAL PROPERTIES

A computer controlled UV/VIS/NIR spectrometer (Perkin Elmer Lambda 900) was used for optical transmission measurements of the films in the range of 400–2,000 nm. Optical transmission is a very complex function and strongly dependent on the absorption coefficient α. The transmission and reflection spectra recorded in the entire spectral range under study were used for the calculation of α in the region of high absorption by[21]:

$$T = \frac{(1-R)^2 \exp(-\alpha d)}{1 - R^2 \exp(-2\alpha d)}, \tag{3}$$

where d is the film thickness, and R and T are the reflection and transmission coefficients, respectively. The optical absorption edge of amorphous semiconductors is generally accepted to be classified into three regions: (i) a weak absorption tail, which originates from defects; (ii) the Urbach region where $\alpha \leq 10^4$ cm^{-1} due to structural disorder; and (iii) a region with $\alpha > 10^4$ cm^{-1} stipulated by electronic interband transitions.

The fundamental absorption edge in most amorphous semiconductors follows an exponential law. Above the exponential tails, the absorption r has been observed to obey the equation[22]

$$\alpha h v = B \, (h v - E_g^{opt})^m, \tag{4}$$

where $\alpha h v$ is the absorption coefficient at the angular frequency $\omega = 2\pi v$, B a constant and m is an exponent with values of 1/2, 3/2, 2 or 3, depending on the nature of the electronic transitions responsible for the absorption. The value of the exponent m depends on the mechanism of carrier transitions. The range, in which this equation is valid, is very small; hence it is difficult to determine the exact value of m (Ref. 23). The linear $(\alpha h v)^{1/2}$ vs. $h v$ plots shown in Figure 9 indicate that the absorption mechanism is stipulated by non-direct electronic transitions. In the case of non-direct interband transitions, which occur in the chalcogenide glasses studied, m has a value of 2.

The optical band gap E_g was determined by extrapolation of the linear part of the relation $(\alpha h v)^{1/2} = f(h v)$ to $\alpha = 0$, according to the Tauc procedure.[24] Figure 10 illustrates the variations of the optical band gap as a function of the B, Ga or Tl content. Generally, for the doped Ge–Se glasses E_g enlarges; the larger the atomic radius of the dopant, the higher is the increase of the band gap.

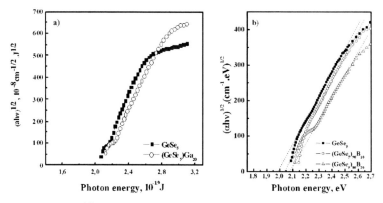

Figure 9. $(\alpha h v)^{1/2}$ versus $h v$ for (GeSe$_5$)–Ga (left) and (GeSe$_5$)–B (right) films.

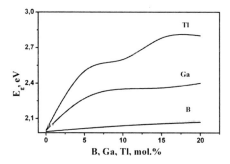

Figure 10. Optical band gap versus boron, gallium or thallium content.

4. Discussion on the Chalcogenide Structure and Properties

These film properties can be discussed in terms of the composition and structure. The structural network of chalcogenide glasses, both in bulk and thin film form, is composed of covalently bonded twofold-coordinated chalcogen atoms forming backbone chains cross-linked by three- or four-fold-coordinated atoms. For $GeSe_4$ a topological threshold is observed at a mean coordination number $Z = 2.40$ (Ref. 25), as result of a structural transition of the glassy network from a flexible (for $Z < 2.40$) to a rigid state (for $Z > 2.40$), which influences properties of these materials[26] such as the appearance of a slight tensile stress in the as-deposited $GeSe_4$ films. A further increase of the germanium fraction, i.e. of atoms with a larger radius ($R_a(Ge) = 1.27$ Å, $R_a(Se) = 1.17$ Å)[27] causes a change of the sign and the magnitude of the stress. As a result, freshly deposited GeSe, $GeSe_2$ and $GeSe_3$ films possess compressive stresses, with the highest value of −50 MPa for $GeSe_2$ (Figure 8a). For these films a reduction of the compressive stress with time observed. The networks of $GeSe_4$ and $GeSe_5$ glasses, which are characterized by a low compactness, include considerable free volume due to the presence of microvoids, which are "frozen" in the structure during the deposition. The selenium-rich $GeSe_5$ films are determined by selenium structural units: bridges, chains and rings.[28] This network is characterized by a comparatively low bonding extent and a low number of constrains per atom, which corresponds to a low internal stress. For this reason the as-prepared $GeSe_5$ films are almost free of stress, as seen in Figure 7 (left).

Additive atoms may enter into the Se chains and rearrange the network of the host Se to satisfy their bonding states and therefore reduce the number of defect states. This leads to an increase of the optical band gap and facilitate the appearance of stress. The introduction of atoms with larger radii ($R_a(Ga) = 1.30$ Å; $R_a(Tl) = 1.90$ Å) causes e.g. the appearance of compressive stress and changes of the optical band gap, while atoms with smaller radii ($R_a(B) = 1.01$ Å)[30] induce tensile stresses (Figure 7 (right)) and a very slight increase of the optical band gap (Figure 9, right). The addition of boron up to 10 at % has almost no influence on the stress in the films and on the optical band gap, while small amounts of gallium and thallium atoms cause substantial stresses and increase the band gap values.

The change in the optical band gap may also be understood in terms of the change in the average bond energy of the systems with boron, thallium or gallium incorporation. The optical band gap is a bond-sensitive property. The systematic replacement of Se–Se bonds (bond energy of 44 kcal mol^{-1}) by Se–Ga (bond energy of 59.5 kcal mol^{-1}) or Se–Tl (110 kcal mol^{-1}) bonds leads to an increase of the average bond energy of the system and consequently of the optical band gap of the thin films.

5. Applications

There are many current and potential applications of chalcogenide glasses and films such as solid electrolytes (batteries, thin film batteries), electro-chemical sensors, photoresists, optical waveguides, diffraction elements, Fresnel lenses, optical memories, holography, and other optical and non-linear optical elements.[29–31] In this section some examples for practical applications of the studied glasses will be presented.

5.1. WAVEGUIDES

The possibilities for practical applications of chalcogenide glasses are defined by the wide region of transparency of the materials from the visible (0.4–0.7 µm) to near (0.7–2.5 µm) and middle infrared (2.5–50 µm) regions. The research in integrated optics is related to the study of the spreading, the transformation into an optical signal, and the amplification of electromagnetic signals by thin film waveguides. Information about the optical losses is of great importance for waveguide evaluation. The optical losses in the studied glasses have been estimated from the spectral depend-ence of the absorption at a certain wavelength of the cut-off region. The transmission losses at $\lambda^{-1} = 900$ cm^{-1} decrease from 15 dB/m for a GeSe$_5$ film to 2.4 dB/m for a (GeSe$_5$)$_{80}$B$_{20}$ thin film. The reduction of the losses is due to phonon absorbance by the boron atoms.

As discussed in the previous section the introduction of boron atoms is associated with some rearrangement in the host glassy structure and the creation of new bonds, which facilitate the light transfer and reduce the optical losses. The concentration dependences of the absorption coefficient and optical transmission losses calculated at 943 cm^{-1} are presented in Figure 11. Analogous dependencies reported for similar glasses verify the applicability of these amorphous thin films as media in integrated optics.[32]

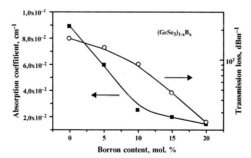

Figure 11. Dependencies of absorption coefficient and optical transmission losses on the boron content in GeSe$_5$–B films.

5.2. OPTICAL STORAGE MEDIA

The chalcogenide glasses exhibit reversible or irreversible photodarkening after prolonged illumination with bandgap light. The optical change is accompanied by structural changes such as volume change,[33] photocrystal-lization[34] or photoamorphisation.[35] Although much effort has been made to understand the mechanism of the photodarkening process, it has not been clarified yet. Holography is an useful tool to study chalcogenide materials as optical storage media.

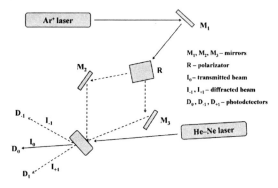

Figure 12. Holographic recording set-up.

The experimental set-up shown in Figure 12 has been utilized in holographic studies of films from the Ge–Se–Ga system. A continuous Ar$^+$ laser with a wavelength of 488 nm in a standard interferometric configuration produces fringes with a spatial modulation of 200 mm^{-1}. The intensity of the interfering beams is controlled by a gradual attenuator. The substrate holder temperature is maintained constant at 20°C by a thermostat. During the recording the diffraction efficiency is measured with a linearly polarized He–Ne laser beam with a wavelength of 632.8 nm. The efficiency is defined as the ratio of the intensity of light transmitted through the sample and the initial intensity of the laser beam. Thin films from the GeSe$_5$ system with different amounts of Ga as additive have been irradiated with an intensity of the recording beam of 3.5×10^3 Wm^{-2}.

The investigations show that the sensitivity of the films practically does not change with the addition of gallium. The magnitude of the diffraction efficiency is three time higher as the gallium concentration increases from 5 to 15 mol % (Figure 13). Further increase of the gallium amount, however, reduces the diffraction efficiency. The variation is most probably associated with structural rearrangements and the ability of the light to reconstruct the glassy skeleton and to cause photoinduced changes of film structure and properties.

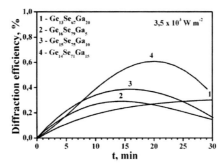

Figure 13. Time dependence of the diffraction efficiency for various compositions.

The photodarkening effect is the sum of at least two independent events: one fast (electronic) and another slow (thermal) one. Both of them may include several processes. The first process represents absorbance changes probably associated with the increased structural disorder and the formation of "wrong" bonds, resulting in a decrease of the bandgap energy, which in turn causes an increase of the refractive index. The second is due to irreversible changes in the density of the glass. Measurements of the glass after exposure show an increased volume in the illuminated region and a permanent decrease of the refractive index. The permanent photodarkening is associated both with increase of the sample volume in the exposed region and also with changes of defect concentration and bonding.

6. Conclusion

The chalcogenide glasses are materials with different applications according to their different properties. Currently many open questions concerning these materials still remain, because the amorphous state is strongly non-equilibrium; therefore many opportunities exist to control the properties over a broad range. Even more difficult is the interpretation of these properties. A comprehensive study of well-known and newly synthesized chalcogenide materials, in bulk and thin film form, will provide a profound knowledge in this field of the materials science.

ACKNOWLEDGEMENT

This research was financially supported by the NATO Collaborative Linkage Grant (CBP.EAP.CLG.982793).

References

1. M. Frumar, T. Wagner, Curr. Opin. Solid State Mater. Sci. 7, 117 (2003).
2. X. Zhang, H. Ma, J. Lucas, Opt. Mater. 25, 85 (2004).
3. K. Kolev, T. Petkova, J. Pirov, Proc. Nanosci. Nanotechnol. 5, 208 (2005).
4. N.F. Mott, E.A. Davis, *Electronic Processes in Non-Crystalline Materials* (Clarendon Press, Oxford, 1979).
5. R. Swanepoel, J. Phys. E 16, 1214 (1983).
6. I. Konstantinov, Private Communication (1989).
7. M. Micoulaut, J.C. Philips, Phys. Rev. B 67, 104204 (2003).
8. M.M. Abdel-Aziz, E.G. Al-Metwally, M. Fadel, H.H. Labib, M.A. Afifi, Thin Solid Films 386, 99 (2001).
9. S. Harshavardhan, M.S. Hegde, Phys. Rev. Lett. 58, 567 (1987).
10. H. Hisakuni, K. Tanaka, Science 270, 974 (1995).
11. P. Nemec, J. Jadelcky, M. Frumar, M. Munzar, M. Jelinek, J. Lancok, J. Non-Cryst. Solids 326 & 327, 53 (2003).
12. I. Sharma, S.K. Tripathi, P.B. Barman, Chalcogenide Lett. 3, 126 (2006).
13. I. Sharma, S.K. Tripathi, P.B. Barman, J. Phys. D: Appl. Phys. 40, 4460 (2007).
14. C. Vodenicharov, Phys. Status Solidi A 29, 233 (1975).
15. C. Vodenicharov, P. Petkov, Thin Solid Films 190, 335 (1990).
16. A. Feltz, C. Kaps, Thin Solid Films 70, 17 (1980).
17. P. Petkov, C. Vodenicharov, C. Kanasirski, Phys. Status Solidi A 168, 447 (1998).
18. P. Petkov, S. Parvanov, Y. Nedeva, E. Kashchieva, Phys. Chem. Glasses 41, 781 (2000).
19. S.R. Ovshinsky, in: B. Chakraverty and D. Kaplan (Eds.), *Proceedings of IX International Conference on Amorphous & Liquid Semiconductors* (Grenoble, France, 1981) pp. 1095.
20. G. Stoney, Proc. Roy. Soc. Lond. A 82, 172 (1909).
21. E. Marquez, A. Bernal-Oliva, J.M. Gonzalez-Leal, R. Prieto-Alcon, J. Non-Cryst. Solids 222, 250 (1997).
22. R. Zallen, *The Physics of Amorphous Solids* (Wiley, New York, 1983).
23. E.A. Davis, in: P.G. Lecomber and J. Mort (Eds.), *Electronic and Structural Properties of Amorphous Semiconductor* (Academic, London/New York, 1973) p. 425.
24. J. Tauc, *Amorphous and Liquid Semiconductors* (Plenum Press, New York, 1974).
25. J.C. Phillips, J. Non-Cryst. Solids 43, 37 (1981).
26. P. Boolchand, W. Bresser, M. Zhang, Y. Wu, J. Wells, R.N. Enzweiler, J. Non-Cryst. Solids 182, 143 (1995).
27. C. Bailar, H.J. Emeleus, R. Nyholm, A.F. Trotman-Dickenson (Eds.), *Comprehensive Inorganic Chemistry* (Pergamon Press, Oxford, 1973).
28. G. Lucovsky, in: P. Grosse (Ed.), *The Physics of Selenium and Tellurium* (Springer, Berlin, 1979).
29. K. Shimakawa, A. Kolobov, S.R. Elliott, Adv. Phys. 44, 475 (1995).
30. T. Wagner, M. Frumar, in: A. Kolobov (Ed.), *Photoinduced Metastability in Amorphous Semiconductors* (Wiley–VCH Verlag, Berlin, 2003).
31. M. Mitkova, in: P. Boolchand (Ed.) *Insulating and Semiconductors* (World Scientific, Singapore, 2000) pp. 814.
32. H.K. Zishan, M. Zulfequar, T. Sharma, M. Husain, Opt. Mater. 6, 139 (1996).
33. J. Cheng, G. Tilloca, J. Zarzycki, J. Non-Cryst. Solids 52, 249 (1982).
34. K. Antoine, H. Jain, J. Li, D.A. Drabold, M. Vlcek, A.C. Miller, J. Non-Cryst. Solids 349, 162 (2004).
35. J. Siegel, A. Schropp J. Solis, C.N. Afonso, M. Wuttig, Appl. Phys. Lett. 84, 2250 (2004).

STUDY OF (As₂Se₃)₁₀₀₋ₓ(AgI)ₓ THIN FILMS PREPARED BY PLD

AND VTE METHODS

T. PETKOVA[*1], V. ILCHEVA[1], C. POPOV[2],
J.P. REITHMAIER[2], G. SOCOL[3], E. AXENTE[3],
I.N. MIHAILESCU[3], P. PETKOV[4], T. HINEVA[4]

[1]*Institute of Electrochemistry and Energy Systems, Bulgarian Academy of Sciences, Acad. G. Bonchev St., bl. 10, 1113 Sofia, Bulgaria*
[2]*Institute of Nanostructure Technologies and Analytics (INA), University of Kassel, Germany*
[3]*National Institute for Lasers, Plasma, and Radiation Physics, Bucharest-Magurele, Romania*
[4]*Departments of Physics, University of Chemical Technology and Metallurgy, Kl.Ohridski Blvd.8, 1756 Sofia, Bulgaria*

Abstract. Amorphous chalcogenide $(As_2Se_3)_{100-x}(AgI)_x$ thin films with x = 5, 10, 15, 20, 25, 30 and 35 mol % have been deposited by pulsed laser deposition (PLD) and vacuum thermal evaporation (VTE). The films were characterized by various techniques with respect to their structure, composition and morphology. The optical properties were studied by transmission spectroscopy; the optical band gap E_g was determined from Tauc plots and as E_g^{04} in the strong absorption region (by $\alpha \geq 10^4$ cm^{-1}) from the relationship $\alpha = f$ (hv). The variation of the band gap is discussed with respect to the influence of the AgI content and the methods of film preparation.

Keywords: chalcogenide thin films; pulsed laser deposition; vacuum thermal evaporation; optical properties

---

[*]tpetkova@bas.bg

1. Introduction

Chalcogenide glasses based on S, Se, and Te have many unique optical properties, which can find a variety of applications.[1] These glasses are very promising materials in fiber optics and waveguiding devices in integrated optics since they exhibit a high transparency in the infrared region, especially at the telecommunication wavelengths of 1.3 and 1.55 μm.[2] For many applications thin films are required due to the necessity of device miniaturization.

There are many methods for film preparation, e.g. spin coating, evaporation of material at low pressures and high temperatures in vacuum, magnetron sputtering, CVD methods, etc. The deposition of chalcogenide thin films on larger areas with complex composition, the stoichiometry desired, sufficient homogeneity, good adhesion to the substrate, and other physico-chemical properties necessary for present and future applications is, however, a difficult task; often classic deposition methods cannot be used. A relatively new deposition process is pulsed laser deposition (PLD). The structure, composition and properties of PLD films can be different from those of films prepared by other methods or from those of the bulk (target) glasses – thus new materials are prepared.

In this paper we report on a comparative study of $(As_2Se_3)_{100-x}(AgI)_x$ films prepared by vacuum thermal evaporation (VTE) and PLD with respect to their optical properties.

2. Experimental

The bulk materials with compositions of $(As_2Se_3)_{100-x}(AgI)_x$ have been prepared by a two-stage synthesis: (i) preparation of binary As_2Se_3 from elemental As and Se with 5N purity; (ii) synthesis of quasi-binary alloys As_2Se_3–AgI from the As_2Se_3 previously obtained and commercial AgI (Alfa Aesar). The scheme of the processes was identical during both stages: the respective amounts of the initial substances were placed in quartz ampoules, evacuated down to ~10^{-3} Pa and heated in a rotary furnace. The temperature was kept constant at the melting point of each component; the melt was continuously stirred for better homogenization. The glasses were then obtained by quenching in a mixture of ice and water, a procedure typical for the preparation chalcogenide materials.[3]

Thin films were deposited by VTE of the respective bulk materials placed in a tantalum crucible in a Leybold LB 370 vacuum set-up, with a residual gas pressure of 1.33×10^{-4} Pa. The experiments were carried out at a constant distance substrate-evaporator; the evaporation temperatures in the range of 850–1,000 K were monitored with a Ni–Ni/Cr thermocouple. The silicon and glass substrates were rotated with a constant rate during the deposition to avoid non-uniformities in the film thickness; they were not additionally heated. The targets for PLD were prepared from the synthesized glasses by milling, pressing and sintering. The deposition of thin $(As_2Se_3)_{100-x}AgI_x$ films was achieved by ablation of the targets from the respective bulk materials with a KrF* excimer laser source ($\lambda = 248$ nm, $\tau_{FWHM} = 25$ ns). The incidence fluence was set to 6.6 J/cm^2; for the deposition of each film 3,000 laser pulses were applied. The films were deposited in vacuum (4×10^{-4} Pa) on silicon and glass substrates, kept at room temperature and placed parallel 5 cm from the target.

After the deposition, the films were characterised in order to investigate the influence of the process parameters and the glass composition on their properties. The structure was studied by X-ray diffraction (XRD) with a DRON UM1 diffractometer using CuK_α excitation radiation within the range of $2\theta = 20-65°$. Energy dispersive X-ray (EDX) analysis of the film composition was performed with a scanning electron microscope (SEM, JEOL JSM-840A) equipped with Link Analytical spectrometer (AN10000), operated at an accelerating voltage of 20 kV. The oxygen content in the films was monitored by wavelength dispersive spectroscopy (WDS) using an accelerating voltage of 10 kV and a beam current of 300 nA. SEM investigations with a Hitachi S4000 apparatus revealed the film morphology. From SEM cross-section (XSEM) micrographs the film thickness was evaluated. The surface topography was studied by atomic force microscopy (AFM, NanoScope II) in tapping mode. A computer controlled UV/VIS/NIR spectrometer (Perkin Elmer Lambda 900) was used for the measurement of the optical transmission of the films in the range of 300–3,300 nm.

3. Results and Discussion

The amorphous nature of the $(As_2Se_3)_{100-x}AgI_x$ films has been proven by the XRD studies. The diffraction patterns of the films deposited by both techniques showed broad halos, without any sharp peaks belonging to crystalline phases of any component of the film composition.

The composition of the VTE and PLD films were found close to the composition of the respective bulk materials with some peculiarities, namely: (i) the ratio As/Se keeps constant within the accuracy of the EDX method; (ii) a loss of dopant (AgI) is observed; (iii) the higher the dopant concentration in the target materials, the larger the difference of the film composition to that of the bulk samples. The deficiency of AgI in the films is most probably caused by the difference of the enthalpy of deposition of the evaporated fragments. The dissociation of the starting material during the evaporation or laser ablation generates fragments with various masses and velocities (As–Se, silver, iodine, etc.), which are transported to substrate.

The enthalpy of deposition, on which the quantity of each fragment deposited onto the substrate depends, is different for the various fragments. No oxygen was detected in the films by WDS within the accuracy of the method (±1%). Top-view SEM pictures of films prepared by both methods reveal the uniform and homogeneous character of their surfaces (Figure 1). Neither defects (liquation) nor traces of an initial nucleation are visible.

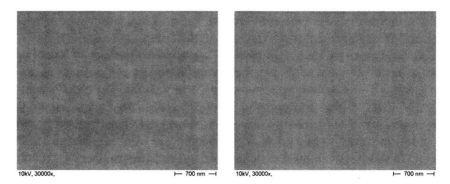

Figure 1. Typical top-view SEM micrographs of VTE (left) and PLD (right).

An analysis of AFM images showed the influence of the dopant (AgI) concentration on the root mean-square (rms) roughness; the latter increases almost linearly with the dopant concentration. The roughness of PLD films was lower than those of the VTE films. For example, the rms roughness of $(As_2Se_3)_{95}(AgI)_5$ films was 0.379 nm for VTE but 0.153 nm for PLD.

The transmission spectra of the films show very pronounced absorption in the region from 300–600 nm; at wavelengths above 700 nm, the films are transparent. The absorption edge is shifted towards lower wavelengths as the AgI concentration increases. This trend is due to the incorporation of the semiconducting AgI with a narrow band gap into the disordered glassy network of the As_2Se_3.

The optical band gap energy was determined by two methods:

- By means of the well-known Tauc procedure[4] in the range of strong absorption ($\alpha \leq 10^4$ cm^{-1}):

$$\alpha\, h\nu = B\, (h\nu - E_g)^2$$

where $h\nu$ is the photon energy and B the slope at the Tauc edge. E_g was determined by extrapolation of the linear part of the relation $(\alpha h\nu)^{1/2} = f\, (h\nu)$ to $\alpha = 0$.
- Taking it as E_g^{04}, i.e. the energy for which $\alpha = 10^4$ cm^{-1} or higher.[4]

The assessment of the optical band gap values as a function of the AgI content and the preparation method are presented in Figure 2. The values for E_g and $E_g{}^{04}$ range from 1.58 to 1.80 eV and from 1.75 to 2.10 eV, respectively. The values for PLD samples are lower, and there are only marginal changes with the AgI content. The small decrease of E$_g$, observed at a composition of 20 mol % AgI could be due to structural trans- formations in the glass network. In general the optical band gap increases as the AgI content grows. The difference of 0.2 eV, which is typical for the band gap calculations by both methods, is also preserved.

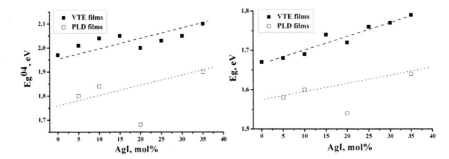

Figure 2. Optical band gaps E$_g{}^{04}$ (left) and E$_g$ (right) of (As$_2$Se$_3$)$_{100-x}$(AgI)$_x$ VTE and PLD films (lines are drawn to guide the eye).

The addition of the narrow-gap silver iodide is expected to reduce the optical band gap. Therefore, the observed widening of the gap opens the question about the appearance of new structures or transformations of the existing. One possible explanation is that AgI creates new levels in the forbidden zone, causing the mechanism of electron transfer to change due to the substantial changes in the electronic structure of the sample.

The optical constants *n* (refractive index) and *k* (extinction coefficient) of the films were calculated from the transmission spectra, using: (i) the Swanepoel method,[5] (ii) a computer program developed by Konstantinov[6] and (iii) the Sellmeier equation which gives an empirical relationship between the refractive index *n* and the wavelength λ for a particular trans- parent medium. The spectral dependences *n* (λ) are presented in Figure 3.

The dependence of the refractive index *n* on the AgI content determined at $\lambda = 1,000$ nm shows that *n* varies in the range 2.4–2.8 for the two types of films.

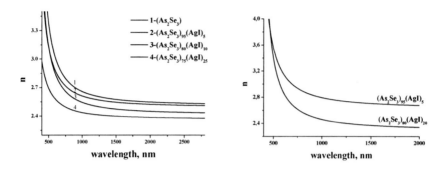

Figure 3. Spectral dependences n (λ). Left: VTE; right: PLD films.

4. Conclusions

The results obtained in this study reveal the good dielectric properties of films from the As_2Se_3–(AgI) system.

The most important conclusions drawn from the investigation of the optical properties of the films can be summarized as follows:

- Films obtained by the PLD method are smoother than VTE films.
- The roughness increases with increasing AgI content.
- Marginal changes of the optical band gap and refractive index with the AgI content are associated with structural changes due to the AgI incorporation.

ACKNOWLEDGEMENT

This research was supported by the NATO Collaborative Linkage Grant (CBP.EAP.CLG.982793). V.I. gratefully acknowledges the financial support of the World Federation of Scientists.

References

1. M. Frumar and T. Wagner, Curr. Opin. Solid State Mater. Sci. 7, 117 (2003).
2. X. Zhang, H. Ma, and J. Lucas, Opt. Mater. 25, 85 (2004).
3. K. Kolev, T. Petkova, and J. Pirov, Proc. Nanosci. Nanotechnol. 5, 208 (2005).
4. N.F. Mott and E.A. Davis, *Electronic Processes in Non-Crystalline Materials* (Clarendon Press, Oxford, 1979).
5. R. Swanepoel, J. Phys. E 16, 1214 (1983).
6. I. Konstantinov, Private Communication (1989).

PHYSICO-CHEMICAL CHARACTERIZATION

OF NANOSTRUCTURED As–Se–Ag GLASSY MATERIALS

V. ILCHEVA[*1], T. PETKOVA[1], D. ROUSSEV[2],
P. PETKOV[2]
[1]*Institute of Electrochemistry and Energy Systems, Bulgarian
Academy of Sciences, Acad. G. Bonchev St., bl. 1113 Sofia,
Bulgaria*
[2]*Departments of Physics, University of Chemical Technology
and Metallurgy, Kl.Ohridski Blvd. 8, 1756 Sofia, Bulgaria*

Abstract. Bulk glasses from the systems $(AsSe)_{1-x}Ag_x$ and $(As_2Se_3)_{1-x}Ag_x$ have been prepared by the melt-quenching technique. Basic physico-chemical characteristics like density, microhardness, and compactness have been investigated. The correlation between the composition and the properties of the AsSeAg glasses is established and comprehensively discussed.

Keywords: chalcogenide glasses; physico-chemical properties

1. Introduction

Multicomponent chalcogenide glasses are promising materials for IR optics and microelectronics. Their properties make them suitable for applications as solid state electrolytes, optical elements and sensors.[1,2]

Arsenic-based chalcogenides are semiconductors with applications in the production of microcircuits, e.g. as resist in photo- and electron beam lithography due to the effects of photodarkening and photodoping.[3]

The introduction of metal atoms like silver into the chalcogenide matrix is often used to vary the glass properties in a desired direction. Accordingly we decide to present the results of a physico-chemical characterization of chalcogenide glasses from the As–Se–Ag system as a function of the silver content. The aim of our work was to study the modification of the physico-chemical behavior and the structure due to the introduction of silver to the

---

[*] Vania_ilcheva@yahoo.com

J.P. Reithmaier et al. (eds.), *Nanostructured Materials for Advanced Technological Applications,*   335
© Springer Science + Business Media B.V. 2009

chalcogenide matrix. The experimentally derived parameters are compared with theoretically calculated ones and elucidated in terms of the Phillips-Thorpe theory.[4]

2. Experimental

Bulk $(AsSe)_{1-x}Ag_x$ and $(As_2Se_3)_{1-x}Ag_x$ glasses (x = 0, 5, 10, 15, 20, 25 mol %) were prepared by the melt-quenching technique. Appropriate amounts of the starting elements were sealed in evacuated quartz ampoules and heated with a constant heating rate of 3 $Kmin^{-1}$ up to a temperature of 1,300 K. The glasses were obtained after quenching in a water–ice mixture.

The amorphous character of the obtained alloys was proven by the X-ray diffraction (XRD) technique by means of a Philips Analyzer APD-15 with Cu Kα radiation.

The density d of the samples was measured by a picnometric method using toluene as immersion fluid; it was calculated by the formula described elsewhere.[5] The molar volume V_m of the samples was calculated from the density data. The compactness δ was obtained from the experimental and theoretical density values by the formula

$$\delta = \left(\sum_i \frac{c_i A_i}{\rho_i} - \sum_i \frac{c_i A_i}{\rho} \right)\left(\sum_i \frac{c_i A_i}{\rho} \right)^{-1} \tag{1}$$

where x_i, A_i and ρ_i are the atomic fraction, atomic weight and density of the components. The microhardness HV was measured by the Knoop method.

The coordination numbers of the components were estimated by the (8-N) rule where N is the number of electrons in the outer shell of the atom. The coordination numbers of As, Se and Ag elements are $Z_{As} = 3$, $Z_{Se} = 2$ and $Z_{Ag} = 3$, respectively. The average coordination number Z of the glasses was estimated according to the formula given elsewhere.[6] The number of con-strains per atom N_{co} were calculated by the equation developed by Thorpe[7]:

$$N_{co} = N_a + N_b = Z/2 + (2Z - 3), \tag{2}$$

where N_a and N_b are the radial and the axial bond strengths, respectively.

3. Results and Discussion

XRD spectra of the bulk samples with x = 5 and 25 at % Ag are presented in Figure 1. They exhibit amorphous plateaus but lack sharp diffraction peaks, which is evidence of the amorphous character of the samples.

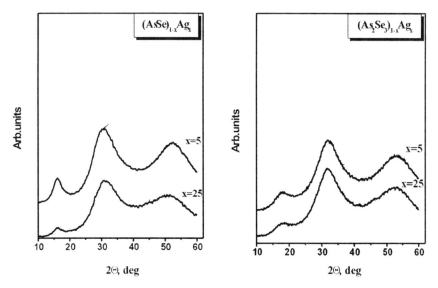

Figure 1. X-ray diffraction patterns: (a) $(AsSe)_{1-x}Ag_x$; (b) $(As_2Se_3)_{1-x}Ag_x$.

TABLE 1. Theoretical calculated and experimentally obtained physico-chemical parameters.

Composition	Z	HV, kpmm^{-2}	δ	N$_{CO}$
AsSe	2.5	97		3.25
$(AsSe)_{95}Ag_5$	2.525	115	−0.224	3.3
$(AsSe)_{90}Ag_{10}$	2.55	128	−0.222	3.38
$(AsSe)_{85}Ag_{15}$	2.575	145	−0.217	3.44
$(AsSe)_{80}Ag_{20}$	2.6	173	−0.168	3.5
$(AsSe)_{75}Ag_{25}$	2.625	192	−0.142	3.56
As_2Se_3	2.4	140		3
$(As_2Se_3)_{95}Ag_5$	2.43	146	−0.175	3.075
$(As_2Se_3)_{90}Ag_{10}$	2.46	150	−0.127	3.15
$(As_2Se_3)_{85}Ag_{15}$	2.49	162	−0.115	3.225
$(As_2Se_3)_{80}Ag_{20}$	2.52	178	−0.098	3.3
$(As_2Se_3)_{75}Ag_{25}$	2.55	180	−0.018	3.375

The values of the experimentally and theoretically established physico-chemical parameters of the glasses under study are presented in Table 1. Information about the glass structure can be derived from the density results since their values depend on the structure and atomic masses of the components constructing the alloy.[8] The density values obtained are in the range of 3.95–4.75×10^3 kgm^{-3}. The dependence of the density on the silver content is presented in Figure 2. The smooth increase of the density with the introduction of silver is associated with the higher density of silver (1.049×10^4 kg.m^{-3}) as compared to the other components As and Se (with densities 5.73×10^3 kgm^{-3} and 4.82×10^3 kgm^{-3}, respectively). The atomic radius and atomic weight of Ag are much larger as compared to the other two elements; therefore the introduction of silver atoms is related to the considerable changes in the properties of the glasses. The ordinary trend of the dependence of the density on the composition is assigned to the random distribution of the Ag atoms in the glassy host matrix.

The compositional dependence of the molar volume is presented in Figure 3. The trend in the dependency of V_m follows the variation of the density. The addition of Ag to the As–Se host matrix causes a decrease of the molar volume. This reduction of the molar volume values is expected owing to the inversely proportional relation between the molar volume and the experimentally obtained density of the samples. The differences in the slope of the curves of density and molar volume are due to the differences of the atomic weight of the components.

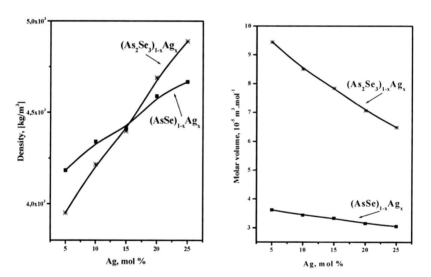

Figure 2 (left). Density dependence on the composition of the As–Se–Ag glasses.

Figure 3 (right). Compositional dependence of the molar volume of the As–Se–Ag glasses.

From the experimental density values the compactness of the glasses has been derived; the results are presented in Table 1. The values are in direct proportion to the density of the glasses i.e. a more compact structure is formed when a higher content of silver atoms is added.

The average coordination number is an important characteristic of amorphous chalcogenide materials, indicating the average number of bonds per atom, which must be broken to obtain fluidity. On the basis of topological arguments, counting constraints and the degrees of freedom, Phillips and Thorpe[4] suggested that glasses with $Z < 2.40$ consist of rigid regions whose volume fraction is too small to be fully connected. In this case the rigid regions are immersed in a 'floppy' matrix. Glasses with $Z = 2.40$ are unique because their number of constrains is equal to the number of degrees of freedom; the floppy and the rigid regions are individually connected matrices with a maximum number of connections.

The values calculated according to Ref. 7 are in the range of 2.4–2.63. When $Z > 2.40$, the solid has continuously connected rigid regions with floppy regions inter-dispersed and may be termed an 'amorphous' solid.

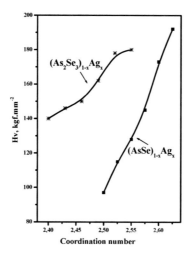

Figure 4. Variation of HV as a function of Z in As–Se–Ag glasses.

The values of N_{co} for the $(AsSe)_{1-x}Ag_x$ and $(As_2Se_3)_{1-x}Ag_x$ systems are also included in Table 1. For the ideal glass, $N_{co} = N_d = 3$, where the mechanical stability of the network is optimized.[9] It is obvious that N_{co} increases with the addition of silver though the values are not very much higher and close to the value of 3. The microhardness HV of the glasses studied is within the range of 97–180 kpmm^{-2}; it exhibits a proportional dependence on the composition of the glasses as revealed in Table 1, and as a function of Z as shown in Figure 4. The variations in HV(x) are reasonable in view of the fact that $HV_{Ag} > HV_{As, Se}$. The absence of any deflections in the microhardness relation at $Z = 2.4$ indicates the absence of 'rigidity percolation' in the glasses.

4. Conclusions

The materials obtained exhibit homogeneous and glassy structures in a wide compositional region. The density results show that the silver atoms are well incorporated in the chalcogenide matrix and uniformly distributed.

ACKNOWLEDGEMENT

The authors thank the Bulgarian Ministry of Education and Science for supporting this research under the contracts VUF 05 /2005 and NATO CBP.EAP.CLG.982793.

References

1. V. Weidenhof, I. Friedrich, S. Ziegler and M. Wuttig J. Appl. Phys. 89, 3168 (2001).
2. M. Mitkova, M. Kozicki J. Non-Cryst. Solids 299–302, 1023 (2002).
3. K. Mietzsch and A.G. Fitzgerald, Appl. Surf. Sci. 162–163, 464 (2000).
4. J.C. Phillips, M.F. Thorpe, Solid-State Commun. 52, 699 (1985).
5. T. Petkova, P. Petkov, S. Vassilev, Y. Nedeva, Surf. Interf. Anal. 36, 880 (2004).
6. S. Elliott (Ed.), *Physics of Amorphous Materials*, Pitman Press (Bath, 2000).
7. M. Thorpe J. Non-Cryst. Solids 57, 355 (1983).
8. M. Mitkova, V. Boev, Mater. Lett. 20, 195–201 (1994).
9. A. Giridhar, S. Mahadevan, J. Non-Cryst. Sol. 134, 94 (1991).

ATOMIC STRUCTURE OF As$_{34}$Se$_{51}$Ag$_{15}$ AND As$_{34}$Te$_{51}$Ag$_{15}$ GLASSES STUDIED WITH XRD, ND AND EXAFS AND MODELED WITH RMC

I. KABAN[*1], W. HOYER[1], P. JÓVÁRI[2], T. PETKOVA[3],
A. STOILOVA[4], A. SCHÖPS[5], J. BEDNARCIK[5],
B. BEUNEU[6]

[1]*Institute of Physics, Chemnitz University of Technology,
D-09107 Chemnitz, Germany*
[2]*Research Institute for Solid State Physics and Optics, H-1525
Budapest, POB 49, Hungary*
[3]*Institute of Electrochemistry and Energy Systems,
G. Bonchev St. Bl.10, 1113 Sofia, Bulgaria*
[4]*University of Chemical Technology and Metallurgy,
Department of Physics, 8Kl Ohridsky Blvd., 1756 Sofia,
Bulgaria*
[5]*Hamburger Synchrotronstrahlungslabor HASYLAB am
Deutschen Elektronen-Synchrotron DESY Notkestrasse 85,
D-22603 Hamburg, Germany*
[6]*Laboratoire Léon Brillouin CEA-SACLAY, 91191 Gif sur
Yvette Cedex France*

Abstract. As$_{34}$Se$_{51}$Ag$_{15}$ and As$_{34}$Te$_{51}$Ag$_{15}$ glasses have been studied with high energy X-ray diffraction, neutron diffraction and extended X-ray absorption spectroscopy. The experimental data were modeled simultaneously with the reverse Monte Carlo simulation method. The combination of these independent measurements together with the application of some plausible physical constraints allowed the separation of the partial pair correlation functions and the estimation of the coordination numbers. An analysis of the results revealed that Ag–Se/Te bonds are preferred to Ag–As ones. The chemically ordered structure of As$_2$Se$_3$ remains essentially intact in As$_{34}$Se$_{51}$Ag$_{15}$, while for As$_{34}$Te$_{51}$Ag$_{15}$ alloying with Ag results in an increase of As–As and Te–Te bonds.

[*] ivan.kaban@physik.tu-chemnitz.de

J.P. Reithmaier et al. (eds.), *Nanostructured Materials for Advanced Technological Applications*, 341
© Springer Science + Business Media B.V. 2009

Keywords: chalcogenide glasses; As_2Se_3; As_2Te_3; silver; atomic structure

1. Introduction

Arsenic triselenide (As_2Se_3) and arsenic tritelluride (As_2Te_3) in the glassy state are semiconductors. However, their electrical properties change upon addition of Ag.[1-5] The conductivity of glassy $(As_2Se_3)_{100-x}Ag_x$ with 6–8 at % Ag is electronic; glasses with 8–15 at % Ag possess a mixed electronic–ionic conductivity while the ionic component dominates for higher Ag contents.[5] The conductivity of $(As_2Te_3)_{100-x}Ag_x$ glasses is essentially electronic in nature, and the ionic component is small.[5] Other physical characteristics like the molar volume, the glass transition temperature, and the crystallization temperature also change remarkably with a variation of the Ag concentration in $(As_2Se_3)_{100-x}Ag_x$ and $(As_2Te_3)_{100-x}Ag_x$ glasses.[5-11]

The scientific interest in the conduction mechanism of these materials originates from their practical application as solid electrolytes in electrochemical devices. Ogusu et al.[9,10] reported that As_2Se_3–Ag glasses are characterized by a very high non-linear refractive index (2,000 times greater than that of fused silica). Such non-linear glasses are promising materials for all-optical switching devices at the telecommunication wavelengths.

To understand these materials, improve their physical properties, and exploit them more efficiently, a more profound knowledge of their atomic structure is needed. However, at the present time much more is known about the physical properties of chalcogenide glasses than about their structure. In many cases, assumptions about the atomic distribution after addition of a third component to As_2Se_3 and As_2Te_3 binaries are derived from composition dependences of the physical properties and not from structural measurements. Also, the few experimental structural studies existing give rather discrepant values of the coordination numbers, which results in different structural conceptions.

Usuki et al.[12] studied the structure of $(Ag_2Te)_x(AsTe)_{1-x}$ glasses with X-ray diffraction (XRD) and extended X-ray absorption spectroscopy (EXAFS). They found that the average coordination number of As is close to three, and each As atom has one As and two Te neighbors. The coordination number of Te increases from ~1.9–3.6 with increasing Ag content. At the same time, a significant number of Ag atoms is assumed to be tetrahedrally coordinated to Te atoms. The authors described the structure of $(Ag_2Te)_x(AsTe)_{1-x}$ glasses as a pseudo-binary homogeneous mixture of $As_2Te_{4/2}$ pyramidal network units and percolated $AgTe_4$ tetrahedra.

Mastelaro et al.[13] investigated the local structure of $(Ag_2Se)_x(AsSe)_{1-x}$ glasses by means of EXAFS at the three K absorption edges. They found that each As atom has three Se neighbors in all alloys studied, while the number of Se–As bonds decrease from 2 for AsSe to 0.8 for $AsSe_3Ag_3$. The number of Se–Ag bonds increases continuously with the addition of Ag to AsSe and reaches ~2 in $AsSe_3Ag_3$ glasses. Each Ag atom was established to have two Se neighbours in the glasses with x between 0.18 and 0.43.

Benmore and Salmon[14] studied g-$Ag_2As_3Se_4$ (g = glassy), and Liu and Salmon[15] investigated g-$AgAsTe_2$ by neutron diffraction (ND) with isotopic substitutions. They concluded that the short-range order of the network former (AsSe and As_2Te_3, respectively) is not destroyed by alloying with a network modifier (Ag_2Se and Ag_2Te, respectively), while elements of intermediate range order of the network modifier are retained.

The partial coordination numbers reported in the above studies are collected in Tables 1 and 2. It can be seen that the main differences concern the silver–chalcogen (Ag–Se, Ag–Te) correlations. Besides, there are discrepancies in the As–Se coordination number reported: $N_{AsSe} = 3.0$ in Ref. 13 but $N_{AsSe} = 2.0$ in Ref. 14. Also, the rather high Ag–Ag coordination numbers (2.7–2.8) in $Ag_2As_3Se_4$ and $AgAsTe_2$ glasses[14,15] give rise to questions.

TABLE 1. The nearest neighbor distances r_{ij} and coordination numbers N_{ij} for amorphous $As_{34}Se_{51}Ag_{15}$ obtained with unconstrained RMC simulations (Se–Se bonding is forbidden). r_{ij} and N_{ij} from Refs. 13 and 14 are cited for comparison.

Pairs	$As_{34}Se_{51}Ag_{15}$ this study		$(Ag_2Se)_{0.18}(AsSe)_{0.82}$ Ref. 13		$Ag_2As_3Se_4$ Ref. 14
	r_{ij} (Å)	N_{ij}	r_{ij} (Å)	N_{ij}	N_{ij}
As–As	2.40	0.23	–	0	2.0
As–Se	2.41	3.06	2.42	3.0	2.0
As–Ag	2.68	0.33	–	–	–
Se–As	2.41	2.04	2.40	1.5	–
Se–Se	–	0	–	0	–
Se–Ag	2.61	0.57	2.59	1.0	–
Ag–As	2.68	0.74	–	–	1.0
Ag–Se	2.61	1.92	2.55	1.9	4,0
Ag–Ag	2.98	0.44	–	–	2.7
As–X	–	3.62	–	–	–
Se–X	–	2.61	–	–	–
Ag–X	–	3.1	–	–	–

TABLE 2. The nearest neighbour distances r_{ij} and coordination numbers N_{ij} for amorphous $As_{34}Te_{51}Ag_{15}$ obtained with RMC simulations (As–Ag bonding is forbidden). r_{ij} and N_{ij} from Refs. 12 and 15 are cited for comparison.

Pairs	$As_{34}Te_{51}Ag_{15}$ this study		$(Ag_2Te)_{0.2}(AsTe)_{0.8}$ Ref. 12		$AgAsTe_2$ Ref. 15
	r_{ij} (Å)	N_{ij}	r_{ij} (Å)	N_{ij}	N_{ij}
As–As	2.48	2.12	2.50	0.94	2.1
As–Te	2.59	1.03	2.66	1.89	2.1
As–Ag	–	0	–	0	–
Te–As	2.59	0.68	–	–	–
Te–Te	2.75	1.65	–	–	–
Te–Ag	2.78	0.71	–	–	–
Ag–As	–	0	–	0.95	–
Ag–Te	2.78	2.42	2.85	3.79	2.9
Ag–Ag	2.81	1.11	–	–	2.8
As–X	–	3.15	–	2.83	–
Te–X	–	3.04	–	3.03	–
Ag–X	–	3.53	–	–	–

In the present work, we studied $As_{34}Se_{51}Ag_{15}$ and $As_{34}Te_{51}Ag_{15}$ glasses with XRD, ND and EXAFS, and modeled the experimental data simultaneously by means of RMC. This combination of independent measurements, together with the application of some plausible physical constraints, allowed the separation of partial pair correlation functions and the determination of the corresponding coordination numbers. This enabled in due course to make conclusions about the local structure of the glasses studied. Also, it is interesting to see the differences between the structures of ternary glasses when one chalcogenide is substituted by another, like Se and Te, in our case.

2. Experimental Details

The samples were prepared in evacuated ($\sim 10^{-3}$ Pa) and sealed quartz ampoules by conventional synthesis in a rotary furnace. At the first stage, binary As_2Se_3 and As_2Te_3 were synthesized from high purity (99.999%) As, Se, and Te. Then, the proper quantities of either As_2Se_3 or As_2Te_3 were mixed with Ag (99.99%) and heated. The molten alloys were kept at 1,200 K and stirred for 2 h. Finally, they were quenched in an ice–water mixture.

The X-ray diffraction experiments were carried out at the BW5 experimental station at HASYLAB (DESY, Hamburg). The samples were filled into thin walled (0.02 mm) quartz capillaries of 2.0 mm in diameter. The energy of the incident beam was 99.8 keV, its size 1×4 mm^2. The raw data were corrected for background, polarization, detector dead-times, and variations in detector solid angle.

The EXAFS measurements were carried out at the Ag, As and Te K-edges at the beamline X of HASYLAB in transmission mode. The samples were finely ground, mixed with cellulose and pressed into tablets. The sample quantity in the tablets was adjusted to the composition of the sample and to the selected edge. The transmission of the samples was about 1/e. EXAFS spectra were obtained with steps of 0.5 eV in the vicinity of the absorption edge. The measuring time was k-weighted during the collection of the signal.

The neutron diffraction was carried out at the 7C2 diffractometer (LLB, France). The samples were filled into thin walled (0.1 mm) vanadium containers of 5 mm diameter. The raw data were corrected for detector efficiency, empty instrument background, scattering from the sample holder, multiple scattering, and absorption.

The experimental XRD and ND structure factors, and the EXAFS spectra for As$_{34}$Se$_{51}$Ag$_{15}$ and As$_{34}$Te$_{51}$Ag$_{15}$ glasses are plotted in Figures 1 and 2.

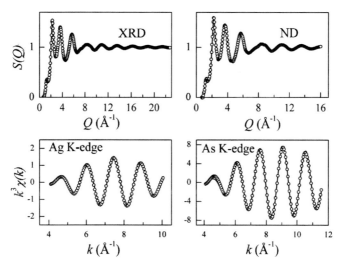

Figure 1. XRD and ND structure factors, and EXAFS spectra for a (As$_2$Se$_3$)$_{85}$Ag$_{15}$ glass. *Circles:* measurement. *Lines:* data obtained by simultaneous RMC simulation of the experimental XRD, ND and EXAFS data without Se–Se bonding.

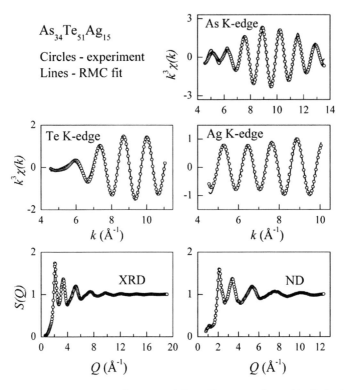

$As_{34}Te_{51}Ag_{15}$

Circles - experiment
Lines - RMC fit

Figure 2. XRD and ND structure factors, and EXAFS spectra for a $(As_2Te_3)_{85}Ag_{15}$ glass. *Circles:* measurement. *Lines: data* obtained by simultaneous RMC simulation of the experimental XRD, ND and EXAFS data without As–Ag bonding.

3. RMC-Modeling

It has been shown in a number of works that the reverse Monte Carlo simulation technique (RMC)[16] is a very useful tool that enables the generation of large three-dimensional structural models of multi-component alloys consistent with the available experimental information. For details of the reverse Monte Carlo simulation method we refer to two recent papers.[17,18]

The calculations in the present work were performed with the new RMCPP code.[19] The simulation boxes contained 12,000 atoms. The number densities of 0.0403 $Å^{-3}$ for $As_{34}Se_{51}Ag_{15}$ and 0.0340 $Å^{-3}$ for $As_{34}Te_{51}Ag_{15}$ used throughout the simulation runs were calculated from the mass densities given in Ref. 5. The backscattering amplitudes needed to calculate the model EXAFS curves from the pair distribution functions were calculated by the FEFF8.4 program.[20]

4. Results and Discussion

4.1. As$_2$Se$_3$ AND As$_{34}$Se$_{51}$Ag$_{15}$ GLASSES

The chemical bonding in crystalline As$_2$Se$_3$ is exclusively heteropolar: As–As and Se–Se direct bonding is lacking. Usually, the structure of As$_2$Se$_3$ glasses is regarded as a random network of corner-shared pyramidal AsSe$_{3/2}$ layers, which are held together through weak intermolecular forces. There are some indications that a small proportion of 'wrong' As–As and Se–Se bonds might exist in As$_2$Se$_3$ thin amorphous films.[21] However, it is very difficult to distinguish between As and Se in an experiment because of the similarity of the scattering properties of As and Se (concerning X-ray and neutron diffraction as well as EXAFS).

It has been suggested by Zha et al.[11] that the creation of covalent Ag–Se bonds upon Ag-doping in As$_2$Se$_3$ glasses should result in the formation of defect As–As bonds due to the shortage of Se. This suggestion has been proven by modeling of the experimental data with RMC. The quality of the EXAFS fits (especially of the Ag K-edge) is notably improved when As–As bonding is allowed, which indicates that As–As bonding is a real feature. The number of homopolar As–As bonds in g-As$_{34}$Se$_{51}$Ag$_{15}$ is small (N_{AsAs} = 0.23 ± 0.12). The most probable interatomic distance found (r_{AsAs} = 2.40 Å) equals to the sum of the covalent radii. However, it should also be mentioned that the uncertainty of this value is rather high because the As–As distance is very close to the pronounced As–Se peak.

It was assumed in a previous structural study[22] that there are no Ag–As bonds in As$_{40}$Se$_{40}$Ag$_{20}$ and (AsSe)$_{65}$(AgI)$_{35}$ glasses. A thorough investigation of the present experimental data suggests that this is not so obvious for g-As$_{34}$Se$_{51}$Ag$_{15}$. There is a small but remarkable improvement in the fit of the Ag K-edge EXAFS measurement if the Ag–As cut off distance is decreased to 2.5 Å. However, the number of As–Ag bonds is small (N_{AsAg} = 0.33 ± 0.16).

Fits obtained by simultaneous modeling of the four independent measurements (without Se–Se bonding) are shown in Figure 1. The agreement between model curves and experimental data is excellent. The partial pair correlation functions, the most probable nearest neighbor distances, and the coordination numbers corresponding to the atomic configuration generated by RMC are given in Figure 3 and Table 1. The bond lengths found in the present work agrees well with those obtained in the EXAFS study of Mastelaro et al.[13]

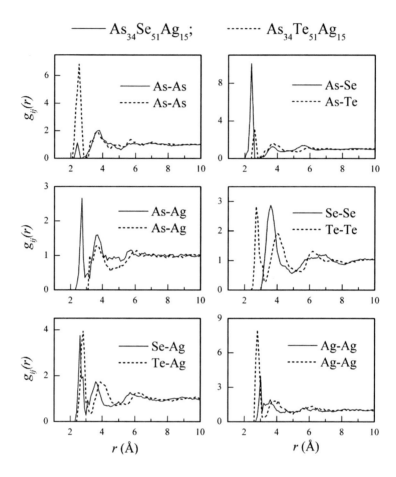

Figure 3. Partial pair distribution functions for glassy $As_{34}Se_{51}Ag_{15}$ and $As_{34}Te_{51}Ag_{15}$ obtained with RMC. For the details of the simulations see the text.

It is noteworthy that the As–Se coordination number in glassy $As_{34}Se_{51}Ag_{15}$ is 3.06 ± 0.23 and the Se-As coordination number 2.04 ± 0.15 without applying any constraint. The mean interatomic distance between As and Se atoms is very close to the sum of their covalent radii. This result strongly suggests that for the composition investigated, the structural $AsSe_{3/2}$ units formed by threefold coordinated As and twofold coordinated Se atoms (like those in binary As_2Se_3 glass) remain intact. The introduction of 15 at % Ag mainly results in the formation of Ag–Se bonds ($N_{AgSe} = 1.92 \pm 0.20$). This coordination number practically coincides with the result of Mastelaro et al.[13] who found $N_{AgSe} = 1.9 \pm 0.2$ for an alloy of a close composition $(Ag_2Se)_{0.18}(AsSe)_{0.82}$. Also, a proportion of Ag atoms is bonded to As ($N_{AgAs} = 0.74 \pm 0.15$), and each Ag atom has on average 0.44 ± 0.22 Ag neighbors.

4.2. As_2Te_3 AND $As_{34}Te_{51}Ag_{15}$ GLASSES

It was shown by Ma et al.[23] that homopolar As–As and Te–Te bonding plays an important role in binary As–Te glasses. This is also supported by some recent results of Jóvári et al.[24] who established that $N_{AsAs} \approx 1.7$, $N_{AsTe} \approx 1.2$, and $N_{TeTe} \approx 1.3$ in g-As_2Te_3. As it can be expected that Ag bonds preferentially to Te, it is reasonable to assume that As–As bonding can be found in glassy $As_{34}Te_{51}Ag_{15}$ as well.

The experimental XRD, ND and EXAFS data were fitted simultaneously both, with and without As–Ag bonding. It was established by the RMC modeling that the average number of As–Ag correlations is negligibly small if they are allowed ($N_{AsAg} = 0.14$; this value is certainly at the limit of the sensitivity of our approach.). Furthermore, unlike the $As_{34}Se_{51}Ag_{15}$ glass, the quality of the As and Ag K-edge fits for g-$As_{34}Te_{51}Ag_{15}$ is not improved when As–Ag bonding is allowed. Ag is mostly connected to Te ($N_{AgTe} = 2.42 \pm 0.25$). Also, each Ag atom has on average 1.11 ± 0.12 Ag neighbors. In view of the high number of Ag–Te bonds, it is remarkable that also Te–Te bonding is significant ($N_{TeTe} = 1.65 \pm 0.17$).

Our results show that As-As bonding is as significant in the ternary g-$As_{34}Te_{51}Ag_{15}$ ($N_{AsAs} \approx 2.1$) as in As–Te glasses. On the average, each As atom participates in 3.15 ± 0.32 As–As and As-Te bonds, while the sum of Te–Te and Te–As coordination numbers is 2.33 ± 0.24. The latter is higher than the average coordination number of Te found for glassy As_2Te_3 ($<N_{Te}> \approx 2.1$).[21]

The homopolar As–As bond length (2.48 Å) is found close to the values obtained in Refs. 12, 23, 24 (2.45–2.50 Å). Similarly to As–Te binary glasses,[24] the As–Te nearest neighbour distance obtained with RMC (2.59 Å) is slightly shorter than that reported by Usuki et al.[12] for $(Ag_2Te)_x(AsTe)_{1-x}$ glasses (2.66 Å).

Fits obtained by the simultaneous modeling of the five independent measurements for a $As_{34}Te_{51}Ag_{15}$ glass without As–Ag bonding are shown in Figure 2. The partial pair distribution functions corresponding to the model configuration are shown in Figure 3. The most probable nearest neighbor distances and the coordination numbers are presented in Table 1.

5. Summary

As_2Se_3 glass is chemically ordered; in other words homopolar bonding is not significant. The results of the present study show that addition of 15 at % Ag to As_2Se_3 mainly results in the formation of Se–Ag bonding. "Wrong" As–As bonds appear due to the shortage of Se. A small number of Ag–Ag bonds is also established. Within the experimental uncertainties, the

As–Se coordination number in glassy $As_{34}Se_{51}Ag_{15}$ is three, and the Se–As coordination number is two. This means that the chemically ordered structural $AsSe_{3/2}$ units remain intact.

Unlike As_2Se_3, glassy As_2Te_3 is characterized by a high number of homopolar bonds, which means that Ag does not have to compete with As for Te neighbors. As a result, the decrease of As–Te bonding is compensated by the increase of As–As bonds. This way, As remains threefold coordinated. According to our model, TeTe bonding increases upon adding Ag. This somewhat surprising result can be understood if we suppose that Te atoms tend to group around Ag atoms. As a result the total coordination number of Te increases from 2.06 ± 0.21 in g-As_2Te_3 to 3.04 ± 0.31 in g-$As_{34}Te_{51}Ag_{15}$.

References

1. V. L. Vainov and S. K. Novoselov, *Izvestiya Vysshikh Uchebnykh Zavedenii, Fizika* 8, 138 (1977).
2. A. Giridhar and S. Mahadevan, *J. Non-Cryst. Solids* 197, 228 (1996).
3. M. M. Hafiz, A. H. Moharram, and A. A. Abu-Sehly, *Appl. Surf. Sci.* 115, 203 (1997).
4. M. Kitao, T. Ishikawa, and S. Yamada, *J. Non-Cryst. Solids* 79, 205 (1986).
5. Z. U. Borisova, *Glassy Semiconductors*, Plenum Press, New York, 1981, p. 505.
6. S. Mahadevan and A. Giridhar, *J. Non-Cryst. Solids* 197, 219 (1996).
7. M. Kitao, C. Gotoh, and S. Yamada, *J. Mater. Sci.* 30, 3521 (1995).
8. K. Ogusu, T. Kumagai, Y. Fujimori, and M. Kitao, *J. Non-Cryst. Solids* 324, 118 (2003).
9. K. Ogusu, J. Yamasaki, S. Maeda, M. Kitao, and M. Minakata, *Opt. Lett.* 29, 265 (2004).
10. K. Ogusu, S. Maeda, M. Kitao, H. Li, and M. Minakata, *J. Non-Cryst. Solids* 347, 159 (2004).
11. C. Zha, A. Smith, A. Prasad, R. Wang, S. Madden, and B. Luther-Davies, *J. Nonlinear Opt. Phys. Mater.* 16, 49 (2007).
12. T. Usuki, O. Uemura, S. Konno, Y. Kameda, and M. Sakurai, *J. Non-Cryst. Solids* 293-295, 799 (2001).
13. V. Mastelaro, S. Benazeth, H. Dexpert, A. Ibanez, and R. Ollitrault-Fichet, *J. Non-Cryst. Solids* 151, 1 (1992).
14. C.J. Benmore and P.S. Salmon, *Phys. Rev. Let.* 73, 264 (1994).
15. J. Liu and P. S. Salmon, *Europhys. Lett.* 39, 521 (1997).
16. R. L. McGreevy and L. Pusztai, *Mol. Simulat.* 1, 359 (1988).
17. R. L. McGreevy, *J. Phys.: Condens. Mat.* 13, R877 (2001).
18. G. Evrard and L. Pusztai, *J. Phys.: Condens. Mat.* 17, S1 (2005).
19. R. L. McGreevy, M.A. Howe, and J.D. Wicks, RMCA version 3, 1993; available at <http://www.isis.rl.ac.uk/RMC/rmca.htm>.
20. A. L. Ankudinov, B. Ravel, J. J. Rehr, and S. D. Conradson, *Phys. Rev. B* 58, 7565 (1998).

21. M. Popescu, F. Sava, N. Aldea, Xie Yaning, Hu Tiandou, Liu Tao, M. Leonovici, *Chalcogenide Letters* 2, 71 (2005).
22. T. Petkova, P. Petkov, P. Jóvári, I. Kaban, W. Hoyer, A. Schöps, A. Webb, B. Beuneu, *J. Non-Cryst. Solids* 353, 2045 (2007).
23. Q. Ma, D. Raoux and S. Benazeth, *Phys. Rev. B* 48, 16332–16346 (1993).
24. P. Jóvári, S. N. Yannopoulos, I. Kaban, A. Kalampounias, I. Lishchynskyy, B. Beuneu, O. Kostadinova, E. Welter, and A. Schöps, unpublished results (2008).

THERMAL BEHAVIOR OF NOVEL (GeS$_2$)$_{1-x}$(AgI)$_x$ GLASSES

B. MONCHEV[1], T. PETKOVA[*1], P. PETKOV[2], J. PHILIP[3]
[1]*Institute of Electrochemistry and Energy Systems, Bulgarian Academy of Sciences, Acad. G. Bonchev St. bl. 10, 1113 Sofia, Bulgaria*
[2]*Departments of Physics, University of Chemical Technology and Metallurgy, Kl. Ohridski Blvd. 8, 1756 Sofia, Bulgaria*
[3]*Sophisticated Test and Instrumentation Center, Cochin University of Science and Technology, Cochin - 682 022, Kerala, India*

Abstract. Novel chalcohalide glasses have been developed. The amorphous nature of the synthesized Ge–S–AgI materials has been proven by means of X-ray diffraction. Differential scanning calorimetry has been used to investigate the thermal characteristics of the materials to derive information about the glassy structure. The glass transition (T_g), the onset of crystallization (T_{on}), and the peak crystallization (T_{cr}) and melting (T_m) temperatures have been determined in the temperature range of 293–750 K. The criteria for thermal stability (ΔT) and the glass-forming ability (H_{gl} and S) have been calculated.

Keywords: chalcohalide glasses; thermal characteristics; criteria

1. Introduction

The thermal features of glassy materials are often essential for elucidation of some structural aspects of the materials. In general, the introduction of a third component into a host matrix, as in our case silver iodide into GeS$_2$ glass, could change the thermal stability of the alloy. Differential scanning calorimetry (DSC) is a most powerful method commonly applied for thermal

---
[*]tpetkova@bas.bg

J.P. Reithmaier et al. (eds.), *Nanostructured Materials for Advanced Technological Applications,*   353
© Springer Science + Business Media B.V. 2009

investigation of inorganic disordered materials.[1] The basic reasons for the utilization of this method are its simplicity and flexibility for a wide range of heating rates.

Investigation of the thermal stability of a material against crystallization is an important problem from a technological point of view,[2] for example for the realization of optical fibers.

2. Experimental

Bulk glasses from the system $(GeS_2)_{100-x}(AgI)_x$, where x = 5, 10, 15, 20 mol %, were prepared by direct monotemperature synthesis in a rotary furnace. The glassy state of the samples was studied by X-ray diffraction (APD-15 Philips 2139 with CuKα radiation and a Ni filter).[1]

The non-isothermal calorimetric measurements of the Ge–S–AgI substances studied were performed using a Diamond DSC (Perkin Elmer GmbH) experimental set-up with a temperature accuracy of ±1 K. The samples, which ware pressed and sealed in aluminum pans, were scanned with a heating rate of 15 K/min.

The thermal stability and the criteria of the glass-formation ability were calculated. The difference between the onset of crystallization and the glass transition temperature represents the Dietzel criterion ΔT.[4] The values of the well-known criteria of Hruby (H_{gl}) and Saad and Poulain (S) were calculated by the expressions given in Refs. 5 and 6, respectively.

3. Results and Discussion

A typical thermogram of a sample with 15 mol % AgI with well-resolved thermal transitions is shown in Figure 1. The thermal characteristics of all samples under study are presented in Table 1. The first endo peak is situated in the range of 420–440 K. In the literature, the β–α silver iodide transition is reported to occur at 420 K.[7] Most probably the peaks observed in the temperature range of 420–440 K are due to the glass transition which is, however, partly covered by the β–α AgI transition. With the increase of the AgI content the position of this first peak is shifted toward lower values as a consequence of the stronger effect of the AgI transition. In addition, T_{on} could be extrapolated from the recorded thermograms. The second endo peak in the temperature range of 603–628 K is due to melting of the material as displayed in Table 1.

The dependency of the experimentally obtained thermal characteristics can be discussed in view of the structure of the materials. The overall bond energy decreases with the addition of silver iodide; alongside the transition temperature changes proportionally to this decrease. The fraction of strong Ge–S bond (551 kJ/mol) diminishes with the introduction of AgI, and the weaker Ag–I (234 kJ/mol)[8] bonds form a considerable amount of the bonds present in the Ge–S–AgI glasses studied. Thus, T_g is decreasing. This consideration is supported by recent results on the microhardness of these materials[3] and the general idea of a network plasticizing effect of AgI in Ge–S based glassy materials.[9]

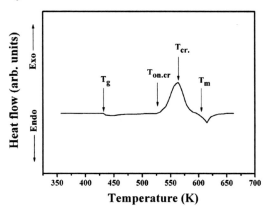

Figure 1. Typical thermogram of a glassy $Ge_{28}S_{57}(AgI)_{15}$ sample.

On other the hand such compositional dependences of the glass transition temperature has already been reported for similar glasses,[8,9] when a "softening" third component is introduced into the main chalcogen network. Seddon et al.[1] and Heo et al.[10] reported that iodine atoms play a role of network terminators in the glassy structure resulting in the formation of rathes-hort –I–S–Ge structures, while the basic Ge–S backbone becomes fragmented in Ge–S–I glasses. In the glasses studied T_g decreases with the introduction of third component (AgI) which could also be attributed to the terminating role of iodine. The shortening of the network structure with increasing of the halogen atomic mass and concentration leads to a considerable decrease of the overall bond energy and hence of T_g.

TABLE 1. Characteristic temperatures of the $(GeS_2)_{1-x}(AgI)_x$ samples.

Composition	T_g (K)	T_{on} (K)	T_{cr} (K)	T_m (K)	ΔT (K)	H_{gl}	S (K)
$Ge_{32}S_{63}(AgI)_5$	443	589	615	628	146	3.74	8.57
$Ge_{30}S_{60}(AgI)_{10}$	435	562	587	620	127	2.19	7.30
$Ge_{28}S_{57}(AgI)_{15}$	430	538	562	614	108	1.42	6.03
$Ge_{27}S_{53}(AgI)_{20}$	421	523	547	603	102	1.27	5.81

The criteria describing the thermal stability, glass-formation ability and fiber-drawing behavior of the chalcohalide materials studied were calculated and summarized in Table 1. Relatively high values reveal a good glass forming tendency and stability of the materials.[1] It should be noticed that glasses with a low AgI content are more stable, probably because they are close to the energetically stable stoichiometric GeS_2 glass. It can be concluded that modifying the Ge–S glassy network with silver iodide has a significant impact on the glass-formation ability; higher AgI concentrations suppress the glass-forming ability and reduce the thermal stability of the materials obtained.

4. Conclusions

The basic thermal features of the studied Ge–S–AgI glasses were presented. The correlation between the thermal characteristics and composition was established. The thermal parameters and glass forming stability criteria decrease as AgI content increases. The specific "softening role" of the AgI salt on the glassy structure in respect to thermal properties was evaluated.

ACKNOWLEDGEMENTS

This research was supported by the NATO Reintegration Grant (EAP.RIG.982374) to which the authors are deeply indebted.

References

1. A. Seddon, M. Hemingway, *J. Non-Cryst. Solids*, 161 323 (1993).
2. V. Shiryaev, J.-L. Adam, X. Zhang, M. Churbanov, *Solid State Sci.*, 7 209 (2005).
3. B. Monchev, T. Petkova, P. Petkov, S. Vassilev, *J. Mater. Sci.*, 4 (23) 4836 (2007).
4. A. Dietzel, *Glastech. Ber.*, 22 41 (1968).
5. A. Hruby, Czech. *J. Phys. B*, 22 1187 (1972).
6. M. Saad, M. Poulain, *Mater. Sci. Forum*, 19–20 11 (1987).
7. V. Boev, M. Mitkova, E. Lefterova, T. Wágner, S. Kasap, M. Vlček, *J. Non-Cryst. Solids*, 266–269 867 (2000).
8. F.M. Daniel, V.M. Stirling, *Dictionary of inorganic compounds*, Chapman & Hall, New York/London (1992).
9. A. Ibanez, E. Philippot, S. Benazeth, H. Dexpert, *J. Non-Cryst. Solids*, 127 25 (1991).
10. J. Heo, J. Mackenzie, *J. Non-Cryst. Solids*, 111 29 (1989).

MICROSTRUCTURAL, MORPHOLOGICAL AND OPTICAL CHARACTERIZATION OF As$_2$Se$_3$–As$_2$Te$_3$–Sb$_2$Te$_3$ AMORPHOUS LAYERS

S. BOYCHEVA[*1], V. VASSILEV[2],
T. HRISTOVA-VASILEVA[2], A.K. SYTCHKOVA[3],
J.P. REITHMAIER[4]
[1]*Department of Thermal and Nuclear Engineering, Technical University of Sofia, 8 Kl. Ohridsky Blvd., 1000 Sofia, Bulgaria*
[2]*University of Chemical Technology and Metallurgy, 8 Kl. Ohridsky Blvd., 1756 Sofia, Bulgaria*
[3]*ENEA, Via Anguillarese 301, 00123 Rome, Italy*
[4]*University of Kassel, INA, Heinrich-Plett-St. 40, 34132 Kassel, Germany*

Abstract. Amorphous thin films from the As$_2$Se$_3$–As$_2$Te$_3$–Sb$_2$Te$_3$ system were prepared by thermal vacuum evaporation from the corresponding bulk glasses. The films were characterized with respect to their microstructure, morphology and surface topology. Optical transmission and reflection VIS and NIR spectra were measured and mathematically simulated using a combination of dispersion models to derive the optical constants.

Keywords: amorphous thin films; optical constants; chalcogenide glasses

1. Introduction

Recently, chalcogenide glasses (ChG) have been considered as advanced optical materials due to their unique optical properties.[1] The most important of them from a practical point of view are As-based chalcogenides and their multilayer structures.

[*] sylvia_boycheva @yahoo.com

J.P. Reithmaier et al. (eds.), *Nanostructured Materials for Advanced Technological Applications,* 357
© Springer Science + Business Media B.V. 2009

Multicomponent glasses from the $As_2Se_3–As_2Te_3–Sb_2Te_3$ system have attracted our interest due to the combination of two As-containing chalcogenides and the presence of large polarizable Te atoms.

The glass-formation in the pseudo-ternary $As_2Se_3–As_2Te_3–Sb_2Te_3$ system, as well as the main physicochemical properties of the glasses obtained have been studied preliminarily.[2]

2. Experimental

Thin $As_2Se_3–As_2Te_3–Sb_2Te_3$ films were prepared by thermal vacuum evaporation of the corresponding glassy bulk materials in a high vacuum set-up. Deposition conditions which guarantee the preparation of amorphous, uniform and homogeneous films, when starting from sublimating complex materials, were established.[3] The morphology and microstructure of the films were examined by a scanning electron microscope (SEM Hitachi S-4000), the surface roughness by an atomic force microscope (AFM Nano-Scope II) in a tapping mode.

Optical transmission and reflection measurements were carried out in the spectral range from 300–2,500 nm with an UV/VIS/NIR double channel spectrophotometer (Perkin Elmer Lambda 900). The experimental spectra were mathematically simulated using a combination of dispersion models which enable us to determine the film thickness, its refractive index n, the extinction coefficient k, and thus the optical band gap of the films $\Delta E_{g,opt}$.

3. Results and Discussions

The compositions selected for thin film deposition are given in Table 1. SEM investigations revealed the uniform and featureless amorphous nature of the layers, both in the depth and on the surface, as can be seen in a typical cross-section SEM micrograph in Figure 1a. Surface topology studies by AFM (Figure 1b) yielded low rms surface roughnesses, as given in Table 1.

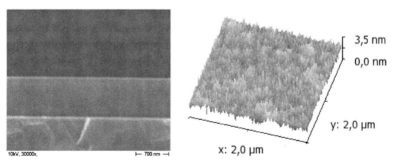

Figure 1. Microstructure, morphology and topography of a $(As_2Se_3)_{81}(As_2Te_3)_9 (Sb_2Te_3)_{10}$ thin film: (a) cross-section SEM micrograph (left); (b) AFM image (right).

TABLE 1. Composition, thickness, rms roughness and optical band gap of $(As_2Se_3)_x(As_2Te_3)_y(Sb_2Te_3)_z$ thin films, $(x + y + z = 100$ mol %).

Composition, mol %			Thickness,	rms roughness,	$\Delta E_{g,opt},$
As₂Se₃	As₂Te₃	Sb₂Te₃	nm	nm	eV
81	9	10	811	0.31	1.460
63	27	10	1,103	1.20	1.210
56	24	20	703	2.45	1.145

The experimental transmittance and reflectance spectra (Figure 2) were used as an input for a fitting procedure to determine $n(\lambda)$ and $k(\lambda)$. An example of the fitting results of the experimental spectra is given in Figure 3.

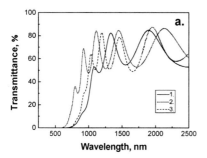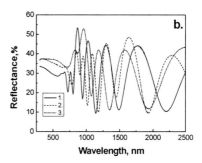

Figure 2. Transmittance (a) and reflectance (b) spectra of: 1: $(As_2Se_3)_{81}(As_2Te_3)_9 (Sb_2Te_3)_{10}$; 2: $(As_2Se_3)_{63}(As_2Te_3)_{27}(Sb_2Te_3)_{10}$; 3: $(As_2Se_3)_{56}(As_2Te_3)_{24}(Sb_2Te_3)_{20}$ films.

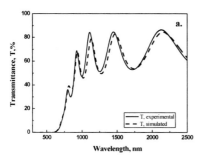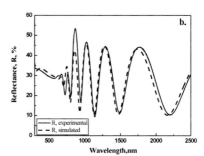

Figure 3. Experimental and simulated transmittance (a) and reflectance (b) spectra of a $(As_2Se_3)_{81}(As_2Te_3)_9(Sb_2Te_3)_{10}$ thin film.

The fitting procedure was started from the Cauchy model for the calculation of n, adding an exponential VIS dispersion for k, from which a first approximation for the film thickness was obtained. Series of iterations in the frame of arbitrary dispersions for both n and k were made for refining, and inhomogeneities in the bulk of the films were added to the model. Examples are presented in Figure 4.

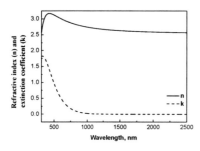 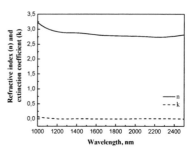

Figure 4. Dispersion of the refractive index (n) and the extinction coefficient (k) of (a) $(As_2Se_3)_{81}(As_2Te_3)_9(Sb_2Te_3)_{10}$ and (b) $(As_2Se_3)_{56}(As_2Te_3)_{24}(Sb_2Te3)_{20}$ thin films.

4. Conclusion

Amorphous thin films from the As_2Se_3–As_2Te_3–Sb_2Te_3 system with low surface roughness were prepared by thermal vacuum evaporation from the corresponding bulk glasses. The film with a composition of $(As_2Se_3)_{81}$ $(As_2Te_3)_9(Sb_2Te_3)_{10}$ has a high refractive index ($2.58 < n < 3.18$) with a strong spectral dispersion at $\lambda < 700$ nm.

ACKNOWLEDGEMENT

The authors thanks to NATO support under CBP.EAP.RIG 981850 and the NSF, Ministry of Education, Science and Technology of R Bulgaria by a grant No.1503/05.

References

1. B. Bureau, X.H. Zhang, F. Smektala, J.-L. Adam, J. Troles, H.-li Ma, C. Boussard-Pledel, J. Lukcas, P. Lucas, D. Le Coq, M. R. Riley, and J. H. Simmons, J. Non-Cryst. Solids 345–346, 276 (2004).
2. T. Hristova-Vasileva, V. Vassilev, L. Aljihmani, and S. Boycheva, J. Phys. Chem. Solids 69, 2540 (2008).
3. V. Vassilev, C. Popov, S. Boycheva, L. Aljihmani, P. Petkov, K. Kolev, and B. Monchev, Mater. Lett. 58, 3802 (2004).

NANOCOLLOIDAL SOLUTIONS OF As–S GLASSES AND THEIR RELATION TO THE SURFACE MORPHOLOGY OF SPIN-COATED AMORPHOUS FILMS

T. KOHOUTEK[1], T. WAGNER[1], M. FRUMAR[1],
A. CHRISSANTHOPOULOS[2], O. KOSTADINOVA[2],
S.N. YANNOPOULOS[*2]
[1]*Department of Inorganic Chemistry, Faculty of Chemical-Technology, University of Pardubice, Legion's sq. 565, 53210, Czech Republic*
[2]*Foundation for Research and Technology Hellas, Institute of Chemical Engineering and High Temperature Chemical Processes, FORTH/ICE-HT, P.O. Box 1414, GR-26 504 Patras, Greece*

Abstract. Amorphous chalcogenide (As–S) thin films can be fabricated by the spin-coating technique from appropriate solutions of the corresponding glasses. Such films exhibit a grainy texture, which is presumably related to the cluster size in the solution. In this paper we report on a possible relation between the grain size of the surface of spin-coated $As_{33}S_{67}$ chalcogenide thin films and the cluster size of the glass in butylamine solutions, using atomic force microscopy and dynamic light scattering, respectively. A novel athermal photo-aggregation effect in the liquid state is reported also.

Keywords: dynamic light scattering; nanocolloidal chalcogenide solutions; spin-coated chalcogenide films

1. Introduction

Thin films of amorphous semiconductors are today essential for modern electronic devices. Several planar waveguides are realized by thin films of chalcogenide glasses which meet the requirements for all-optical signal processing. The major methods usually applied for thin film deposition are

---

[*] sny@iceht.forth.gr.

J.P. Reithmaier et al. (eds.), *Nanostructured Materials for Advanced Technological Applications*, 361
© Springer Science + Business Media B.V. 2009

thermal evaporation, magnetron sputtering, and pulsed laser deposition. The main disadvantages of these methods include the poor control of film compositions and the high cost of film fabrication. On the other hand, using wet-chemistry methods one can deposit thin/ultrathin films by spin coating of a material solution. The fabrication of metal chalcogenides thin films of technological importance has recently been reported[1]. However, it has long ago been demonstrated that thin films of arsenic chalcogenides can be prepared by the spin-coating method from solutions of chalcogenide glasses in organic solvents[2]. Arsenic sulphide glasses can readily be dissolved in organic solvents such as amines in the concentration range of $1-10^3$ mg ml^{-1}.

Studies towards understanding the mechanism of dissolution of chalcogenides in amines have been reported[2] where the presence of chalcogenide clusters with dimensions of several nanometers was proposed. Further studies focused on the surface morphology of spin-coated As–S chalcogenide thin films deposited from amine solutions which revealed a nanoscale grainy character as measured by atomic force microscopy (AFM, Figure 1).[3] Taking in mind that the film structure is usually fragmented due to the existence of large grains – a problem for applications – it would be important to ask if there is any particular relation between the cluster size in the solution of the chalcogenide glasses and the grain size in the film morphology. The present work aims to elucidate this issue by combining results from various experimental techniques, including dynamic light scattering (DLS), optical absorption spectroscopy, and AFM.

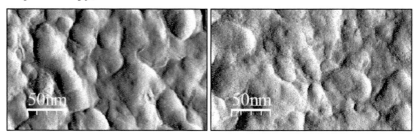

Figure 1. AFM tapping mode images of the surface morphology of an as-deposited spin-coated As$_{33}$S$_{67}$ film with a thickness of 1,700 nm with a typical grain size of about 50 nm (left) of the same film after stabilization at 90°C for 1 h (right).

2. Experimental

Chalcogenide glass solutions with three different concentrations $c_1 = 16.6$, $c_2 = 83$, and $c_3 = 166$ mg/ml of As$_{33}$S$_{67}$ were obtained by dissolution of the bulk glasses in butylamine (BA). Thin films were prepared on silica glass substrates by spinning the solutions with concentrations c_1 and c_2. Dust-free solutions for the DLS experiments were prepared by passing the As$_{33}$S$_{67}$/BA solutions through 0.2 μm PTFE filters into pyrex tubes. Electronic absorption spectra of all solutions were recorded in the region of 200–800 nm in order to select off-resonant laser wavelengths for the DLS studies, as well as to check possible changes of the concentrations of the samples after filtering the solutions. The following laser wavelengths were used: 496, 488, 514.5, 632.8, and 671 nm. AFM images of the film surfaces were recorded over an area of 250×250 nm^2 in high-resolution tapping mode to reveal the surface morphology of the films. More details can be found elsewhere.[4]

3. Results and Discussion

The energy gap E_g of the nanocolloidal solutions of As$_{33}$S$_{67}$ particles were estimated from optical absorption spectra. The concentration dependence of E_g is shown in the inset of Figure 2. The strong increase of E_g at low concentrations indicates the onset of quantum confinement and the quantum-dot character of the particles in solution. Usually, the band gap of the solutions decreases towards that of As$_{33}$S$_{67}$ bulk glass (~2.42 eV) denoted by the horizontal dashed line in the inset. The analysis showed that the solutions c_1 and c_2 are monodispersed, whilst solution c_3 exhibits a bimodal distribution of the particle size. The particle sizes R_h were estimated by the Stokes-Einstein relation (which relates R_h and the diffusion coefficient) and were found to depend almost linearly on the concentration in the range from 1 nm for c_1 to ~4 nm for c_3.[4]

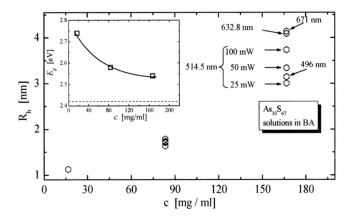

Figure 2. Spectral and concentration dependence of As$_{33}$S$_{67}$ glass cluster sizes in BA solutions exhibiting an almost linear increase with the concentration. The inset shows the concentration dependence of E_g for the samples studies; the solid line is a guide to the eye while the dashed horizontal line marks the E_g the bulk glass.

While studying the spectral dependence of the cluster size an interesting new effect was observed. In particular, a systematic *athermal photo-aggregation* phenomenon was found to occur with increasing the power density of the incident light as shown in Figure 2. Increasing the power by a factor of 2 and 4 results in an increase of the hydrodynamic radius by ~11% and ~25%, respectively.[4] These differences in R_h are beyond the experimental error. A thermal origin of the effect caused by laser radiation absorption can be excluded since the opposite behavior of R_h would be expected in this case.

Figure 1 reveals that the typical surface grain sizes of the films were about 25 and 50 nm for films prepared from solutions c_1 and c_2, respectively. A substantial smoothening of the grainy character of the surface morphology took place after thermal stabilization of the films. We consider that the grainy character of the surface morphology of as-deposited spin-coated thin films is related to the agglomeration of chalcogenide clusters during the solvent evaporation from the films at the time of their deposition. This agglomeration of chalcogenide clusters is to be expected in view of the concentration dependence of the cluster size as revealed by DLS. In particular, during the first stages of film formation the continuously evaporation of the solvent leads to a progressively increasing concentration of the chalcogenide material and hence to an increase of the cluster size.

4. Conclusions

A simple method for thin film fabrication of amorphous chalcogenides involving spin-coating from appropriate solutions has been used to fabricate $As_{33}S_{67}$ films. The grainy surface structure was found to be related to the cluster size of the dissolved chalcogenide material in the parent solutions. Grains are formed by agglomeration of clusters during solvent evaporation from the films at the stage of their preparation. Hydrodynamic cluster radii between 1 and 4.2 nm were found. A novel athermal photo-aggregation process was observed for increased laser power densities.

References

1. D. J. Milliron, S. Raoux, R. Shelby, J. Jordan-Sweet, Nat. Mater. 6, 352 (2007).
2. G. C. Chern, I. Lauks, J. Appl. Phys. 53, 6979 (1983).
3. T. Kohoutek, T. Wagner, J. Orava, M. Krbal, T. Mates, A. Fejfar, S.O. Kasap, M. Frumar, J. Non-Cryst. Solids 353, 1437 (2007).
4. T. Kohoutek, T. Wagner, M. Frumar, A. Chrissanthopoulos, O. Kostadinova, S. N. Yannopoulos, J. Appl. Phys. 103, 063511 (2008).

NANOSTRUCTURAL CHARACTERIZATION OF AMORPHOUS CHALCOGENIDES BY X-RAY DIFFRACTION AND POSITRON ANNIHILATION TECHNIQUES

T. KAVETSKYY[*1,2], O. SHPOTYUK[1,2], I. KABAN[3], W. HOYER[3], J. FILIPECKI[2]

[1]Institute of Materials, Scientific Research Company "Carat", Stryjska 202, 79031 Lviv, Ukraine
[2]Institute of Physics, Jan Dlugosz University of Czestochowa, Armii Krajowej 13/15, 42201 Czestochowa, Poland
[3]Institute of Physics, Chemnitz University of Technology, Reichenhainer 70, D-09107 Chemnitz, Germany

Abstract. Nanostructural features of binary As–Se(Te), pseudo-binary $As_2S(Se)_3$–$Sb_2S(Se)_3$, ternary stoichiometric $As(Sb)_2S_3$–GeS_2 and non-stoichiometric $As(Sb)_2S_3$–Ge_2S_3 chalcogenide glasses are studied by using conventional X-ray diffraction, high-energy synchrotron X-ray diffraction and positron annihilation lifetime spectroscopy (PALS). Compositional trends of the position of the first sharp diffraction peak (FSDP) (size of nanovoids) are analyzed within the void-based model as well as cation-correlation approaches along with the PALS results. The dominant cation correlations responsible for the FSDP occurrence and sizes of nanovoids are established in dependence on the type of the glass structure.

Keywords: chalcogenide glasses; first sharp diffraction peak; positron annihilation lifetime spectroscopy; nanostructure; nanovoids

1. Introduction

The first sharp diffraction peak (FSDP) of chalcogenide glasses has been intensively studied during the last 2 decades. Its relation with internal structural nanovoids within the void-based model[1] helps to understand the

---

[*]kavetskyy@yahoo.com

J.P. Reithmaier et al. (eds.), *Nanostructured Materials for Advanced Technological Applications*, 365
© Springer Science + Business Media B.V. 2009

nanostructure of glassy materials. In the framework of this model, it was shown for tetrahedral GeX_2-type glasses that the position of the FSDP Q_1 is linked with the nearest-neighbour cation–cation distance or cation-void distance d through the expression[1]

$$Q_1 = 3\pi/2d. \tag{1}$$

A similar relation has been established for pyramidal As_2X_3-type glasses[2,3]

$$Q_1 = 2.3\pi/D, \tag{2}$$

where D ($\equiv d$) is the diameter of the nanovoids. Naturally, the question arises about a possible relation between the first sharp diffraction peak and the nanostructure of mixed pyramidal-tetrahedral type glasses.

Jensen and co-workers[4] suggested that positron annihilation lifetime spectroscopy (PALS) can be a sensitive tool to test the void-based model of the FSDP of covalent chalcogenide glasses. Based on theoretical calculations, they established a relation between the positron lifetime τ and the open volume (positron traps) V for g-As_2Se_3 (g = glassy) which can be expressed as follows (where τ is given in ns and V is in Å^3):

$$\tau \cong 0.240 + 0.0013 \cdot V. \tag{3}$$

Assuming that (i) the lifetime is independent of the number of vacancies for positrons trapped at hypothetical vacancy chains[5] and (ii) di-vacancies and tetra-vacancies have very similar lifetimes,[6] Eq. (3) can also be used for other glassy chalcogenides with structures like that in g-As_2Se_3. Hence, the size of nanovoids determined from FSDP and PALS data can be compared.

The present work is aimed to examine the compositional behaviour of the FSDP and the size nanovoids for binary As–Se(Te), pseudo-binary As_2S $(Se)_3$–$Sb_2S(Se)_3$, ternary stoichiometric $As(Sb)_2S_3$–GeS_2, and non-stoichiometric $As(Sb)_2S_3$–Ge_2S_3 glasses by means of conventional X-ray diffraction (XRD), high-energy synchrotron X-ray diffraction (HESXRD), and PALS.

2. Experimental

Bulk chalcogenide glasses were prepared by the melt-quenching technique. Conventional XRD studies were carried out with a HZG-4a diffractometer (Cu K_α-radiation). HESXRD experiments were performed at the experimental station BW5 at HASYLAB, DESY (Hamburg, Germany). The scattered intensity was measured between 0.5 and 18 Å^{-1}. The raw data were corrected for detector dead-time, polarization, absorption and variation of the detector solid angle. PALS measurements for selected glass compositions were done by using a conventional fast–fast coincidence system with an ORTEC spectrometer and radioactive ^{22}Na isotopes with an activity of

0.74 MBq as a positron source, placed between two sandwiched samples. All experiments were carried out at room temperature.

3. Results and Discussion

The main features of the FSDP (position and intensity) for the chalcogenide glasses investigated exhibit remarkable changes with the alloy composition as shown in Figure 1 for some samples from the As_xSe_{100-x} and $Ge_xAs_{40-x}S_{60}$ systems. This suggests that the nanovoid size, which is directly related to the FSDP, varies with the composition. The question, however, remains, which type of structural units is responsible for such changes.

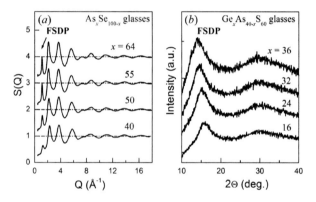

Figure 1. (*a*) Experimental structure factors $S(Q)$ obtained with HESXRD and (*b*) FSDP patterns obtained with conventional XRD for some glasses from the As_xSe_{100-x} and $Ge_xAs_{40-x}S_{60}$ systems. The curves are shifted for clarity.

It can be seen from Figure 1a that the FSDP intensity remarkably increases with increasing As concentration in As–Se glasses. A similar dependence was observed for glasses from the binary As–Te and pseudo-binary As_2S $(Se)_3$–$Sb_2S(Se)_3$ systems (not shown). Such a behaviour has also been found by other researches (see e.g. the work of Bychkov et al.[7] on binary seleni-des). Our findings prove that for binary and pseudo-binary pyramidal (P-type) glasses the As cations are responsible for the FSDP occurrence.

The intensity of the FSDP for mixed pyramidal-tetrahedral (P-T-type) ternary stoichiometric As_2S_3–GeS_2 and Sb_2S_3–GeS_2, as well as for non-stoichiometric As_2S_3–Ge_2S_3 and Sb_2S_3–Ge_2S_3 glasses exhibits a strong dependence on the Ge concentration (see Figure 1b). This suggests that the nature of the FSDP is related to the Ge content and its distribution in these materials. It has been shown[7] that Ge–Ge cation–cation correlations are responsible for the increase of the FSDP intensity in tetrahedral (T-type) Ge–Se glasses.

Taking into account that the size of nanovoids in chalcogenides is related to the FSDP position by Eqs. (1) and (2), Q_1 has been determined from the XRD data for the alloys studied (Figure 2a–c). The data obtained for the As–Se and As–Te systems are in good agreement with the results of Bychkov et al.[7,8] The FSDP position for binary tetrahedral (T-type) g-GeS_2 and g-$GeSe_2$ with a Ge content of 33.3 at % is known to be in the range of 0.98–1.04 Å^{-1} (Figure 2d)[9–12]. It is noteworthy that the FSDP for ternary P–T-type glasses with 30–35 at % Ge is also situated at 1.01–1.04 Å^{-1} (Figure 2c). This is a further argument for the dominant role of Ge cations in the occurrence of the FSDP in amorphous P–T-type chalcogenides. It was also established that the value of Q_1 for $Ge_{23.5}As_{11.8}S_{64.7}$ and $Ge_{23.5}Sb_{11.8}S_{64.7}$ is the same although As and Sb differ by their scattering power as well as their atomic size. The same picture is observed for $Ge_{28.125}As_{6.25}S_{65.625}$ and $Ge_{28.125}Sb_{6.25}S_{65.625}$ alloys. These compositions are marked in Figure 2c. A strong correlation between Q_1 and the Ge concentration exists also for non-stoichiometric As_2S_3–Ge_2S_3 and Sb_2S_3–Ge_2S_3 glasses. It can therefore be concluded that the FSDP position for P–T type amorphous chalcogenides is determined exclusively by Ge, while As and Sb cations do not affect the FSDP occurrence.

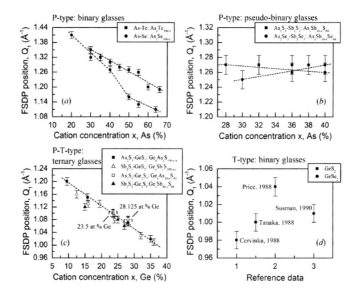

Figure 2. FSDP position as a function of the cation concentration in chalcogenides for the binary, pseudo-binary, and ternary systems studied (*a*, *b*, *c*), and for GeS(Se)$_2$ glasses taken from literature (*d*) with different types of structure: pyramidal (P-type), mixed pyramidal-tetrahedral (P–T-type), and tetrahedral (T-type). The dashed lines are drawn to guide the eye.

The dependence of the FSDP position on the Ge concentration for mixed pyramidal-tetrahedral type glasses suggests that the size of nanovoids can be determined within the void-based model by applying Eq. (1) developed for simple tetrahedral AX_2-type structures.

The volumes of nanovoids for glasses of various types determined from the XRD and PALS data with Eqs. (1)–(3) are given in Table 1. A good agreement between the sizes obtained from different experiments is observed. The volume decreases from ~100 Å^3 for P-type glasses to ~55 Å^3 for T-type glasses and further to 30–50 Å^3 for P–T-type glasses. This can be interpreted as change of the nanovoid structure from three atomic vacancies for P-type to di- and/or mono-vacancies for T- and P–T-type glasses.

TABLE 1. FSDP and PALS data (position Q_1, open-volume related lifetime τ_2) and the corresponding values of the nanovoid volume V obtained from (1)–(3) for different glasses.

Glass system	FSDP, Q_1 (Å^{-1}) ±0.01	V_{FSDP} (Å^3) ±5	PALS, τ_2 (ns) ±0.01	V_{PALS} (Å^3) ±5
P-type (pyramidal type of structure)				
As–Se	1.41–1.11	70–144	0.36–0.39[13]	92–115
As–Te	1.35–1.19	80–117	–	–
As$_2$S$_3$	1.26	98	0.37	100
As$_2$Se$_3$	1.25–1.27	101–96	0.37	100
As$_2$Te$_3$	1.30	90	–	–
As$_2$S$_3$–Sb$_2$S$_3$	1.26–1.27	98	–	–
As$_2$Se$_3$–Sb$_2$Se$_3$	1.25–1.27	101–96	–	–
T-type (tetrahedral type of structure)				
GeS$_2$	0.98–1.04[9–11]	58–49	–	–
GeSe$_2$	1.01[12]	53	–	–
Ge$_2$S$_3$	–	–	0.42[14]	56*
P–T-type (pyramidal-tetrahedral type of structure)				
As$_2$S$_3$–GeS$_2$	1.20–1.07	32–45	–	–
Sb$_2$S$_3$–GeS$_2$	1.16–1.08	35–43	0.40	48*
As$_2$S$_3$–Ge$_2$S$_3$	1.13–1.01	38–53	0.39	44*
Sb$_2$S$_3$–Ge$_2$S$_3$	1.12–1.02	39–52	0.40	48*

*The data are obtained with the empirical relation for tetrahedral glasses $\tau \cong \tau_1 + k \cdot V$, where $\tau_1 = 0.28$ ns is the bulk positron lifetime and the coefficient $k = 0.0025$.

4.   Conclusions

Nanostructural features of binary As–Se(Te), pseudo-binary As$_2$S(Se)$_3$–Sb$_2$S(Se)$_3$, ternary stoichiometric As(Sb)$_2$S$_3$–GeS$_2$, and non-stoichiometric As(Sb)$_2$S$_3$–Ge$_2$S$_3$ chalcogenide glasses are characterized by strong composition dependences of the FSDP position and intensity. It was found that As cations are responsible for the FSDP occurrence in glass structures based on a pyramidal configuration, whereas Ge cations play a dominant role for glasses with a structure constituted by a mixture of pyramidal and tetrahedral configurations. The size of nanovoids obtained from FSDP and PALS data show a good agreement, their type changes from three atomic vacancies for P-glasses to di- and/or mono-vacancies for T- and P–T-glasses.

ACKNOWLEDGMENTS

The authors are indebted to Pál Jóvári (Research Institute for Solid State Physics and Optics, Budapest, Hungary) for the help with experimental data treatment. T.K. gratefully acknowledges a grant by INTAS for Young Scientist Fellowship (Ref. No. 05-109-5323) and his stay in Chemnitz University of Technology (Chemnitz, Germany) and Jan Dlugosz University of Czestochowa (Czestochowa, Poland). T.K. and I.K. also acknowledge Deutsches Elektronen-Synchrotron DESY for support of the experiments performed at HASYLAB (Hamburg, Germany).

References

1. S.R. Elliott, *Phys. Rev. Lett.* 67(6), 711–714 (1991).
2. T.S. Kavetskyy and O.I. Shpotyuk, *J. Optoelectron. Adv. Mater.* 7(5), 2267–2273 (2005).
3. O. Shpotyuk, A. Kozdras, T. Kavetskyy, and J. Filipecki, *J. Non-Cryst. Solids* 352, 700-703 (2006).
4. K.O. Jensen, P.S. Salmon, I.T. Penfold, and P.G. Coleman, *J. Non-Cryst. Solids* 170, 57–64 (1994).
5. T.E.M. Staab, M. Haugk, A. Sieck, T. Frauenheim, and H.S. Leiper, *Physica B* 273–274, 501–504 (1999).
6. D.V. Makhov and L.J. Lewis, *Phys. Rev. Lett.* 92(25), 255504 (2004).
7. E. Bychkov, C.J. Benmore, and D.L. Price, *Phys. Rev. B* 72, 172107 (2005).
8. E. Bychkov and M. Miloshova, ISIS Experimental Report (Rutherford Appleton Laboratory) No 13701, 2003, <http://www.isis.rl.ac.uk/isis2003/reports/13701.PDF>.
9. L. Cervinka, *J. Non-Cryst. Solids* 106, 291–300 (1988).
10. K. Tanaka, *Phil. Mag. Lett.* 57(3), 183–187 (1988).
11. D.L. Price, S.C. Moss, R. Reijers, M-L Saboungi, and S. Susman, *J. Phys.C: Solid State Phys.* 21, L1069–L1072 (1988).
12. S. Susman, K.L. Volin, D.G. Mantague, and D.L. Price, *J. Non-Cryst. Solids* 125, 168–180 (1990).
13. O.K. Alekseeva, V.I. Mihajlov, and V.P. Shantarovich, *Phys. Status Solidi A* 48, K169–K173 (1978).
14. B.V. Kobrin and V.P. Shantarovich, *Phys. Status Solidi A* 83, 159–164 (1984).

HETEROGENEOUS STRUCTURES IN THE SYSTEMS

B_2O_3–MoO_3–CoO AND B_2O_3–MoO_3–Fe_2O_3

D. ILIEVA[*1], E. KASHCHIEVA[1], Y. DIMITRIEV[1],
R. YORDANOVA[2]

[1] Department of Physics, University of Chemical Technology
and Metallurgy, 8, Kl. Ohridski Blvd., 1756 Sofia
[2] Institute of General and Inorganic Chemistry, BAS, Bl. 11
Acad G. Bonchev St. 1113 Sofia

Abstract. The phase separation regions of the systems B_2O_3–MoO_3–CoO and B_2O_3–MoO_3–Fe_2O_3 were determined. The presence of different types of heterogeneous structures composed of glassy and crystalline phases was proven. In compositions located on the immiscibility-crystallization boundaries, complex micro-aggregates were observed by TEM. The crystalline phases were identified by XRD analysis. IR spectra of homogeneous glasses outside the immiscibility gap possess an intensive band at about 1100 cm^{-1} and bands with lower intensities around 860–680 and 470 cm^{-1}. It is suggested that the amorphous network is build up by non-associated BO_3 and BO_4 groups together with MoO_6 and CoO_6 units.

Keywords: glass; heterogeneity; metastable immiscibility; crystallization

1. Introduction

The study of the liquid phase separation in multi component systems allows to synthesize homogeneous transparent materials as well as to create compositions permitting in various degrees some additional initiation of immiscibility processes. It has been confirmed that the large range of metastable immiscibility in the binary system B_2O_3–MoO_3 influences the structural evolution in several three-component systems, such as B_2O_3–WO_3–MoO_3,

* darjailieva@abv.bg

J.P. Reithmaier et al. (eds.), *Nanostructured Materials for Advanced Technological Applications,* 371
© Springer Science + Business Media B.V. 2009

B_2O_3–CuO–MoO_3, and B_2O_3–MnO–MoO_3.[1] The boron atoms participate in the network mainly as BO_3 complexes, dominating in a wide concentration interval. Droplet-like and more complex microstructures are obtained during the cooling of the melts. The reason for their appearance is that transition atom ions do not stimulate the transformation of B_3O_6 (BO_3) to BO_4 units.[2–4] Probably the formation of BO_4 groups is coupled with a stabilization of the glass, the improvement of its chemical resistance and a decrease of the trend towards phase separation.[5, 6]

The purpose of the present work is to study the influence of CoO and Fe_2O_3 on the immiscibility gap and on the formation of heterogeneous structures. These formations give the opportunity to suppose that some selected compositions could be used as matrices for the production of new nanostructured materials.

2. Experimental

Bathes prepared from chemically pure H_3BO_3, MoO_3, Co_3O_4 and Fe_2O_3 were homogenized and melted in air in porcelain or alumina crucibles for 10–15 min at 1,200°C, the maximal melting temperature. Slow cooling with a rate of 100 K/min was performed by leaving the melts in the hot crucible outside the furnace in order to allow the development of the phase separation process. The nano- and micro-scale heterogeneous formations were observed by a Philips TEM-EM-400 using the C + Pt replica method from freshly fractured glassy surfaces. The structure of the glasses was investigated by IR spectroscopy with a Nicolet-320 FTIR spectrometer. The XRD analysis was done with a Bruker D5005 diffractometer with Cu Kα radiation.

3. Results and Discussion

Figure 1 shows the Gibbs triangles for the three-component systems B_2O_3–CoO–MoO_3 and B_2O_3–Fe_2O_3–MoO_3. In the system containing CoO, two immiscibility regions superpose: the first one is the metastable phase separation in the binary system B_2O_3–MoO_3, the other one the stable phase separation in the system B_2O_3–CoO. For this reason, a large number of compositions are included in the liquid phase separation range of the ternary system. At slow cooling of the melts, stable glasses were obtained near the B_2O_3–CoO side at about 30–40 mol % B_2O_3 and 60–70 mol % CoO. The crystallization ability increases at a MoO_3 content above 20 mol %.

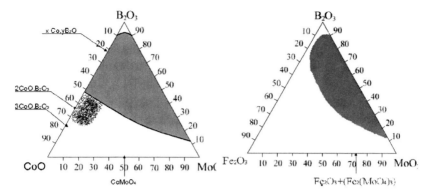

Figure 1. Left: Gibbs triangle of the B$_2$O$_3$–CoO–MoO$_3$ system. Right: Gibbs triangle of the B$_2$O$_3$–Fe$_2$O$_3$–MoO$_3$ system.

No glasses were obtained in the system B$_2$O$_3$–Fe$_2$O$_3$–MoO$_3$. In this case, the phase diagram of the binary system B$_2$O$_3$–MoO$_3$ plays also an important role, and phase separation is found for many samples. The gap includes compositions containing between 25 and 90 mol % B$_2$O$_3$ and between 5 and 80 mol % MoO$_3$. Increase of the Fe$_2$O$_3$ content leads to an increase of the melting temperature. TEM investigations show different cases of micro-heterogeneous structures, related to metastable and stable immiscibility and crystallization processes. The structural evolution was followed for samples with a constant MoO$_3$ content of 10 mol %. In Figure 2 (left) the micro-structure of a 70B$_2$O$_3$,10MoO$_3$,20CoO two-phase glass is illustrated as corroded amorphous matrix with glassy immiscibility drops. A sample with a composition of 60B$_2$O$_3$,10MoO$_3$,30CoO (not shown graphically) pos-sesses also two layers. The lower one is amorphous with several dispersed heterogeneities. The upper layer is hydrated with an additional droplet-like micro-phase separation.

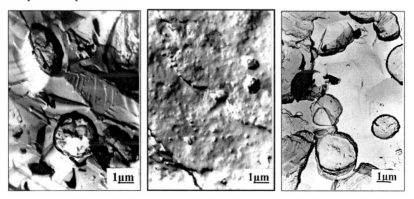

Figure 2. TEM micrographs of left: 70B$_2$O$_3$,10MoO$_3$,20CoO; middle: 40B$_2$O$_3$,20MoO$_3$, 40CoO; right: 30B$_2$O$_3$,30MoO$_3$,40CoO.

The glass-formation region could be reached by following the same line, i.e. decreasing the B_2O_3 content. Visually the samples are homogeneous black glasses. It is interesting to follow the microheterogeneous structures for compositions at the immiscibility-crystallization boundary as such the selected sample with a composition of $40B_2O_3,20MoO_3,40CoO$ (Figure 2, middle). The matrix contains fine-scale compact crystals, whereas droplet-like immiscibility takes place during the melt cooling of a $30B_2O_3,30MoO_3,40$ CoO sample (Figure 2, left).

The drops are rich in B_2O_3. TEM studies of the $B_2O_3-Fe_2O_3-MoO_3$ samples show crystalline formations, dispersed in the borate matrix, which are formed even at lower B_2O_3-contents, about 30–40 mol %. The samples situated inside the phase separation range, show similar microheterogeneous structures (Figures 3, 4).

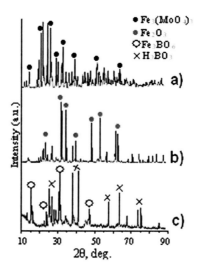

Figure 3. XRD patterns of crystallized samples: (a) $20B_2O_3,70\ MoO_3,10\ Fe_2O_3$; (b) $40B_2O_3,10\ MoO_3,50\ Fe_2O_3$; (c) $70B_2O_3,10\ MoO_3,20\ Fe_2O_3$.

XRD analysis was performed for selected compositions. The results confirm the visual and microscopic observations for the evolution of the crystallization processes. Orthorhombic crystals of Fe_3BO_6 were detected in a sample with a composition of $70B_2O_3,10MoO_3,20Fe_2O_3$, combined with a high H_3BO_3 content. At higher MoO_3 contents, the sample with composition of $20B_2O_3,70MoO_3,10Fe_2O_3$ is separated into two layers: the upper one is rich in B_2O_3, while the lower is crystalline with $Fe_2(MoO_4)_3$ as main crystal phase without the presence of H_3BO_3.

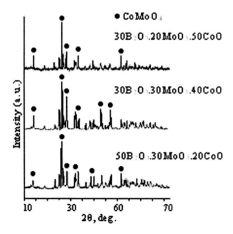

Figure 4. XRD patterns of crystallized samples in the B_2O_3–CoO–MoO_3 system.

The main crystalline phase in the system with the participation of CoO is $CoMoO_4$. Boric acid (H_3BO_3) is not found in materials containing less than 60 mol % B_2O_3, such as $30B_2O_3,30MoO_3,40CoO$; $50B_2O_3,30MoO_3$, 20CoO and $30B_2O_3,20MoO_3,50CoO$. IR spectra (Figure 5) show strong bands at 1,200–1,120, 880, 640 and 550 cm^{-1}, which are typical for H_3BO_3 for the sample with a composition of $70B_2O_3,10MoO_3,20CoO$. Their intensity decreases with increasing CoO content. The absorption maximum at 1200 cm^{-1} (550 cm^{-1}) completely disappears for homogeneous glasses ($40B_2O_3$, $10MoO_3,50CoO$ and $30B_2O_3,10MoO_3,60CoO$) outside the immiscibility gap which is related with the higher corrosion stability of CoO-riched glasses. Their characteristic bands with high intensities are located at about 1,100 cm^{-1} same and with smaller intensities, at about 880, 680 and 470 cm^{-1}. It is thus suggested that the amorphous network is build up by non-associated BO_3 and BO_4 groups, and MoO_6 and CoO_6 units.

4. Conclusions

Different heterogeneous structures are observed in the established range of liquid phase separation in the systems B_2O_3–MoO_3–CoO and B_2O_3–MoO_3–Fe_2O_3. One phase glasses are obtained for slow cooling of the melts in the system B_2O_3–MoO_3–CoO between 30–40 mol % B_2O_3 and 50–60 mol % CoO or 10 mol % MoO_3. CoO does not stimulate the formation of superstructural units.

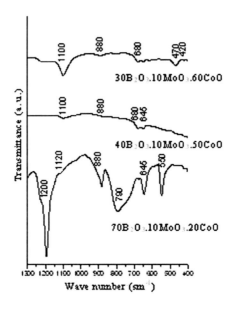

Figure 5. IR-spectra of glasses in the B_2O_3–CoO–MoO_3 system.

ACKNOWLEDGMENTS

The work was performed with the financial support of the Bulgarian National Scientific Foundation Contract BY-TH-102/2005 (806-UCTM) and by UCTM – Contract 10520/2008.

References

1. Y. Dimitriev, E. Kashchieva, D. Ilieva, R. Iordanova, Phys. Chem. Glasses 47, 435 (2006).
2. Y. Dimitriev, E. Kashchieva, R. Iordanova, G. Tyuliev, Phys. Chem. Glasses 44, 155 (2003).
3. E. Kashchieva, M. Pankova, Y. Dimitriev, Ceramics Silikaty 3, 111 (2001).
4. R. Iordanova, Y. Dimitriev, E. Kashchieva, D. Klissurski, Ceramics Silikaty 3, 115 (2001).
5. E. Milyukov, S. Kasymova, *Immiscible Melts and Glasses* (Fan, Tashkent, 1981) (in Russia).
6. B. Varshal, *Two Phase Glasses: Structure, Properties and Application* (Nauka, Leningrad, 1991) (in Russian).

RARE-EARTH DOPED FLUOROCHLOROZIRCONATE (FCZ) GLASSES AND GLASS-CERAMICS: SELECTED THERMAL PROPERTIES AND X-RAY LUMINESCENCE IN SAMARIUM DOPED FCZ

D. TONCHEV[*1], G. BELEV[1], S. PANIGRAHI[1], C. VAROY[2],
A. EDGAR[2], H. VON SEGGERN[3], S.O. KASAP[1]
[1]*Department of Electrical and Computer Engineering,
University of Saskatchewan, Saskatoon, SK, S7N 5A9, Canada*
[2]*School of Chemical and Physical Sciences, Victoria
University of Wellington, Wellington, New Zealand*
[3]*Institute of Materials Science, Technische Universität
Darmstadt, D-64287 Darmstadt, Germany*

Abstract: Recently, a new class of rare-earth (RE) doped fluorochloro-zirconate (FCZ) glass-ceramics has been developed and studied with the goal of using these materials in indirect conversion x-ray imaging applications. We report selected thermal and x-ray induced luminescence (XL) properties of rare-earth (EuF$_2$ and SmF$_3$) doped fluorochlorozirconate glasses and glass-ceramics, and highlight some of their interesting properties. Sm^{3+}-doped FCZ glasses exhibit XL even in the unannealed state, while annealing at a temperature of 250°C in a reducing atmosphere (5% H$_2$ and 95% Ar) significantly enhances the XL efficiency.

Keywords: rare-earth doped fluorochlorozirconate glass-ceramics; nanocrystals; thermal properties; DSC; TMDSC; X-ray digital imaging

1. Introduction

Most of the currently used indirect digital X-ray detector systems utilize crystalline materials that exhibit storage or scintillation effects. Such detector materials have relatively high energy conversion efficiencies but their

—————
[*] Dan.Tonchev@Usask.Ca

J.P. Reithmaier et al. (eds.), *Nanostructured Materials for Advanced Technological Applications*,   377
© Springer Science + Business Media B.V. 2009

spatial resolution is limited due to lateral spread of light during the readout process. Recently, a new class of rare-earth (RE) doped fluorochlorozirconate glass-ceramics has been developed and studied with the goal of improving the spatial resolution of X-ray images. It has been already shown that the image resolution in such doped glass-ceramics can exceed the resolution of crystalline materials by a factor of 5, when used as a storage phosphor, and at least by a factor of 40 for scintillation purposes.[1,2]

The X-ray luminescence in the most extensively studied glass-ceramic system with Eu^{2+} doping depends strongly on the annealing temperature of these glasses. As was shown in previous papers, the annealing of the glass encourages the formation of $BaCl_2$ nanocrystals in the glass matrix and the luminescence of Eu^{2+} ions in $BaCl_2$ crystals is responsible for both, scintillation and storage phosphor properties.[3] It is therefore necessary to have a good knowledge of the critical temperatures, where phase transformations take place in this glass system, as a function of composition.

In this work, we have examined the glass transition, crystallization and heat capacity C_p of rare-earth doped FCZ glasses by conventional differential scanning calorimetry (DSC) and temperature modulated DSC (TMDSC) experiments. We have used the TMDSC method to obtain more precise results for the glass transition phenomena since it is well known that this technique has a number of distinct advantages, especially in the study of glass transitions in various glasses.[4,5]

We have found that, in contrast to Eu-doped FCZ glasses, Sm-doped FCZ glasses show no storage phosphor effect. They exhibit x-ray luminescence from the excitation of Sm^{3+} ions even without annealing the glass to form previously identified $BaCl_2$ nanocrystals. However, annealing the glass at a temperature that is slightly higher than the temperature of the first of the two crystallization peaks appearing in the thermograms of the material and in a reducing gas atmosphere (5% H_2 and 95% Ar) improves the XL by a factor of almost 7.

2. Experimental Procedure

FCZ glasses with a typical composition of 53.5% ZrF_4 + 3% LaF_3 + 3% AlF_3 + 20% NaF + 0.3% InF_3 + 20% (BaF_2 and $BaCl_2$) + 0.2% (EuF_2 or SmF_3) (in molar percentages) have been prepared by conventional melt-quenching techniques.[6] Different rare-earth concentrations were achieved by substituting rare-earth fluorides for the appropriate quantity of BaF_2. The substitution of BaF_2 by $BaCl_2$ allows us to vary the relative amount of Cl R_{Cl}, which is defined as the ratio of the Cl concentration C_{Cl} to the overall concentration of halogens, i.e. $R_{Cl} = C_{Cl}/(C_{Cl} + C_F)$. Commercial grade high purity SmF_3 and EuF_2 were used as dopants. The amount of rare earth doping was in the range of 0.05–2 mol %. DSC and TMDSC experiments were performed as described previously[4] with DSC Q100 and 2910 thermal analysis systems from TA Instruments. Heat capacity (reversing heat flow) measurements were carried out by running the TMDSC experiments with an underlying heating rate of 1 K/min while using a modulation amplitude of $\pm 1°C$ and a modulation period of 60 s.

The set-up used for X-ray luminescence measurements is presented in Figure 1. The X-ray source used in this work is a standard GENDEX, GX-1000 dental unit. The X-ray tube has a tungsten anode and a 2.7 mm Al filter. The sample under investigation is positioned in a 2 in. PTFE integrating sphere (IS200, Thorlabs), which collects all light emitted from the sample and couples it into a PMMA optical fiber of 2 mm diameter. The light exiting the fiber is coupled through a collimator to a photomultiplier tube (PMT, R508, Hamamatsu) that covers the spectral range of 200–850 nm. The detector arrangement allows the insertion of a wide variety of optical filters between the end of the fiber and the photomultiplier tube. The Keithley 6514 electrometer allows measurements in current and charge mode. We use a Tektronix TDS 210 digital oscilloscope connected to the current pre-amplifier of the electrometer to record the fast changes in the photocurrent. All the equipment is controlled by a personal computer through an IEEE 488 interface (GPIB). The X-ray tube and the integrating sphere are mounted in a lead chamber. The throughput of the system was calibrated using a set of light emitting diodes in the range of 430–680 nm. The latter

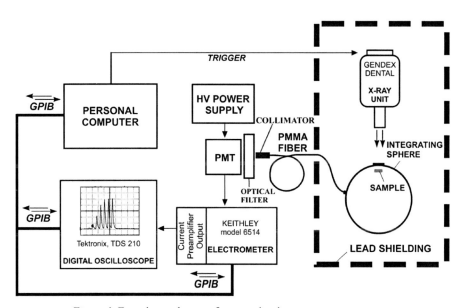

Figure 1. Experimental set-up for x-ray luminescence measurements.

were coupled to an optical fiber; the other end of which was positioned in the place of the sample. The actual power entering the integrating sphere through the optical fiber during the calibration procedure was measured with an Industrial Fiber Optics power meter.

3. Results and Discussion

Figure 2 shows typical conventional DSC heating scans for FCZ samples up to 320°C for different relative Cl contents, all doped with 0.2% SmF_3. The SmF_3–FCZ glasses with R_{Cl} = 0.05 and 0.07 are characterized by a single, wide, exothermic crystallization peak in the temperature range up to 350°C.

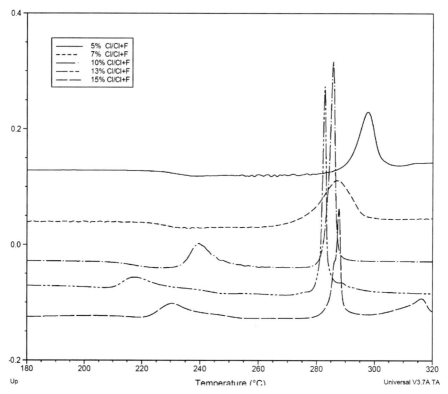

Figure 2. DSC thermograms of 0.2% SmF_3-doped FCZ glasses with different amounts of halogens.

The latter glasses have been observed in TMDSC scans to exhibit a clear single glass transition phenomenon as a sigmoidal step-like change in the heat capacity C_p from the reversing heat flow; one can also observe the glass transition for these two glasses in the conventional DSC in Figure 2 in the range of 230–250°C. The remaining FCZ samples, corresponding to

R_{Cl} = 10%, 13% and 15%, evince two crystallization peaks. The first has a smaller enthalpy and occurs at a lower temperature, while the second, which is rather narrow, has a larger enthalpy and occurs at a higher temperature. (There are other exothermic and endothermic peaks at higher temperatures, which are not discussed in this paper.) The glass transition events in these glasses are somewhat obscured by the crystallization phenomena. The DSC thermograms for EuF_2 doped FCZ glasses are similar and not shown.

Figure 3 shows two TMDSC thermograms for R_{Cl} = 5% and 13%, in which C_p is shown as a function of the sample temperature. For the R_{Cl} = 5% sample, there is only one glass transition at 242°C. For the R_{Cl} = 13% sample, one can clearly identify two sigmoidal-type step-like changes of C_p, one at a lower temperature (211°C) just before the $BaCl_2$ crystals formation and the other at a higher temperature (251°C). The low-temperature C_p change is followed immediately by a slight drop of C_p, where $BaCl_2$ crystals formation takes place.[3] Glasses with R_{Cl} = 10% and 15% show a similar behavior.

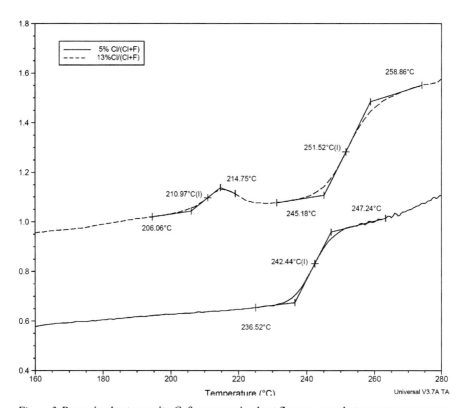

Figure 3. Reversing heat capacity C_p from reversing heat flow vs. sample temperature measurements, where (I) (as in 210.97°C (I)) represents the glass transition temperature from the inflection point in the C_p curve in the T_g-region.

Both, the low and high-temperature C_p transitions for the 13% sample in Figure 3 are characteristic features of glass transition phenomena, which are typically observed for FCZ glasses. Further, the low-temperature C_p transition occurs at a temperature that is expected from an extrapolation of the T_g vs R_{Cl} behavior from the values for $R_{Cl} = 5\%$ and 7% to higher R_{Cl} values. The observation of two possible glass transitions points at the possibility of having two phases in the glasses with higher relative Cl contents. Similarly separated C_p changes in TMDSC thermograms have been used to identify different phases in other glasses.[7] Indeed, Edgar et al., using molecular dynamics,[8,9] have proposed that FCZ glasses with these compositions can be phase separated. The heat capacity vs. composition dependence (not shown) exhibits a distinct change of its behavior. A significant drop of C_p is observed when going from a relative Cl content of 7–10%, which is also indicative of a significant change of the structure around this composition. Further studies of phase separations in this glass system would be useful to confirm the present observations. In particular, it would be useful to rule out that the first step-like change in C_p vs. T for the 13% sample in Figure 3 is not simply due to a change of the heat capacity or a measurement artifact, due to the formation of $BaCl_2$ crystallites in the glass matrix.

Figure 4 shows the dependence of the crystallization temperature T_c, defined (in this work) as the temperature at which the crystallization exotherm reaches its maximum, on the Cl content. The crystallization temperature of

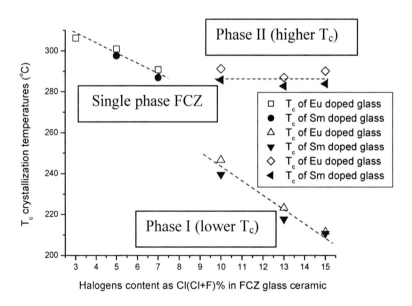

Figure 4. Crystallization temperatures vs. the relative Cl content R_{Cl}.

the bulk glass initially decreases with R_{Cl} and then remains constant in the two-phase region, whereas the crystallization temperature related to the formation of $BaCl_2$ crystals decreases with R_{Cl}. Figures 3 and 4 together are useful in choosing the annealing temperature for the formation of nanocrystals in these rare-earth doped FCZ glasses.

We have also examined the dependence of the heat capacity and crystallization temperature (of both phases) on the rare-earth content, and found very little dependence. The heat capacity tends to decrease slightly, the crystallization temperature for the bulk glass to increase slightly with the EuF_2 and SmF_3 content. The crystallization temperature related to the formation of $BaCl_2$ crystals shows more scattering but, within the experimental errors, there was no significant dependence on the rare-earth content.

The luminescence properties of FCZ:Eu glasses have been extensively studied in the past (e.g. Ref. 10) and will not be discussed further in this work. On the other hand, there is not much data on Sm-doped FCZ glasses. We have examined the X-ray luminescence in FCZ:Sm glasses with a relative Cl content of 13%. This choice was based on previous works[1,2] where it was shown that significant storage phosphor and scintillation effects were observed in this concentration range for Eu-doped FCZs. It can be seen from Figure 2 that when the relative Cl content is lower than 10%, the FCZ glasses do not exhibit the $BaCl_2$ crystallization exotherm. Further, glasses with high Cl contents tend to be more hygroscopic. Thus the sample with a relative Cl content of 13% seems a reasonable choice.

The optical absorption spectrum of Sm-doped FCZ glasses shows typical Sm^{3+} absorption peaks, which suggests that the Sm ions in the glass are mainly in the Sm^{3+} state. The absorption spectrum does not change with the Sm content nor with annealing at 250°C in N_2 and H_2/Ar gas atmospheres, which was quite surprising since FCZ:Eu needs to be annealed to form $BaCl_2$ nanocrystals to activate the Eu^{2+}-ions. (Eu^{2+} ions in the glass matrix do not fluoresce[11]).

We have examined the x-ray induced luminescence (XL) of FCZ:Sm as a function of the Sm-content. Our results show that FCZ:Sm glasses work in the pure scintillator mode inasmuch as we were unable to observe any storage effect in these materials; unlike FCZ:Eu glasses which can be made to function as a storage phosphor or as a scintillator, depending on the annealing temperature, which determines the $BaCl_2$ crystals formed.[3]

Figure 5 shows the dependence of the XL yield (total collected charge per unit exposure) on the Sm-content. The measured XL yield is the integration of the photodetector current over all the light emitted in the four major PL peaks of the Sm^{3+} ion (562, 596, 642 and 706 nm). (There are two minor peaks at 898 and 942 nm, which were excluded and do not significantly affect the XL yield and efficiency measurements.) The dependence of the XL yield on the Sm content is not linear, especially at high Sm concentrations. Although it may be possible to attribute the origin of the non-linearity to luminescence quenching,[12] the optically pumped luminescence yield depends linearly on the Sm content. For a given amount of Sm-doping, the XL yield was found to depend linearly on the x-ray exposure up to the highest exposures used, 5 R, which is an essential property of a good scintillator. The decay of the XL signal upon cessation of the excitation provides a means of determining the XL decay time, which was estimated to be ~4 ms.

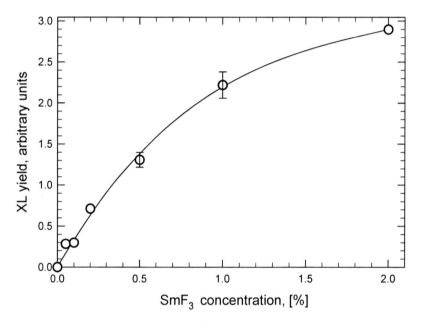

Figure 5. X-ray luminescence signal (XL) from as-prepared FCZ glasses doped with different amount of samarium (Sm^{3+} was introduced as SmF_3).

FCZ:Eu glass phosphors depend on suitable annealing of the glass and the formation of $BaCl_2$ nanocrystals for the activation of the Eu^{2+} ions. We have also examined the effects of annealing on the XL yield for FCZ:Sm glasses. The annealing was done at 250°C for 30 min since, at this temperature, according to Figure 3 and Ref. 3 $BaCl_2$ nanocrystals should have formed in the sample over this duration of time. 250°C is around the high-temperature glass transition region (Figure 3) but below the bulk glass crystallization temperature (Figure 4). Annealing the FCZ:Sm samples in N_2 atmosphere only increased the XL yield by a factor of about 1.6, while annealing in a reducing gas atmosphere (5% H_2 + 95% Ar) increased the yield by a factor of nearly ~7, a very significant increase in the XL yield. While the formation of $BaCl_2$ nanocrystals is essential in FCZ:Eu glasses to activate the Eu-ions, in FCZ:Sm glasses, on the other hand, annealing seems to enhance the XL by suitably eliminating non-radiative recombination paths (e.g. reducing defects) or by additional activation of more Sm^{3+} ions in the glass-ceramic. The fact that there is no marked change in the optically excited PL spectra with annealing, and given the large increase in the XL yield in an H_2 containing atmosphere, it is more likely that annealing diminishes non-radiative paths competing with the luminescent decay. Further work is needed to clarify the exact nature of the effects of annealing.

We have also measured the absolute energy conversion efficiency of a FCZ glass doped with 0.2% Sm, which had been annealed in an H_2/Ar atmosphere for 30 min at 250°C, and found this efficiency to be about 0.16%. This value is typical for glass scintillators and, of course, substantially lower than those of 10–20% observed for crystalline scintillators. However, activator doped glass based scintillators provide other advantages, the most important of which are the possibility of an increased resolution and a tailoring of the activators and the matrix composition for dual energy imaging.

4. Summary and Conclusions

We have reported selected thermal and x-ray induced luminescence properties of rare-earth (Eu and Sm) doped fluorochlorozirconate glasses, in which the relative Cl content (with respect to the total halogen content) was varied. The FCZ glasses with a relative Cl content of 5% and 7% exhibit a single glass transition and crystallization peak and are likely to be single-phased. The glasses with a relative Cl content of 5%, 13% and 15% exhibit two crystallization peaks; the lower temperature crystallization peak is strongly related to the formation of $BaCl_2$ nanocrystals. In addition, the latter glasses also most probably exhibit two glass transition regions as observed in reversing heat flow TMDSC scans. The glasses with a relative Cl content of 5%, 13% and 15% may be two-phase glasses. The thermal properties are not significantly affected by the addition of 2% EuF_2 or SmF_3. In contrast to FCZ:Eu glass-ceramics, FCZ:Sm glasses and glass-ceramics with a relative Cl content of 13% do not exhibit any storage phosphor effect. Further, FCZ:Sm glasses evince X-ray induced luminescence arising from the incorporation of Sm as Sm^{3+} in the glass even in the unannealed state. Annealing FCZ:Sm in a reducing atmosphere (5% H_2 and 95% Ar) at 250°C for 30 min, above the lower of the two crystallization temperatures and in the higher temperature glass transition region, increases the XL significantly, nearly by a factor of 7. The conversion efficiency of the FZC:0.2% Sm material annealed under the conditions above described was measured to be about 0.16%.

ACKNOWLEDGEMENTS

The authors are grateful to the New Zealand Foundation for Research, Science and Technology, and the Natural Sciences and Engineering Research Council (NSERC) for financial support.

References

1. Edgar, G.V.M. Williams, S. Schweizer, J.-M. Spaeth, Curr. Appl. Phys. 6, 399 (2006).
2. G. Chen, J. Johnson, R. Weber, R. Nishikawa, S. Schweizer, P. Newman, D. MacFarlane, J. Non-Cryst. Solids 352, 610 (2006).
3. S. Schweizer, L. Hobbs, M. Secu, J.-M. Spaeth, A. Edgar, G.V.M. Williams, J. Hamlin, J. Appl. Phys. 97, 083522 (2005).
4. S.O. Kasap, D. Tonchev, J. Mater. Res. 16, 2399 (2001).
5. D. Tonchev, S.O. Kasap, in: C.A. Brebbia and W.P. De Wilde (Eds.), *High Performance Structures and Materials II* (WIT Press, Southampton, UK, 2004), ISBN: 1-85312-717-5, 223.
6. Edgar, G.V.M. Williams, M. Secu, S. Schweizer, J.-M. Spaeth, Radiat. Meas. 38, 413, (2004).
7. T. Wagner, S.O. Kasap, M. Vlcek, A. Sklenar, A. Stronski, J. Mater. Sci. 33, 5581 (1998).
8. S.C. Hendy, A. Edgar, J. Non-Cryst. Solids 352, 415 (2006).
9. A. Edgar, S.C. Hendy, D.A. Bayliss, Phys. Chem. G.: Eur. J. Glass. Sci. Technol. A, 47, 254 (2006).
10. S. Hobbs, L., Secu, M., Spaeth, J.-M., Edgar, A., G.V.M. Williams, App. Phys. Lett. 83, 449 (2003).
11. D.R. MacFarlane., P.J. Newman., J.D. Cashion, A. Edgar, J. Non-Cryst. Solids 256–257, 53 (1999).
12. P. Hackenschmied, H. Lia, E. Epelbaum, R. Fasbender, M. Batentschuk, A. Winnacker, Radiat. Meas. 33, 669 (2001).

6. APPLICATIONS OF NANOSTRUCTURED MATERIALS

6.1. ELECTROCHEMICAL APPLICATIONS

MIXED NANO-SCALED ELECTRODE MATERIALS
FOR HYDROGEN EVOLUTION

P. PAUNOVIĆ[*]
*University St. Cyril and Methodius, 1000 Skopje, Faculty
of Technology and Metallurgy, FYR Macedonia*

Abstract. The role and importance of a hydrogen economy is given within the energetic and environmental problems facing mankind. The electrocatalytic activity of metals is quoted as well as physical and chemical pathways to improve the performance of electrode materials. Some characteristics of advanced nano-structured hypo-hyper *d*-electrocatalysts deposited on multi-walled carbon nanotubes (MWCNTs) are shown. Non-platinum electrocatalysts approach the activity of the conventional Pt/Vulcan XC-72, while one containing only 20% Pt in the metallic phase exceeds it by far.

Keywords: hydrogen economy; hypo-hyper *d*-electrocatalysts; nano-structured materials; multiwalled carbon nanotubes MWCNTs

1. Introduction

Fossil fuels were and still are the most convenient and available energy sources. But the unclosed loop of their exploitation, i.e. extraction, transportation, processing and use, makes them an environmentally unfriendly and unsustainable source of energy. The products of fossil fuels combustion, CO_2 and CO, do not revert to the source, but instead, go into the atmosphere, accompanied by other gases, and cause serious pollution. This pollution is worldwide known as "green-house effect"; its impact on environment and humanity is well-known.[1] On the other side, the non-renewable nature and enormous consumption rate of fossil fuels contribute to their rapid total exhaust. The final price of the present life-quality, including many material and technological goods and pleasures, is global environmental disturbance

---

*[*]pericap@tmf.ukim.edu.mk

J.P. Reithmaier et al. (eds.), *Nanostructured Materials for Advanced Technological Applications,*   391
© Springer Science + Business Media B.V. 2009

with long-term consequences as well as total exhausting of the most convenient source of energy, fossil fuels. So, our civilization will be facing serious energy crisis in near future.

As result of this situation, modern science is facing the challenge to find alternative sources of energy which can satisfy the global energy needs, but are also environmentally friendly. The question therefore is: **what fuel will be the main energy source in the future?** The answer can be found by an analysis of the history of fossil fuels with respect to the hydrogen to carbon ratio (HCR). A Rise of hydrogen contribution in fuels is obvious: starting with wood which is low in carbon content, coal with HCR near 50:50, oil or gasoline with 67:33 and finally natural gas with HCR of 75:25. So, further extrapolation reveals that pure **hydrogen is a logical alternative**. In this context, in the last few decades, a hydrogen economy as a system of hydrogen production/conversion to energy has been shown perspective.

The concept **"hydrogen economy"** is a closed loop of hydrogen production, transportation to energy centers or to electricity conversion sites, and conversion to electricity by fuel cells. The term "hydrogen economy" was coined by J. O'M. Bockris in the early seventies of the last century.[2] The main goal of its development is the establishment of a global energy system, in which hydrogen will be produced from available renewable energy sources, converted to electricity by fuel cells and further applied in transportation, residential, industrial and other sectors instead of fossil fuels. In the last few decades, the growth of the hydrogen economy system affected fast development of many scientific fields and mobilized a large part of the world scientific community. Hydrogen economy is a **interdisciplinary scientific and engineering branch** that includes electrochemistry, particularly electrocatalysis (investigation of hydrogen evolution reaction/oxidation in different media), material science (producing advanced electrode materials for hydrogen evolution/oxidation reactions), polymer science (development of polymer membranes as solid electrolytes for hydrogen electrolysers/fuel cells), mechanical engineering (design of appropriate electrolysers/fuel cells), etc. Recently, the globally most attractive scientific area are nanomaterials and nanotechnologies (development of nanotechnologies for producing nano-scaled electrode materials for hydrogen evolution/oxidation, using carbon nanotubes as support material, etc.).

Hydrogen is not only fuel, but also energy carrier of the future which could provide a painless transition from fossil fuels to the "hydrogen era". As a fuel, hydrogen has many advantages as compared to fossil fuels, as: (i) a clean fuel with high a caloric value (100% H_2); (ii) a closed loop fuel vs. product: H_2O is the source for production but also the product of combustion; (iii) production from/to electricity with high efficiency (50–60%); (iv) conversion to energy by different ways (combustion, electrochemical conversion, hydriding); (v) storage and transport possible in all aggregate states (gas, liquid and metal-hydrides) and (vi) environmental friendliness.

One of the most important parts of the "hydrogen economy" loop are electrochemical hydrogen generation and conversion to energy. So, the electrode material on which hydrogen evolution/oxidation occurs is very important, since it can reduce the cost of hydrogen production or increase the efficiency of conversion to electricity. The higher the catalytic activity of the electrode material, the lower the overpotential and the energy consumption and finally the more cost-effective the hydrogen production/conversion.

2.   Enhancing the Electrocatalytic Activity of Pure Metals

A heterogeneous catalytic process includes adsorption phenomena and the formation of intermediates between reacting species and substrate atoms/ ions. Thus, the adsorptive characteristics of electrode materials are of great importance for their electrocatalytic behavior. According to P. Sabatier's principle,[3] an electrode material shows optimal electrocatalytic activity, when the adsorption strength of the electrode to H-adatom has an intermediate value. A high strength implies a high activation energy for further electrochemical transformation and desorption, and is not favorable for the purpose. A weaker bond leads to premature desorption of the intermediar from the electrode surface, so no electrochemical transformation occurs.

The best individual catalysts for hydrogen evolution reactions (HER) are Pt, Pd and Ni with an average electronic configuration of d.[8] But is this the maximum that could be achieved in electrocatalysis for HER? The performance of pure bulk metals is not sufficient for cost-effective hydrogen evolution/oxidation. Is it possible to enhance the electrocatalytic properties of individual metals? How to achieve that? The answer to this question is confirmative. There are two basic approaches to improve the electrocatalytic activity of pure bulk metals. The first is based on an increase of the real surface area of the electrode material (physical approach) where the rise of the catalytic activity is a result of the so called "size-effect". The second one alludes an increase of the intrinsic activity through alloying of metals (chemical approach), the corresponding effect is called "intrinsic effect".

2.1.   PHYSICAL APPROACH

In this context, the question arises how to increase the electrode activity, i.e. the hydrogen electrolyser/fuel cell efficiency, using the same quantity of some metal, e.g. platinum. The unique path to solve this problem was to increase the real surface area of the electrode vs. its geometric surface (roughness factor S_R/S_G) or vs. its mass (specific surface area, m^2g^{-1}).

So, involving a porous medium as electrode provides a considerable increase (10^3–10^4 times) of the limiting current density i_L and consequently of the power density. Improving these parameters makes fuel cells practically useful. In porous electrodes the diffusion layer thickness is 10^{-7} to 10^{-4} cm instead of 10^{-2} cm of a planar electrode. Involving porous electrodes for successful and economical work of fuel cells represents the beginning of **modern electrochemistry**.

Increase of the real surface area of porous electrodes can be achieved by reducing the grain size of the electrocatalytic material. So, a maximum surface area can be reached by using **nano-scaled electrocatalysts**. Thus, **nanotechnologies** are priceless tools of modern electrocatalysis.

In modern electrolysers, two-phase porous electrodes with proton exchange membranes (PEM) were introduced. These membranes have a bifunctional role: as a separator of cathodic and anodic areas, and as a solid electrolyte, which can selectively transport cations across the cell junction. Using polymer membranes, there is no concern connected with liquid electrolytes as e.g. corrosion, watertight compartment and safety issues. Also, the membranes are chemically resistant and long-term stable. Recently, the mostly used membranes are those based on a polyfluorosulfonic acid/PTFE copolymer, commercially named Nafion® PFSA.[4] Its operating temperature is below 100°C. In order to extent the operating temperatures even up to 350°C, polybenzimidazole (PBI) membranes can be used.[5]

2.2. CHEMICAL APPROACH

The need for active, stable and cheaper electrocatalysts motivated intensive research, resulting in the development of multicomponent catalysts. The attempt of M. Jakšić et al.[6] to alloy a good individual catalyst (Ni) with a metal behaving as a poor individual catalyst (Mo) resulted in a synergetic effect, the i.e. activity of the non-platinum catalyst produced was comparable to that of platinum. It was predicted and proven that a combination of *d*- metals with dissimilar electronic character exhibits a pronounced synergetic effect.[7] As a result, combinations of hyper *d*-electronic transition metals (good individual electrocatalysts) with hypo *d*-electronic transition metals (poor individual catalysts) become the subject of scientific and technical interest. A **"new era" of modern electrocatalysis** was opened.

The phenomena of synergetic effects of hypo-hyper *d*-electrocatalysts can be explained by Jakšić's interpretation of the Brewer resonance bond-valence theory of intermetallic compounds.[7] It offers a better understanding and thermodynamic prediction of the stability and properties of intermetallic compounds as most promising electrocatalytic materials for HER.

Hypo and hyper d-components may be not only in the elemental state, but also in higher oxidation states. Depending on the valence state of the hypo and hyper d-components, we can distinguish these electrocatalysts into five main groups as shown in Figure 1.

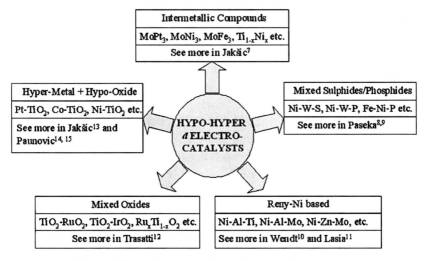

Figure 1. The main groups of hypo-hyper d-electrocatalysts.

3. Advanced Electrode Materials

The most widely used electrode material for hydrogen electrolysers/fuel cells is nano-scaled platinum deposited on nano-sized carbon black (Vulcan XC-72). But there are several barriers limiting the commercial application of Pt, e.g. the low abundance and the high costs. According to the analysis of Lee et al.,[16] the present resources are estimated to cover ~20% only of the automotive industry needs. What about the remaining 80%? What about future electricity needs that is considerably higher than in automotive industry? So, one of the imperatives of material scientists and electrochemists is to reduce or even to replace Pt from fuel cells/hydrogen electrolysers.

The main activity of the present author and his group in the last few years is the production, characterization and modification of **non-platinum hypo-hyper d-electrocatalysts** for hydrogen evolution/oxidation. Nano-scaled electrocatalysts were prepared by the solgel method. A metallic phase (10%) was deposited on carbon black (Vulcan XC-72). In order to improve the intrinsic catalytic activity, TiO_2 (anatase) was added as hypo d-phase. The hypo d-component has a bifunctional role, as a catalyst support (with the carbon phase) and by contributing to the catalysts overall synergetic effect by the so-called strong metal-support interaction (SMSI).

The unique role of the anatase form of titania in the improvement of the catalytic activity in both heterogeneous chemical catalysis and electrocatalysis has been noticed by other authors.[17,18]

As metal phase Ni and Co or CoNi (1:1 atomic ratio) were used. Why Co and Ni? Ni is a non-noble metal of the platinum group, while Co is near Ni in the periodic table, having similar properties. Instead of the conventional carbon Vulcan XC-72 substrate, multiwalled carbon nanotubes (MWCNTs) were involved. Their inner structure, geometry, surface area and conductive characteristics are favourable for use as a carbon substrate of nano-scaled electrocatalysts. The conductive properties enable an easier electron exchange with hydrogen protons, which intensifies the formation of adsorbed hydrogen atoms and further hydrogen molecules. Their highly developed surface area enables a better dispersion of active catalytic centres across the catalysts surface. On the other hand, the high **inter- and trans-particle nano porosity** of MWCNTa as well as their geometry (empty cylinders) facilitate the escape of hydrogen molecules from catalysts surface.

Figure 2a polarization curves of non-platinum electrocatalysts (10% Me + 18% TiO_2 + MWCNTs) as well as of a corresponding conventional platinum-based catalyst (10% Pt deposited on MWCNTs by the same sol–gel procedure) are shown. As result of the above modifications, the electrocatalytic activity of the Co-based catalyst approaches that of the platinum one. Ni-based electrocatalysts have shown the lowest activity. The large difference of the activity between Co- and Ni-based electrocatalyst is a result of the relatively large difference of the average particle size of the hyper d-metallic phase as the active catalytic centre. The intrinsic interaction between hyper and hypo d-phase is almost the same in both cases[14,15] as well as the specific surface area of the catalysts as a whole.[19] But the size of the Co metallic phase is ~2 nm, while that of the Ni particles is 15–20 nm. Thus, the surface area of the active catalytic centres is considerably higher

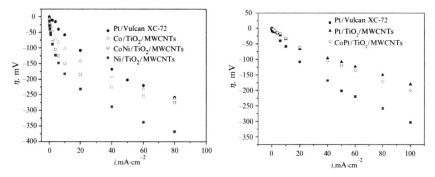

Figure 2. Polarization curves of mixed (a) non-platinum electrocatalysts (b) different Pt-based electrocatalysts.

in case of Co-based catalysts; also the distribution of the metallic phase over the catalyst surface is much better. In the case of the Ni-based catalyst, larger grains and their poor dispersion over the surface a cause lower surface area of the catalyst available for hydrogen evolution/oxidation. CoNi-based catalysts behave very similar to the systems containing Co only. XRD analysis[20] as well as the results presented by other authors[21] indicate that Co in this case exists in two different phases. A minor part forms a Ni–Co solid state solution while the rest continues to exist as pure (amorphous) Co, thus keeping the superior activity typical for the pure Co-system.

But the main goal was to produce superior non-platinum catalysts, exceeding the activity of the most widely used catalyst Pt/Vulcan XC-72. As shown above, the maximal activity of non-platinum (Co-based) catalysts approaches that of Pt-based ones, but does not exceed it. Therefore, a catalyst with a mixed metallic phase was prepared, containing some amount of platinum. Figure 2b shows polarization curves for a catalyst containing 10% CoPt (4:1, at.) and 18% anatase and MWCNTs, compared with corresponding ones of Pt/Vulcan XC-72 and a Pt-based catalyst with an analogous composition (10% Pt + 18% anatase + MWCNTs) produced under the same conditions as the previous one. The catalyst with the mixed metallic phase with only 20% Pt shows an impressive catalytic activity, exceeding that of conventional platinum by almost 100 mV at a reference current density of 60 mA·cm^{-2}. It is even more interesting that this catalyst shows a catalytic activity very close to that of the corresponding catalyst containing 100% Pt as metallic phase. This paradox can be explained by results of XRD and TEM investigations.[22] Namely, the size of the platinum particles in the catalyst with pure Pt as metallic phase is almost three times larger that that of the platinum particles mixed with Co. Thus, Co contributes to lower the size of platinum grains, i.e. to increase the surface area of the metallic phase as active catalytic centres, and consequently to increase the catalytic activity.

4. Conclusion

According to the above discussion it is obvious that hypo-hyper d-electrocatalysts (hypo-oxides + hyper-metals) are very promising in reducing or avoiding Pt in modern electrode materials. In this context a few conclusion can be noted: (i) MWCNTs have a considerable contribution as an extraordinary carbon support; (ii) TiO$_2$ as a hypo d-component has a bifunctional role: as a catalyst support and by increasing the intrinsic catalytic activity; and (iii) Co obtained by sol–gel synthesis using organometallic precursor has a high electrocatalytic activity due to its small grains (~2 nm). Co/TiO$_2$/MWCNTs catalysts mixed with very small amounts of Pt show a superior catalytic activity as compared with that of conventional Pt/Vulcan XC-72.

ACKNOWLEDGEMENT

This paper was supported by and carried out within the Bilateral Project of the Ministry of Education and Science of Macedonia with the Ministry of Science of Bulgaria.

Special thank to Academitian Evgeni Budevski and the staff of the "*Institute of Electrochemistry and Energy Systems*", Bulgarian Academy of Science, Professor Milan Jakšić, University of Belgrade and Professor Svetomir Hadži Jordanov, Faculty of Technology and Metallurgy, Skopje.

References

1. J. O'M. Bockris, T.N. Veziroglu, and D. Smith, *Solar Hydrogen Energy, The Power to Save the Earth* (Macdonald & Co., London, 1991).
2. J. O'M Bockris, Science 176, 1323 (1972).
3. P. Sabatier, *La Catalyse en Chemie Organique* (Librairie Polytechnique, Paris, 1913).
4. S. Banerjee, *Advances in Materials for Proton Exchange Membrane Fuel Cell Systems* (Asilomar Conference Grounds, Pacific Grove, CA, February 23–27 2003).
5. T. Zawodzinski, *Advances in Materials for Proton Exchange Membrane Fuel Cells Systems* (Asilomar Conference Grounds, Pacific Grove, CA, February 23–27 2003).
6. M. Jakšić, V. Komnenić, R. Atanasoski, and R. Adžić, Elektrohimiya, 13, 1355 (1977).
7. M.M. Jakšić, Int. J. Hydr. Energy 12, 727 (1987).
8. I. Paseka, Electrochim. Acta 40, 1633 (1995).
9. I. Paseka and J. Velicka, Electrochim. Acta 42, 237 (1997).
10. S. Rausch and H. Wendt, J. Electrochem. Soc. 143, 2852 (1996).
11. D. Miousse, A. Lasia, and V. Borck, J. Appl. Electrochem. 25, 592 (1995).
12. S. Trasatti, Electrochim. Acta 36, 225 (1991).
13. S.G. Neophytides, S.H. Zaferiatos, and M.M. Jakšić, Chem. Ind. 57, 368 (2003).
14. P. Paunović, O. Popovski, A.T. Dimitrov, D. Slavkov, E. Lefterova, and S. Hadži Jordanov, Electrochim. Acta 52, 1610 (2006).
15. P. Paunović, O. Popovski, A.T. Dimitrov, D. Slavkov, E. Lefterova, and S. Hadži Jordanov, Electrochim. Acta 52, 4640–4648 (2007).
16. K. Lee, J. Zhang, H. Wang, and D.P. Wilkinson, J. Appl. Electrochem. 36, 507 (2006).
17. S.G. Neophytides, S. Zaferiatos, G.D. Papakonstantnou, J.M. Jakšić, F.E. Paloukis, and M.M. Jakšić, Int. J. Hydr. Energy 30, 393 (2005).
18. S.J. Tauster and S.C. Fung, J. Catal. 55, 29 (1978).
19. S. Hadži Jordanov, P. Paunović, O. Popovski, A. Dimitrov, and D. Slavkov, Bull. Chem. Technol. Macedonia 23, 101 (2004).
20. P. Paunović, O. Popovski, S. Hadži Jordanov, A. Dimitrov, and D. Slavkov, J. Serb. Chem. Soc. 71, 149 (2006).
21. Y.L. Soo, G. Kioseoglou, S. Kim, Y.H. Kao, P. Sujatha Devi, J. Parise, R.J. Gambino, and P.I. Gouma, Appl. Phys. Lett. 81, 655 (2002).
22. P. Paunović, Unpublished data within the Bilateral Project: Catalysts and electrode structures for environmentally friendly electrochemical energy converters, with Bulgarian Academy of Sciences, 2006–2008.

SILICON/GRAPHITE NANO-STRUCTURED COMPOSITES FOR A HIGH EFFICIENCY LITHIUM-ION BATTERIES ANODE

T. STANKULOV [*], W. OBRETENOV, B. BANOV,
A. MOMCHILOV, A. TRIFONOVA
*Institute of Electrochemistry and Energy Systems,
Bulgarian Academy of Sciences, Acad. G. Bonchev St., bl. 10,
1113 Sofia, Bulgaria*

Abstract. Si/C composites have been employed in an attempt to overcome the problems of Si-based negative Li-ion electrodes. The composites were produced by coating, followed by a two-step solid state reaction. Electrodes were prepared therefrom by spreading on a Cu foil. The materials were characterized by SEM and XRD. The Si phase in the composite shows an enhanced crystalline structure compared to the pristine Si powder. Electrochemical cycling tests displayed two discharge plateaus at 0.250 and 0.1 V, where lithiated amorphous Si is formed and the entire crystalline phase is completely depleted. During the Li extraction the composite shows a distinct plateau at 0.45 V, where delithiated amorphous Si is formed. The behaviour of the composite was investigated in a 1M $LiPF_6$/EC:EMC (ethylene carbonate:ethyl methyl carbonate, 1:2) electrolyte with and without vinylene carbonate additives. The VC-based electrolyte exhibited a better efficiency, approaching 99.2% after the first few cycles and remaining constant in the next 30 cycles.

Keywords: nano-materials; silicon/carbon composite; anode; lithium ion batteries

1. Introduction

Silicon-based materials have been proposed as candidates to replace the conventional graphite negative electrodes for Li-ion batteries due to their markedly high specific capacity: 3,579 mAh/g for Si vs 372 mAh/g for

---

[*]trifonova@bas.bg

J.P. Reithmaier et al. (eds.), *Nanostructured Materials for Advanced Technological Applications*, 399
© Springer Science + Business Media B.V. 2009

graphite. This arises from the reaction stoichiometry: 3.75 Li atoms per Si atom as opposed to 1 Li atom per 6 carbon atoms.[1,2] Obrovac et al.[3] showed that $Li_{15}Si_4$ (3,579 mAh/g) represents the highest lithiated phase of Si at room temperature, not $Li_{22}Si_5$ (4,200 mAh/g), as found in the thermo-dynamic equilibrium phase diagram. However, the reaction mechanisms of silicon and carbon are rather different. Li intercalates in graphite[4] following a topotactic reaction with minor changes in the host structure (~10%). In contrast, the reaction of Si with lithium produces various alloy-type com-pounds[5,6] and causes enormous volume changes (~280%) during the lithiation/delithiation process. This, together with the low electrical conductivity of silicon, obstruct the commercial use of Si as a negative electrode material. Several approaches have been made to solve these problems. For instance: using a nano-structured material, elastomeric binder[7–9] or formation of composites of Si particles with conductive materials such as different kinds of carbons and graphites, which could also buffer the overall volume change of Li–Si alloys. These composites are a compromise in cycle life and lithium storage capacity. It should be noted that graphite is also electro-chemically active with respect to lithium ion intercalation in the voltage region of 0.2–0.1 V vs. Li/Li^+.

We present a silicon-graphite nano-sized composite prepared by a two-step synthesis process. The morphology of the material obtained, the crys-tallographic parameters of the Si phase observed as well as preliminary electrochemical data are presented.

2. Experimental

2.1. SYNTHESIS OF NANO-SIZED SILICON/GRAPHITE (50/50) POWDER

For the preparation of the composites, nano-sized silicon and graphite powders (Nanostructured and Amorphous Materials, Inc. USA; 80 nm particles size for Si and 55 nm mean diameter for the graphite) were used. Equal weight portions of the two materials were dispersed consequently (first Si, then gra-phite) in a precursor solution, containing methacryloxypropyltrimethoxysilane $C_{10}H_{20}O_5Si$, Merck, Germany). Presumably, the silane induces the formation of chemical bonds between the silicon and graphite particles, resulting in a more effective coating. The mixture was treated in a sonication bath for 20 min to ensure a uniform dispersion. Finally, the composite was heat treated at 700°C for 2 h under N_2 atmosphere.

2.2. ELECTRODE PREPARATION

Prior to the preparation of the electrodes, 85% (w/o) of composite material (Si:Gr 50:50) was mixed together with 2 wt % of a conductive additive (Super P carbon black, Timcal, Switzerland) and 13 wt % of a conductive binder (PVdF) in N-methyl pyrrolidone and homogenized until a suitable viscosity of the slurry formed was obtained. It was spread on a copper foil and vacuum dried at 120°C for 12 h. The active material loading was typically $\sim 2.0 \pm 0.1$ mg/cm^2 for 12 mm electrode diameter. The morphology of the synthesized powder was analyzed by SEM (JEOL Superprobe 733), while the phase composition was characterized by X-ray diffraction (XRD, Philips PW 3040/60) with Cu Kα radiation.

2.3. ELECTROCHEMICAL EXPERIMENT

Galvanostatic tests were carried out in a three-electrode laboratory-type metallic cell (prototype of the 2032 button cell), using a metallic Li foil as both, counter and reference electrode. The electrolyte was 1M LiPF$_6$ dissolved in EC:EMC (1:2 in volume), with and without the addition of 3 wt % vinylene carbonate. The cells were assembled with a Micro-porous PP battery separator (Freundenberger, Germany) and 130 µl electrolyte. All manipulations were carried out in an Ar-filled glove box. Chronopotentiometry was performed in the potential range 5–1,500 mV vs. Li/Li$^+$ with a current of 0.412 mA, corresponding to 0.1 C-rate.[†]

3. Results and Discussion

SEM images of the Si/C composite after annealing are shown in Figure 1. It can be seen that the post-treatment step leads to a substantial agglomeration of the nano-particles, but nevertheless they are still porous (Figure 1a). As can be seen from the high magnification image in Figure 1b, the Si grains are covered approximately uniformly by carbon. The primary graphite particles are quite spherical with a size about 80 nm. The follicle-like formations have a diameter up to 400 nm. This observation suggests that the electrode preparation used ensures good surface linkage between the silicon and graphite particles.

[†]The C-rate is a standard charge-discharge regime for electrochemical power sources, where the current is expressed by means of the electrode capacity.

Figure 1. SEM images of Si/Gr composite after heat-treatment at 700°C for 2 h: (a) overview; (b) high magnification.

The composition of the Si/C composite was identified by XRD-patterns (Figure 2). The basic peaks of graphite and silicon are observed. The diffraction lines of the silicon phase in the composite diffractogram reveal an enhanced crystalline structure as compared to the pristine Si-powder. The heat treatment induces growth of the nanocrystals and variations in the silicon structure.

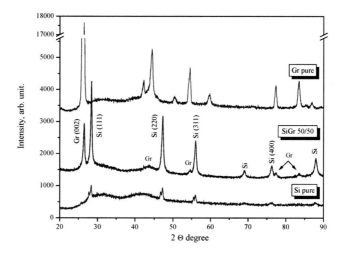

Figure 2. XRD patterns of the original graphite and silicon powders and composite material.

A comparison of the voltage profiles of the composite for both electrolyte compositions tested is presented in Figure 3. It is found that the first lithium insertion starts at ~0.250 V and continues with a pronounced plateau around 0.1 V. The plateaus correspond to two-phase regions, in which lithiated amorphous silicon is formed[10] and all the crystalline silicon is completely depleted. Li and Dahn[6] observed the formation of a $Li_{15}Si_4$ phase by an *in situ* XRD study.

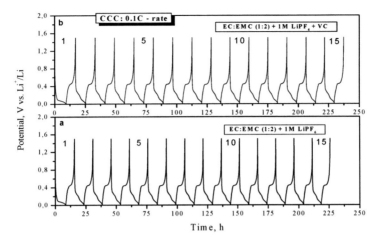

Figure 3. Voltage profiles of a Si/C composite, cycled in: (a) EC:EMC (1:2) + 1M LiPF$_6$ and (b) the same electrolyte with addition of 3 wt % VC.

During the extraction process, the composite sample shows a distinct plateau at 0.45 V where delithiated amorphous silicon is formed. If this phase is lithiated again (second charge), the voltage profile describes two sloping plateaus. The voltage drop below 50 mV is due to re-crystallization of the Li$_{15}$Si$_4$ phase, indicated by its characteristic plateau (0.45 V) during the subsequent extraction. This behaviour holds for both electrolytes but the VC-based electrolyte exhibits a better cycleability and capacity retention. The capacities delivered by the composite electrode are given in Figure 4.

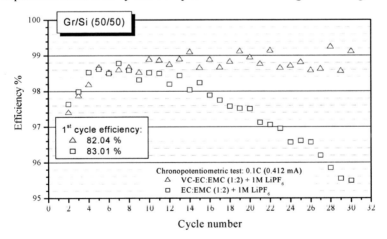

Figure 4. Capacity recovery for the Si/C composite electrodes, cycled in EC- and VC-based electrolytes with a constant current density corresponding to 0.1 C-rate.

The highest efficiencies were those for the VC-based electrolyte, approaching 99.2% after the first few cycles and remaining constant during the next cycles. When a EC-based electrolyte without additive was used, the efficiency during the first cycle was 83%, but reached 98.6% in the successive cycles, then starting to slowly decrease.

4. Conclusions

A two-step method was applied for preparing a Si/Gr nano-composite active mass. The post-treatment step leads to the agglomeration of the Si nanoparticles. The silicon in the composite reveals an enhanced crystalline structure compared to the pristine Si-powder. The highest efficiencies were those for the VC-based electrolyte, which approached 99.2% after the first few cycles and remained constant in the next cycles.

ACKNOWLEDGMENTS

This work was funded by NATO via EAP.RIG.982531.

References

1. B. A. Boukamp, G. C. Lesh, R. A. Huggins, J. Electrochem. Soc., 128, 725 (1981)
2. M. Yoshio, H. Wang, K. Fukuda, T. Umeno, J. Electrochem. Soc., 149, A1598 (2002)
3. M. N. Obrovac, L. Christensen, Electrochem. Solid-State Lett., 7, A93 (2004)
4. M. Winter, J. O. Besenhard, M. E. Spahr, P. Novak, Adv. Mater., 10, 725 (1998)
5. T. D. Hatchard, J. R. Dahn, J. Electrochem. Soc., 151, A838 (2004)
6. Jing Li, J. R. Dahn, J. Electrochem. Soc., 154 (3), A156–A161 (2007)
7. W. R. Liu, M. H. Yang, H. C. Wu, S. M. Chiao, N. L. Wua, Electrochem. Solid-State Lett., 8, A100 (2005)
8. Z. Chen, L. Christensen, J. R. Dahn, Electrochem. Commun., 5, 919 (2003)
9. N. S. Hochgatterer, M. R. Schweiger, S. Koller, P. Reimann, T. Woehrle, C. Wurm, M. Winter, Electrochem. Solid-State Lett., 10 (2), A17–A20 (2007)
10. P. Limthongkul, Y.-I. Jiang, N. J. Dudney, Y.–M. Chiang, J. Power Sources, 119, 604 (2003)

MAGNETRON SPUTTERED THIN FILM CATHODE MATERIALS FOR LITHIUM-ION BATTERIES IN THE SYSTEM Li–Co–O

B. KETTERER[*1], H. VASILCHINA[1], S. ULRICH[1],
M. STÜBER[1], H. LEISTE[1], C. ADELHELM[1], T. KAISER[1],
TH. SCHIMMEL[2]
[1]*Forschungszentrum Karlsruhe, IMF I,*
Hermann-von-Helmholtz-Platz 1, D-76344
Eggenstein-Leopoldshafen, Germany
[2]*Universität Karlsruhe (TH), Institut für Angewandte Physik,*
Wolfgang-Gaede-Straße 1, D-76131 Karlsruhe, Germany

Abstract. R.F. magnetron sputtered $LiCoO_2$ films were prepared at various working gas pressures, substrate biases and annealing temperatures. The film composition and the intrinsic stress were determined. The structure was examined using X-ray diffraction and Raman spectroscopy. Depending on the deposition parameters, $HT-LiCoO_2$ with a hexagonal layered structure and the cubic spinel-related $LT-LiCoO_2$ were identified after annealing. $LT-LiCoO_2$ is assumed to be stabilized due to the high compressive stress in the films. Layers deposited at a working gas pressure of 10 Pa showed a nanocrystalline rocksalt structure with an unordered cation arrangement. With increasing annealing temperature a cation ordering was observed.

Keywords: lithium ion battery; lithium cobalt oxide; thin-films; cathode materials; r.f. magnetron sputtering

1. Introduction

The high specific capacity and high operating voltage of $LiCoO_2$ make this compound an attractive cathode material for Li ion batteries. Three phases of lithium cobalt oxide, $LiCoO_2$, are known.[1–3] The hexagonal high temperature

---

[*]Bernt.Ketterer@imf.fzk.de

phase HT-LiCoO$_2$ with a layered structure and the low temperature phase LT-LiCoO$_2$ with a cubic spinel-related structure are based on the same oxide sublattice, but with a different spatial arrangement of the cations. Both are superstructures of a rocksalt structure isomorphic to NaCl, where the Li and Co atoms randomly occupy the cation sites. The rocksalt phase with completely random cation ordering is not thermodynamically stable, but rather stabilized only kinetically.[4]

2. Experimental Details

Films of 3 μm thickness were deposited on Si substrates by non-reactive r.f. magnetron sputtering of a LiCoO$_2$ target (diameter 7.5 cm) in a Leybold Z550 coating facility at a target power of 200 W with argon gas pressures between 0.15 Pa and 10 Pa. At 10 Pa a negative substrate bias of −15 V was applied. Annealing of the films took place in an argon/oxygen atmosphere of 10 Pa at temperatures between 100°C and 600°C for 3 h. The lithium and cobalt content was determined by atomic emission spectrometry with an inductively coupled plasma using an OPTIMA 4,300 DV from Perkin Elmer. The oxygen content was determined by inert gas fusion analysis. A commercial N/O analyzer TC 600 (Leco, USA) was used. The radius of the curvature of the Si substrates was measured before and after film deposition to determine the film stress following the Stoney equation.[5] Structural information was obtained by micro-Raman spectroscopy at room temperature using a Renishaw-1,000 system equipped with an argon ion laser (excitation wavelength λ_{Raman} = 514.5 nm, power output 21 mW) and by X-ray diffraction (Seifert PAD II, Bragg-Brentano arrangement, Cu Kα1 radiation, λ_{XRD} = 0.154 nm).

3. Results

3.1. FILM DEPOSITION AND PROPERTIES

Thin film cathode materials were deposited by variation of the working gas pressure between 0.15 and 10 Pa. As shown in Figure 1 the elemental composition of the as deposited films noticeable changes with the gas pressure. Between 5 and 10 Pa, the cathode materials nearly reach the stoichio-metric LiCoO$_2$ composition. At 0.15 Pa, the ratio Co/O ≈ 0.53 is almost the same as in LiCoO$_2$, but an overstoichiometric Lithium content is observed.

Stress is induced within the films during deposition as shown in Figure 1. The high compressive stress of about 1.5 GPa at 0.15 Pa lowers with increasing argon gas pressure and even changes to tensile above 5 Pa.

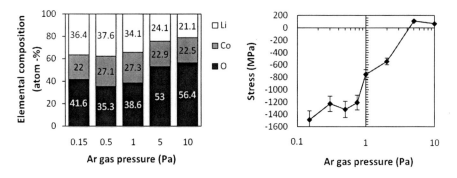

Figure 1. Elemental composition (left) and intrinsic stress (right) as a function of the argon gas pressure.

3.2. X-RAY AND RAMAN-SPECTROSCOPY

Figure 2a shows the results of X-ray diffraction experiments of the as deposited films for different argon gas pressures. Two peaks at $2\theta \approx 38°$ and $2\theta \approx 82°$ are observed at 10 Pa, whereas at 0.15 Pa one peak at $2\theta \approx 64°$ is present, indicating a change of the texture between deposition at high and low pressures. The sample at 0.5 Pa is X-ray amorphous.

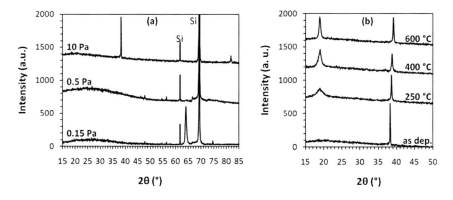

Figure 2. X-ray spectra (a) of the as-deposited films at 0.15, 0.5 and 10 Pa argon gas pressure and (b) of the film deposited at 10 Pa after annealing at 250°C, 400°C and 600°C.

The influence of the heat treatment of the films prepared at 10 Pa is shown in Figure 2b. An additional broad peak develops at $2\theta \approx 19°$ at temperatures above 100°C, which becomes more intense at higher temperatures. A peak shift to higher values of 2θ as well as a decrease of the peak width is observed with increasing annealing temperature.

The Raman spectrum (after annealing at 600°C) at the bottom of Figure 3a) exhibits four bands at 485, 525, 595 and 694 cm⁻¹. As an effect of a substrate bias of −15 V applied (at 10 Pa Argon gas pressure), two of the four band vanish and only two (at 486 and 596 cm⁻¹) remain after heat treatment at 600°C (Figure 3a top). In the Raman spectrum of the film deposited at 0.15 Pa and annealed at 400°C presented in Figure 3b four bands at 448, 482, 588 and 605 cm⁻¹ can be distinguished.

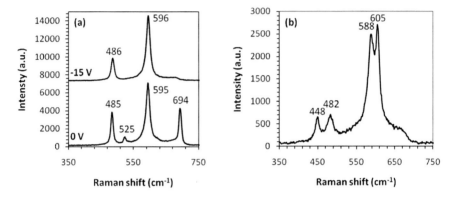

Figure 3. Raman-spectra (a) of films deposited at 10 Pa with and without a substrate bias of -15 V annealed at 600°C and (b) of a film deposited at 0.15 Pa after annealing at 400°C.

4. Discussion

The comparison of the elemental composition and the X-ray diffraction spectra of the as-deposited films show that crystalline films with compositions, which are close to that of stoichiometric $LiCoO_2$, have been obtained at pressures between 5 and 10 Pa. The sample with the largest deviation from the composition of $LiCoO_2$ deposited at 0.5 Pa is indeed X-ray amorphous. Films deposited at 0.15 Pa are crystalline due to an almost stoichiometric ratio Co/O. The two peaks observed at 10 Pa belong to the cubic (222) and (444) reflexes of the metastable $Li_xCo_{1-x}O_2$ rocksalt phase with randomly distributed Li- and Co-cations in an oxygen lattice.[1] The occurrence of the cubic (111) superstructure reflex after annealing at temperatures above 100°C (Figure 2) indicates the beginning of a cation ordering process that leads to the formation of the HT-$LiCoO_2$ superstructure at 600°C. The characteristic E_g and A_{1g} modes of layered $LiCoO_2$ at 485 and 595 cm⁻¹ indicate the formation of HT-$LiCoO_2$ after annealing at 600°C.[6] The two additional bands in the spectrum are assigned to the formation of Co_3O_4.[7] As shown in Figure 4b, the formation of Co_3O_4 is suppressed by the application of a low negative substrate bias of −15 V during the 10 Pa process.

After annealing at 400°C the Raman-spectrum of the film deposited at 0.15 Pa exhibit the characteristic shape known for the cubic spinel-like low temperature phase LT-LiCoO$_2$.[6] The possibility of a transition between the layered and the cubic spinel-like structure as a function of the applied pressure is mentioned by Li.[8] McKenzie showed that an applied biaxial stress causes a hydrostatic pressure[9]. Since the unit cell volume of layered LiCoO$_2$ is slightly larger than that of spinel LT-LiCoO$_2$, an applied pressure may hinder the hexagonal distortion of the HT-LiCoO$_2$ lattice. As the as-deposited rocksalt structure shows a strong spinel (440) orientation, the hexagonal distortion during the formation of HT-LiCoO$_2$ should occur in a direction parallel to the substrate surface. Regarding the high compressive stress of about 1.5 GPa measured in the films at 0.15 Pa we assume that this lattice expansion is suppressed by the directly counteracting biaxial stress field. The effectiveness of stress in stabilizing high pressure phases is well illustrated by the boron nitride system, where the stabilization of the cubic phase c-BN by high compressive stress in thin films – in contrast to the formation of the hexagonal phase h-BN at low stress conditions – is a well-known phenomenon.[9]

5. Conclusions

Three phases of LiCoO$_2$ have been deposited by r.f. magnetron sputtering. As-deposited crystalline thin films with nearly stoichiometric LiCoO$_2$ composition deposited at high argon gas pressures exhibit a rocksalt structure with random cation ordering. Annealing of these films at 600°C leads to the formation of layered HT-LiCoO$_2$. The Raman spectrum of the film deposited at 0.15 Pa argon pressure and annealed at 400°C indicates the formation of the low temperature LT-LiCoO$_2$ phase which is assumed to be stabilized due to high compressive stress in the film.

References

1.   M. Antaya, K. Cearns, J.S. Preston, J. Appl. Phys. 76, 2799 (1994).
2.   E. Antolini, Solid State Ionics 170, 159 (2004).
3.   T. Bak, J. Nowotny, M. Rekas, C.C. Sorrell, S. Sugihara, Ionics 6, 92 (2000).
4.   C. Wolverton, A. Zunger, Phys. Rev. B 57, 2242 (1999).
5.   G. Stoney, Proc. Royal Soc. A 82, 172 (1909).
6.   W. Huang, R. Frech, Solid State Ionics 86–88, 395 (1996).
7.   V.G. Hadjiev, M.N. Iliev, I.V. Vergilov, J. Phys. C: Solid-State Phys. 21, L199 (1988).
8.   W. Li, J.N. Reimers, J.R. Dahn, Phys. Rev. B 49, 826 (1994).
9.   D.R. McKenzie, J. Vac. Sci. Technol. B 11, 1928 (1993).

6.2. DATA STORAGE

PHASE-CHANGE MATERIALS FOR NON-VOLATILE DATA STORAGE

D. LENCER[*], M. WUTTIG
I. Physikalisches Institut (IA), RWTH Aachen University, 52056 Aachen, Germany

Abstract. Phase-change materials comprise a unique portfolio of properties, which enables their use in non-volatile data storage devices. In particular, this material class is characterized by a pronounced contrast of optical and electrical properties between crystalline and amorphous state, while switching between these states proceeds on the timescale of a few nanoseconds.

Keywords: phase-change materials; amorphous; glassy; crystalline; crystallization kinetics; dielectric function; data storage

1. Introduction

In this contribution, we will first address actual applications, showing that device performance depends crucially on the material choice. Therefore, we will then tackle the kinetics of the phase change as exhibited by frequently employed materials. Subsequently, the origin of the contrast will be addressed. A comprehensive study of these important aspects of phase-change materials by means of various experimental techniques will enable us to characterize the generic properties of this exciting material class.

2. The Principle of Phase-Change Recording

In order to store massive amounts of data on a short timescale as demanded by today's information technologies, only those data storage principles are viable that meet the following requirements:

---

[*]lencer@physik.rwth-aachen.de

- Competitive data transfer rates
- Reliability, i.e. sufficient data retention times and guaranteed read/write-cycles
- Scalability
- Non-volatility
- Energy-efficiency
- Economic mass production

Today, mainly two techniques are employed for electronic memories. On the one hand, only the costly DRAM (Dynamic Random Access Memory) provides high data transfer rates, but needs to be constantly refreshed. On the other hand, the Flash Memory dominates the market for non-volatile memories, but offers only slow data transfer rates. Therefore, various novel data storage principles are currently investigated to provide superior memories that hold the potential to combine the strengths of both types. A particular promising approach is phase-change recording, which is already widely employed for data storage in rewritable optical media such as the Compact Disc ReWritable (CD-RW), Digital Versatile Disc (DVD±RW, DVD-RAM) and Blu-Ray Disc (BD-RE). It is commonly referred to as Phase-Change Random Access Memory (PCRAM or PRAM) or Ovonic Unified Memory (OUM).

In order to store binary information, bits of a phase-change material are switched between a crystalline and an amorphous state. As both phases differ in their optical and electrical properties, the state of a phase-change bit can be read by measuring its reflectance or its resistance. Figure 1 explains how to switch between the phases. Heating an amorphous bit above a certain material-specific temperature for a sufficiently long time enables the atoms to relax to their crystalline positions. To switch a crystalline bit back to the amorphous state, it must be rapidly heated above the

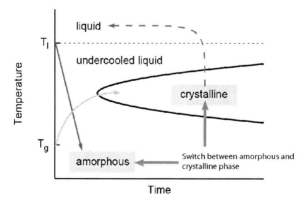

Figure 1. Principles of phase-change recording.

melting temperature. As the surrounding material remains unaffected, there is a huge temperature gradient which results in a quenching of the liquid. If the cooling rate is sufficiently high, the configuration of the liquid is eventually frozen-in once the temperature falls below the so-called glass transition temperature T_g. Within this regime, the atomic mobility is too small to allow for crystallization on a reasonable time scale. Hence, the bit remains in the metastable glassy state.

In optical media as sketched in Figure 2, laser pulses are employed to facilitate these processes, whereas current pulses are used for electrical devices. For the latter, two principle designs have been proposed as shown in Figure 3. Unlike in the case of optical recording, there is a very exciting phenomenon observed for the switching of the amorphous phase, called threshold switching.[1] Figure 4 shows the current–voltage characteristics, i.e. the I–V-curve, of a PCRAM-cell. Starting from the amorphous phase, a high non-ohmic resistivity is exhibited. As switching to the crystalline state necessitates the deposition of a specific amount of energy by Joule heating, such an I–V-characteristic is unfavorable as large voltages would be

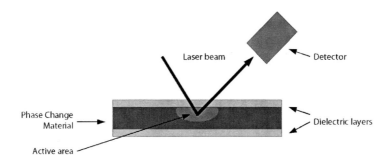

Figure 2. Cross section through a simplified rewritable optical phase-change disc.

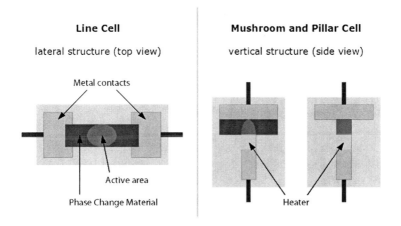

Figure 3. Proposed designs of PCRAM-cells.

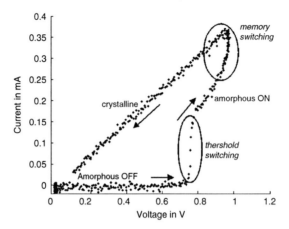

Figure 4. Current–voltage characteristics of a PCRAM-cell exhibiting a threshold-switch. From Ref. 2.

required. However, for many materials known from optical phase-change recording it has been found that the I–V-characteristics change dramatically once a specific electrical field (resp. voltage) is exceeded, the so-called threshold-field (resp. threshold-voltage). While remaining in the amorphous state, the conductivity is enhanced by orders of magnitude, so that crystallization becomes feasible with only moderate voltages applied. Though there are phenomenological descriptions of this effect, its physical origin is not yet sufficiently understood.

The performance of phase-change devices is severely limited by the properties of the phase-change material employed.[2] By the change of its optical and electrical properties, it yields the contrast, but a concurrent density contrast would negatively affect the device endurance. Its crystallization kinetics dictate the time scale of the slowest process involved in phase-change recording, the re-crystallization, and also the data retention time at ambient temperatures. So we see that it is crucial to characterize materials with respect to these aspects and to model their behavior. This shall be the scope of the remainder of this contribution.

3. Known Phase-Change Materials

As of today, various 'generations' of phase-change materials exist, which have all been found primarily by extensive trial-and-error approaches.[2] Most of them are related to the ternary Ge:Sb:Te-phase diagram shown in Figure 5. In particular, pseudo-binary materials composed as $(GeTe)_m(Sb_2Te_3)_n$ ('GST-compounds'), m and n being integer numbers, as well as compounds based on Sb_2Te and Sb (e.g. $Ge_{15}Sb_{85}$) have been shown to exhibit suitable phase-change properties. In literature, GST-compounds have served as prototype-materials and hence have been thoroughly studied. Therefore, we will primarily present the properties of GST-compounds in the following.

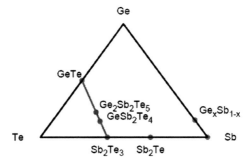

Figure 5. Ternary Ge:Sb:Te-phase diagram (at %) hosting various phase-change materials (dots). From Ref. 3.

4. Phase-Change Kinetics

To gain a deeper insight into the mechanisms of the phase changes, we will provide some background on the kinetics of the phase transformations involved.

4.1. THE GLASS TRANSITION

If a liquid is cooled down, its viscosity will increase, i.e. the atomic mobility will decrease. If the cooling rate is sufficiently high, crystallization of an undercooled liquid is avoided and eventually, the material solidifies in a glassy state once a glass transition temperature is reached. The latter is not a unique property, e.g. it depends on the cooling rate. The glassy state is only a metastable state in internal equilibrium. As amorphization of phase-change materials takes place by quenching below the glass transition temperature, T_g is an important parameter for modeling device behavior.

Theoretically, it can be assessed by means of a simple analysis of bond strengths.[4] To do so, the bond strengths of all possible bonds for a given stoichiometry have to be evaluated, for instance by applying Pauling's equation,

$$H_{AB} = \frac{(H_{AA} + H_{BB})}{2} + 96.14(S_A - S_B)^2 \cdot$$

Here, H_{AB} denotes the bond enthalpy of a heteropolar A–B-bond, which is calculated from the homopolar A–A- and B–B-bond enthalpies as well as from the electronegativities S_A and S_B of elements A and B. The total bond enthalpy of a system, H_{tot}, then results as

$$H_{tot} = \sum n_{AB} H_{AB},$$

where the number of particular A–B-bonds is given by n_{AB}. Assuming the 8-N-rule (N being the valence) to be valid, i.e. that an atom of valence N forms 8-N bonds, we can now evaluate this sum such that starting with the strongest bonds, as many bonds as given by the stoichiometry and the 8-N-rule are formed. From the resulting total bond enthalpy, the glass transition temperature can be determined according to Lankhorst's empirical relation[4]

$$T_g = 3.44 \frac{K}{kJ/g\text{-}atom} H_{tot} - 480K.$$

However, the resulting glass transition temperatures are in poor agreement with the experimental ones (cf. Figure 6), though the general trend is reproduced very well as shown by the linear fit that represents a modification of Lankhorst's empirical relation.

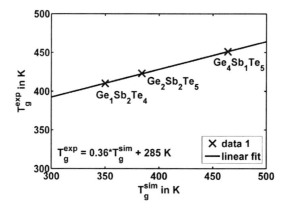

Figure 6. Comparison between experimental and theoretical glass transition temperatures for three phase-change compounds. From Ref. 5.

Experimentally, the glass transition, i.e. the transition from the glass to the undercooled liquid, can be measured by differential scanning calorimetry (DSC), a technique that is capable of monitoring the transition heat associated with a phase-change upon heating the glass. Since the undercooled liquid has more degrees of freedom, its heat capacity is larger, so there would be an endothermic step at T_g (cf. Figure 7).

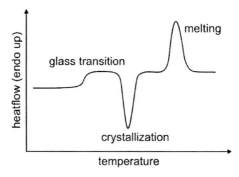

Figure 7. Schematics of a DSC-measurements. From Ref. 5.

However, this is only true if the crystallization takes place at higher temperatures, which is generally not the case for phase-change materials. Nevertheless, we know that 'glass' is not a stable state, it can be relaxed by thermal annealing. As can be seen in Figure 8, this would result in a shift of the glass transition temperature towards lower temperatures. So by pre-annealing the amorphous phase, it was studied whether a separation be-tween glass transition and crystallization could be obtained. Indeed, it was possible to separate crystallization and an endothermic step by some 10 K, though unidentified irreversible processes resulted from the annealing as well. For that reason, we can conclude that for phase-change materials, glass transition and crystallization temperature roughly coincide. This is not surprising, since a strong tendency for fast crystallization is desirable.

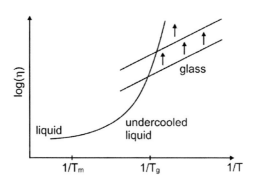

Figure 8. Schematic dependence of the viscosity on temperature. The arrows indicate the effect of relaxation of the glass. From Ref. 5.

4.2. THE MECHANISM OF CRYSTALLIZATION

As mentioned above, cooling a liquid with a finite cooling rate below the melting temperature is possible without solidification, i.e. one can prepare an undercooled liquid, since the formation of crystalline nuclei leads to the formation of crystal-liquid interfaces. This is illustrated by Figure 9. Only for nuclei exceeding a critical size, growth is favored, as the volume contribution wins over its surface counterpart. Mathematically, the dependence of the change in free energy on the cluster radius r can be expressed as

$$\Delta G(r) = V(r) \cdot \Delta G_V + A(r) \cdot \sigma .$$

Here, V and A denote the nuclei volume and surface, ΔG_V the difference in Gibb's free energy per volume between the two phases, i.e. the driving force for nucleation, and σ the interfacial energy.

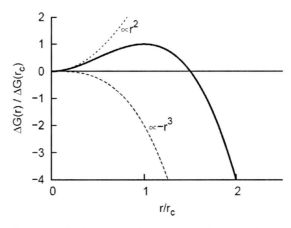

Figure 9. Energetic competition between volume and interface upon the formation of crystalline clusters within a liquid. For the critical cluster size r_c, the difference in Gibb's free energy exhibits a maximum.

After an incubation time, a steady-state-distribution of subcritical clusters will evolve, which may become postcritical due to thermal fluctuations. Both the corresponding temperature-dependent steady-state nucleation rate I^{ss} and the growth velocity u can be evaluated within the classical theory of crystallization.[6] The former is determined to

$$I^{ss} = I_0 \cdot \frac{1}{\eta(T)} \cdot \exp\left(-\frac{16\pi}{3kT} \frac{\sigma^3}{\Delta G_V{}^2} \cdot f(\Theta) \right),$$

while the latter results as

$$u = f_s \cdot \frac{2kT}{\pi\lambda^2} \frac{1}{\eta(T)} \left[1 - \exp\left(-\frac{|\Delta G_V|}{kT} \right) \right].$$

Both quantities depend on the reciprocal of the viscosity η, i.e. scale with the atomic mobility. I_0 and f_s are constant prefactors, λ is a typical length scale. The main contributions to both equations are represented by the exponential functions. Note that the possibility of heterogeneous nucleation, i.e. nucleation at impurities or surfaces, enters the steady-state nucleation rate by the term $f(\Theta)$, Θ being the wetting angle. This contribution accounts for an effective reduction of the interfacial energy case in the presence of such heterogeneous nucleation sites, compared to the homogeneous.

4.3. CRYSTALLIZATION AT LOW TEMPERATURES

The driving force for crystallization of an undercooled liquid is high around the glass transition temperature, but also the viscosity is high, so the atomic mobility is very small. Hence, crystallization proceeds very slowly in this regime. Consequently, one can directly observe crystallization experimentally by a combination of careful isothermal annealing and atomic force-microscopy (AFM).[7] Figure 10 shows AFM-images of a 40 nm thick GeSb$_2$Te$_4$ film on Si, which was sequentially annealed at 368 K. Since the crystalline phase has a higher density than the amorphous one, crystallites appear as depressions of the film thickness. Upon isothermal annealing, both nucleation and growth can clearly be identified as a function of time.

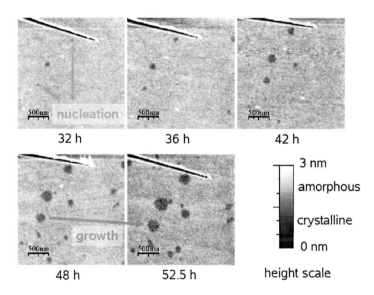

Figure 10. AFM-images (3 × 3 μm) showing nucleation and growth upon isothermal annealing. From Ref. 6.

In particular, the growth velocity of the crystallites can be analysed by measuring the change of their area between subsequent annealing steps. The same procedure can be repeated for various anneal temperatures, as long as the experimental timescale suits the timescale on which crystallization of the film proceeds. From these data, a strict Arrhenius-like dependence of the growth velocity on the temperature is obtained for phase-change materials, with an activation energy for growth E_v (Figure 11).

These results allow for an assessment of data retention properties of phase-change materials. By extrapolating to ambient temperatures, one can chose a material with a small growth velocity to ensure that an amorphous spot does not crystallize within a given period of time. The smaller the bit size gets, the more important becomes this requirement.

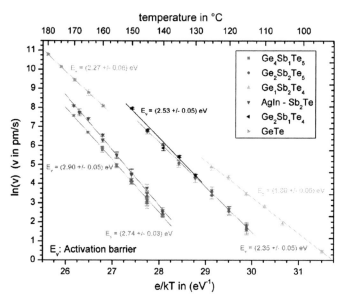

Figure 11. Arrhenius-like dependence of the growth velocity on the temperature. From Ref. 5

4.4. CRYSTALLIZATION AT HIGH TEMPERATURES

At temperatures around the melting temperature, the atomic mobility is high, but the driving force for crystallization is low. Hence, also in this temperature regime nucleation can be investigated on experimentally fairly accessible timescales.[6] To do so, small chunks of phase-change compounds are embedded in B_2O_3-filled crucibles. B_2O_3 is an easy glass former that prevents the samples from evaporating. Furthermore, it may prevent hetero-geneous nucleation since it dissolves impurities and provides a non-crystalline container. The samples are heated at a constant heating rate above the melt-ing temperature by means of a differential thermal analyzer (DTA).

Figure 12 shows the endothermic melting of a Ge_4SbTe_5 sample taking place at around 690°C. The peak area corresponds to the heat of fusion ΔH_f. Also shown in this viewgraph is the corresponding signal upon cooling at a constant cooling rate. As the exothermic peak of crystallization is shifted towards lower temperatures (about 625°C) as compared to the transition temperature upon heating, it is concluded that the sample could be undercooled by more than 60°C. This measurement represents the largest undercooling achieved for this particular material out of a series of analogous experiments. From the amount of undercooling, i.e. the difference between melting temperature $T_m{}^\dagger$ and the temperature at which crystallization happens, T_n, the interfacial energy can be obtained, assuming that crystallization has occurred due to the occurrence of one nucleation event,

$$1 = \frac{V}{\dot{T}} \int_{T_m}^{T_n} I^{ss}(T)\, dT \cdot$$

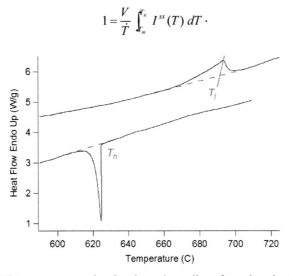

Figure 12. DTA-measurement showing the undercooling of a molten droplet. From Ref. 6.

TABLE 1. Maximal undercooling, heat of fusion and interfacial energy obtained from undercooling experiments. From Ref. 6.

	$T_m\text{-}T_n$ (K)	ΔH_f (J/cm^3)	σ (mJ/m^2)
$Ge_{12}Sb_{88}$	77 ± 1	1,225 ± 10	76 ± 5
$AgIn\text{-}Sb_2Te$	62 ± 1	902 ± 15	55 ± 4
Ge_2SbTe_5	66 ± 1	705 ± 60	47 ± 6
$Ge_2Sb_2Te_5$	43 ± 1	825 ± 40	40 ± 3

---

†Please note that in the following, we will only refer to the melting temperature T_m for the sake of simplicity. However, for materials which depart from eutectic compositions the liquidus temperature T_l is taken into account. The latter is indicated in Figures 12 and 13.

In order to evaluate this equation, supporting measurements or estimates for the quantities entering the steady-state nucleation rate given above have to be obtained; for more information the reader is referred to Ref. 6. Assuming that homogeneous nucleation was observed, $f(\Theta)$ is set to unity. The latter assumption leads to the fact that the obtained interfacial energy is only a lower limit of the real value. The results for four materials are compiled in Table 1.

Concurrently, an upper limit for the steady-state nucleation rate can now be computed (Figure 13). The results obtained agree well with the experimental observation that two classes of phase-change materials can be distinguished. On the one hand there are so-called growth-dominated materials, which tend to crystallize by the growth of only a few nuclei or from a crystalline rim. On the other hand, there are so-called nucleation-dominated materials, which show a tendency to crystallize by the formation of many nuclei. Both cases are illustrated in the inset of Figure 13. Indeed, the interfacial energies obtained provide a clear explanation for the occurrence of these two crystallization mechanisms. The higher the interfacial energy, the lower is the nucleation rate, and eventually growth becomes the dominating process.

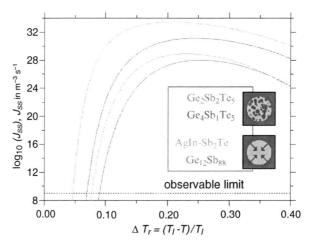

Figure 13. Steady-state nucleation rate (here: J^{ss}) derived from the undercooling experiments versus reduced temperature (zero corresponds to the liquidus temperature T_l). The inset illustrates the resulting dominant mechanism of crystallization. From Ref. 6.

4.5. DEVICE TESTING

So far, we have been concerned with crystallization in the border cases of high and low temperatures. But also the crystallization at intermediate temperatures under operation conditions can be investigated, though the short timescale allows only a comparison between the initial and the final state. A laser set-up ('static tester') for the treatment of thin film samples similar to those used in actual applications is employed for that purpose. In these experiments, bits are switched by the application of laser pulses of different lengths and intensities. Monitoring the reflectance of the bit by a second beam (or alternatively by a second low-intensity pulse of the same laser), a power-time-effect diagram can be derived. An example ($Ge_3Sb_4Te_8$) is shown in Figure 14. From the change in reflectivity, not only suitable process parameters are obtained, but also the minimal time needed for crystallization. Since this is the slowest process involved in phase-change recording and hence limits data transfer rates, this is a very important parameter to characterize. If too much energy is deposited, i.e. for long pulses of high intensities, film ablation is observed. Analogously, also PCRAM-cells can be tested under operation conditions with current pulses replacing the laser pulses.

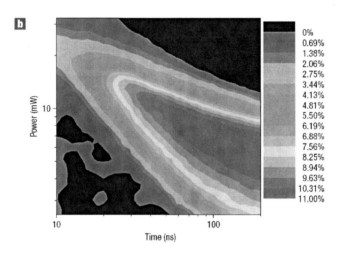

Figure 14. Example of a power-time-effect diagram. From Ref. 9.

5. The Origin of the Optical Contrast

For optical media, phase-change materials have to exhibit a sufficiently large contrast at the wavelength of the employed laser, i.e. 780 nm for CD-RWs, 650 nm for rewritable DVDs and 405 nm for BluRay discs, respectively. While it is easy to measure this contrast, for instance by ellipsometry, no scheme exists to predict whether a particular semiconductor will show it. Figure 15 shows the results of Fourier transform infrared spectroscopy (FTIR) measurements in reflectance mode for both, an ordinary semiconductor, $AgInTe_2$, and a typical phase-change material, Ge_2SbTe_4.[10]

While the reflectance of the former hardly changes upon crystallization, there is a pronounced change for the latter. The maximum reflectance is considerably reduced, caused by a Drude-like contribution. The spacing between the fringes, roughly corresponding to the product of film thickness and index of refraction, is reduced, hinting at a significant change of the polarizability. Finally, the oscillations fade out at lower energies in the crystalline case, so the optical gap is smaller than that of the amorphous phase. On the right hand side of Figure 15, the dielectric functions fitted concurrently to these results and ellipsometry measurements at higher energies are shown.

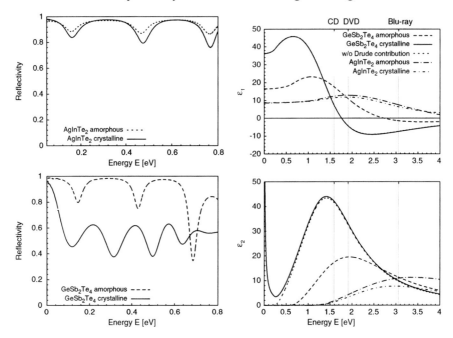

Figure 15. Left: raw data of a FTIR-measurement in reflectance mode for an ordinary covalent semiconductor (top) and a phase change material (bottom). Right: fitted real (top) and imaginary (bottom) part of the dielectric functions. From Ref. 10.

From structural investigations, it has been concluded that phase-change materials have a nearly sixfold coordinated crystalline structure in common, which is slightly (Peierls-like) distorted. From density function theory (DFT) calculations, the occurrence of this coordination has been linked to the average number of valence electrons.[11] Only materials exceeding an average of four valence electrons per atom were found to exhibit both, the afore-mentioned coordination and phase-change properties. The crystalline structures are very similar to those observed in some IV–VI-compounds such as GeTe[12] For these, the relation between electronic and crystallographic structure has been discussed in the past, introducing the model of resonance bonding. These systems have only three p-electrons per atom but form six bonds (slightly altered by the distortions), so there are only half as many electrons present as needed to form saturated bonds. The electronic groundstate can approximately be interpreted as a superposition of (symmetry-related) saturated-bond configurations. This electronic structure is prone to rearrange upon external perturbations, i.e. a change of the mixing coefficients, for instance in order to screen an applied electric field. This yields an explanation for the polarizability enhancement in the crystalline state, i.e. the large value of ε_∞.[13] Since resonant bonding relies on long-range order, the nearly sixfold coordination can only be present in the crystalline state, providing an explanation for the change in chemical bonding between the two phases. Hence, only materials exhibiting similar crystalline properties are considered possible candidates for phase-change applications.

6. Summary

Phase-change materials enable data storage devices due to the optical and electrical contrast between the amorphous and crystalline phase. This contrast stems from the fact that there is a generic change in chemical bonding between the two states. While the amorphous state behaves more like an ordinary covalent semiconductor, the crystalline phase is characterized by a slightly distorted sixfold-coordinated structure. Its electronic configuration can be interpreted within the model of resonant bonding, which is only possible in the presence of long-range order and which provides an explanation for the observed contrast. The kinetics can be investigated in various temperature regimes by specialized experiments. From these measurements, data retention at ambient temperatures, the minimal crystallization time and the dominant crystallization mechanism (nucleation or growth) can be addressed.

ACKNOWLEDGMENTS

We would like to thank the organizers for the kind invitation and the great organization. Furthermore, we would like to thank Dr. Martin Salinga and Dr. Johannes Kalb for providing material for this contribution alongside helpful discussions.

References

1. S.R. Ovshinsky, Phys. Rev. Lett. 21, 1450 (1968).
2. M. Wuttig and N. Yamada, Nat. Mater. 6, 824 (2007).

3.   D. Lencer et al., Nat. Mater. 7, 972 (2008).
4.   M.H.R. Lankhorst, J. Non-Cryst. Solids 297, 210 (2002).
5.   M. Salinga et al., *European Phase Change and Ovonics Science Symposium* (Zermatt, Switzerland, 2007).
6.   J. Kalb, *Crystallization kinetics in antimony and tellurium alloys used for phase change recording*, Ph.D. thesis, RWTH Aachen (2006).
7.   J. Kalb, F. Spaepen, and M. Wuttig, App. Phys. Lett. 84, 5240 (2004).
8.   M. Klein, *Crystallization kinetics in thin films of amorphous tellurium alloys used for phase change recording*, Diploma thesis, RWTH Aachen (2006).
9.   M. Wuttig et al., Nat. Mater. 6, 122 (2007).
10. K. Shportko et al., Nat. Mater. (in print) (2008).
11. M.B. Luo and M. Wuttig, Adv. Mater. 16, 439 (2004).
12. P.B. Littlewood, J. Phys. C 13, 4855 (1980).
13. P.B. Littlewood, J. Phys. C 12, 4459 (1979).

STRUCTURE AND CRYSTALLIZATION BEHAVIOUR
OF (GeTe$_5$)$_{100-x}$Ga$_x$ NANOSIZED THIN FILMS
FOR PHASE-CHANGE APPLICATIONS

P. ILCHEV[*1], P. PETKOV[2], D. WAMWANGI[3],
M. WUTTIG[3]
[1]*Central Laboratory of Photoprocesses "Acad.
J. Malinowski", Bulgarian Academy of Sciences, Acad.
G. Bonchev St., bl. 109, 1113 Sofia, Bulgaria*
[2]*Department of Physics, University of Chemical Technology
and Metallurgy, Kl.Ohridski Blvd.8, 1756 Sofia, Bulgaria*
[3]*Department of Physics 1A, Aachen University of Technology,
14 Sommerfeld St., 56072 Aachen, Germany*

Abstract. Nanosized thin films from the (GeTe$_5$)$_{100-x}$Ga$_x$ system have been investigated to establish the effect of Ga addition on the structure and crystallization kinetics of GeTe$_5$ alloys. XRD measurements and TEM images of a Ge$_{13}$Te$_{67}$Ga$_{20}$ alloy in the as-deposited state and after annealing at 90°C and 150°C have been performed to identify the structural relaxation of the amorphous state. The crystallization proceeds after annealing the alloy at 250°C, which coincides with the phase separation of this alloy to Ga$_2$Te$_3$, GaTe and GaGeTe phases.

Keywords: chalcogenide glasses; thin films; structure; phase transition; phase-change materials

1. Introduction

In recent years research interest in chalcogenide glasses has increased profoundly as a result of their widespread applications in optical and electrical devices. In the area of optical applications alone, Te based chalcogenide

---
* p_ilchev2001@yahoo.com

glasses have been intensively studied due to their infrared transmission abilities,[1] as well for use in optical data storage applications.

Investigations of GeTe in the past 2 decades have shown the suitability of this alloy for optical data storage applications. Pioneer studies on the electrical properties of GeTe by Ovshinsky[2] demonstrated the switchable characteristics of this alloy and hence layed the foundation for further research on GeTe as a potential phase change alloy. The studies of Yamada et al.[3] on $Te_{(1-x)}Ge_x$ phase change films showed that stoichiometric GeTe are most promising for optical data storage due to its fast crystallization time. Research interests on GaGeTe alloys have varied considerably over the years; literature accounts have reported e.g. ferroelectric properties for this system. The investigations of Sripathi et al.[4] on $Ga_{20}Ge_{30}Te_{50}$ revealed high absorption coefficients ($k > 10^4$) in the visible range and a sufficiently high optical contrast ratio of about 0.2 at 830 nm. The work of Bose and Pal[5] on $Ga_{1-x}Ge_xTe$ alloys found that the ferroelectric behavior depends on the Ge concentration. In addition, their results also indicate that the crystal structure and the optical properties largely change with the Ge concentration; a GaTe-like monoclinic structure was found up to x = 0.10. However, for compositions x $\geq$ 0.25, the structure of the alloy changed to the rhombohedral GeTe-like structure. Their results have also shown that the addition of Ge changes the electronic properties of GaGeTe. Regarding the thermal properties, Ga–Ge–Te glasses have been intensively investigated to determine their excess enthalpies in the liquid state. Many thermodynamical data on chalcogenide glasses have been generated in the course of testing topological models.[6] In this work we investigate the effect of Ga addition on the structural, optical and thermal properties of $GeTe_5$ alloys.

2. Experimental

Elemental Ga, Ge and Te have been mixed in the appropriate mass ratios and used to prepare $(GeTe_5)_{1-x}Ga_x$ alloys by the melt-quenching technique. Thin films of $(GeTe_5)_{1-x}Ga_x$ alloys with thicknesses between 80 and 500 nm were deposited on glass substrates using vacuum thermal evaporation. The residual gas pressure of the evaporation chamber was 2.6×10^{-4} Pa.

To identify and distinguish the structural properties of $(GeTe_5)_{1-x}Ga_x$ alloys before and after annealing, grazing incidence XRD investigations were performed using a Philips MRD system. In addition, transmission electron microscopy was used for selected area diffraction studies to determine the nature of the transformations of $Ga_{20}Ge_{13}Te_{67}$. All X-ray measurements were performed at room temperature.

The optical behaviour of amorphous and crystalline $Ge_{13}Te_{67}Ga_{20}$ phases was investigated using variable angle spectroscopic ellipsometry (VASE) at incident angles of 65°, 70° and 75°, respectively.

3. Results and Discussion

The structure of the $(GeTe_5)_{100-x}Ga_x$ films has been determined by performing grazing incidence XRD at room temperature for each sample of the various structural phases. The glancing angle was set to 0.75° for all samples.

Our results establish that for x = 0–20 mol % Ga, the alloy decomposes into GaTe, GaGeTe and Ga_2Te_3 phases. Typical X-ray diffractograms of a $Ge_{13}Te_{67}Ga_{20}$ sample in as-deposited state, and after annealing at 90° and 150°C, respectively, are presented in Figure 1. It is evident from this figure that the two broad halos represent the amorphous phase, which indicates the absence of any long range order. Figure 1 shows that the amorphous phase does not change significantly even upon annealing up to 150°C. Upon further annealing at 250°C, a phase separation becomes evident; the $Ge_{13}Te_{67}$ Ga_{20} alloy separated into GaGeTe, GaTe and Ga_2Te_3 phases as shown in Figure 2.

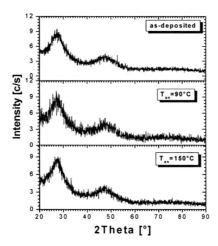

Figure 1. XRD spectra of $Ge_{13}Te_{67}Ga_{20}$ in the as-deposited state and after annealing at 90°C and 150°C.

For the GaGeTe phase, a rhombo-centred hexagonal lattice has been identified; the corresponding lattice parameters have been determined to a = b = 3.967 ± 0.003 Å and c = 33.773 ± 0.029 Å.

Similar calculations for GaTe phase yielded a monoclinic structure with lattice parameters of a = 17.355 ± 0.024 Å, b = 10.442 ± 0.014 Å and c = 4.069 ± 0.024 Å, respectively, after identifying the Bragg reflexes using the PDF card.[7] In addition, the Ga_2Te_3 phase was also identified in the XRD spectrum; it possesses a face centered cubic structure with a lattice parameter of a = b = c = 5.886 ± 0.003 Å.

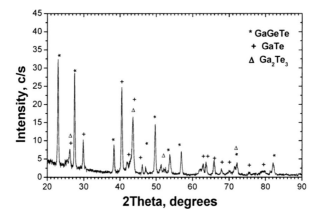

Figure 2. Phase separation of the GaGeTe film after annealing at 250°C into GaGeTe, GaTe and Ga_2Te_3.

Structural investigations using TEM also corroborated the XRD results. Figure 3a, b show the diffuse rings of TEM pictures of an as-deposited alloy as well as of a sample measured after annealing at 90°C. The occurrence of these diffuse rings confirms the amorphous character of these samples; however, the annealing increases the sharpness of these diffraction rings. Further annealing of the sample at 190°C (Figure 3c) and 250°C (Figure 3d) leads to the appearance of diffraction spots. These results supplement our previous observation with XRD that crystallization of the sample occurs at 190°C and completes at 250°C.

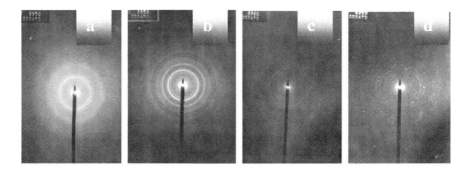

Figure 3. Electron diffraction pattern of a $Ge_{13}Te_{67}Ga_{20}$ film in the as-deposited state (a) and after annealing at 90°C (b) 190°C (c) and 250°C (d).

The optical behaviour of the amorphous and crystalline phases of the Ge$_{13}$Te$_{67}$Ga$_{20}$ alloy were investigated using VASE at incident angles of 65°, 70° and 75°, respectively.

Figure 4 shows the real and imaginary parts of the dielectric function as a function of wavelength for the sample in the as-deposited state and for the same sample after annealing at 100°C and 250°C, respectively. The optical constants have been determined using the Tauc-Lorentz oscillator model and the Drude model to simulate the free carriers in the sample. The intersection of the imaginary part of the dielectric function gives a measure of the band gap of the alloy. For the as-deposited sample, the band gap has been determined to 0.95 eV, which decreases slightly to 0.93 eV after annealing at 100°C.

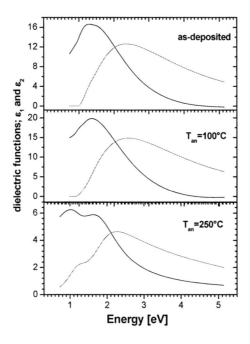

Figure 4. Optical characterization of the Ge$_{13}$ Ga$_{20}$Te$_{67}$ film the in as-deposited state and after annealing at 100°C and 250°C.

The optical constants decrease significantly upon annealing at 250°C due to the separation of the amorphous Ge$_{13}$Te$_{67}$Ga$_{20}$ film into three crystalline phases. The separation possibly leads to the formation of compounds with different optical properties. The optical band gap decreases to 0.56 eV upon crystallization at 250°C.

4. Conclusions

For an alloy with the composition of $Ge_{13}Te_{67}Ga_{20}$ the presence of only two transitions was observed, one amorphous-amorphous and the other amorphous-crystalline. XRD measurements and TEM images of the alloy in the as-deposited state and after annealing at 90°C and 150°C revealed a structural relaxation of the amorphous state. The crystallization takes place after annealing at 250°C; it coincides with a phase separation of this alloy to Ga_2Te_3, GaTe and GaGeTe phases.

The optical characterization of $Ge_{13}Te_{67}Ga_{20}$ films has shown that the optical band gap decreases marginally from 0.95 to 0.93 eV upon annealing at 100°C. On the whole the optical constants do not change very much upon annealing of the sample at 100°C, where the amorphous relaxation is observed. Upon crystallization of the film at 250°C the band gap decreases to 0.56 eV, and a reduction of the optical constants is observed. The decrease of the optical constants is attributed to the phase separation of the amorphous film $Ge_{13}Te_{67}Ga_{20}$ into GaGeTe, Ga_2Te_3 and GaTe crystalline phases, respectively.

ACKNOWLEDGEMENT

The authors thank the Bulgarian Ministry of Education and Science for supporting this research under the contracts VUF 05 /2005 and NATO CBP.EAP.CLG.982793.

References

1. J. Scottmiller, M. Tabak, G. Lucovsky, and A. Ward, J. Non.-Cryst. Solids 4, 80 (1970).
2. S.R. Ovshinsky, Phys. Rev. Lett. 21, 1450 (1968).
3. N. Yamada, MRS. Bull. 21, 48 (1996).
4. Y. Sripathi, L.M. Malhotra, and G.B. Reddy, Thin Solid Films 270, 60 (1995).
5. D.N. Bose and S. Pal, Mater. Res. Bull. 29, 111 (1994).
6. J.C. Philips, J. Non-Cryst. Solids 34, 43 (1979).
7. F. Alapini, J. Solid-State Chem. 28, 309 (1979).

Ag-CONDUCTING CHALCOGENIDE GLASSES: APPLICATIONS IN PROGRAMMABLE METALLIZATION CELLS

A. PRADEL[*], A.A. PIARRISTEGUY
Institut Charles Gerhardt Montpellier UMR 5253 UM2 CNRS ENSCM UM1, Equipe Physicochimie des Matériaux Désordonnés et Poreux, CC 1503, Université Montpellier 2, F-34095 Montpellier Cedex 5, France

Abstract. Recent work performed on the Ag–Ge–Se glass system, a potential candidate to the fabrication of electrical memories, is reviewed in this paper. The combined investigation of the electrical conductivity by complex impedance spectroscopy and the microstructure by field emission scanning electron microscopy and electric force microscopy indicated that phase separation is a key to the understanding of the electrical properties. An ab-initio molecular dynamic simulation along with neutron diffraction experiments performed on the D4 instrument at the Institut Laue Langevin helped to understand the short and intermediate range order of the glasses. Under dynamic conditions, a thermodiffraction study carried out on the D1B instrument at the ILL evidenced the existence of a new Ag_2GeSe_3 phase at high temperatures. This phase partially decomposed with time or when the temperature decreased to $GeSe_2$ and a new $Ag_{10}Ge_3Se_{11}$ phase which can be found in the compound at room temperature.

Keywords: chalcogenide glasses; programmable metallization cells

1. Introduction

The chalcogenide glasses have been proposed as potential component of electrical memories (Programmable Metallization Cells (PMC)). This application is possible owing to the property known as photodiffusion phenomenon

_____
[*]apradel@lpmc.univ-montp2.fr

J.P. Reithmaier et al. (eds.), *Nanostructured Materials for Advanced Technological Applications*,   435
© Springer Science + Business Media B.V. 2009

present in Ge–Se or Ge–S glasses and the ionic conductivity observed in Ag–Ge–Se or Ag–Ge–S glasses. A PMC memory typically comprises a silver-photodoped glassy thin film with a composition of $\sim Ge_{0.25}Se_{0.75}$ placed between two electrodes, a silver one and a nickel one for example. The conductivity of the film is reversibly changed by several orders of magnitude when a weak voltage is applied (~ 0.3 V).[1] To date the explanation of the phenomenon is still controversial.

In this paper, a brief review of the related bulk glasses from the Ag–Ge–Se system, i.e. $Ag_x(Ge_{0.25}Se_{0.75})_{100-x}$ glasses with $x < 30$ at. %, is presented.

2. Sample Preparation

Bulk $Ag_x(Ge_{0.25}Se_{0.75})_{100-x}$ glasses with $x < 30$ at. % were prepared from high-purity (4N) elements by the melt quenching technique. The materials were synthesized by placing the powdered elements in stoichiometric proportions in a cylindrical quartz ampoule. The batches were evacuated to a pressure of $\sim 10^{-5}$ mbar and sealed. After synthesis and homogenization for 7 h at $T = 1,200$ K in a furnace, the ampoules were quenched in a mixture of ice and water to obtain the glassy materials. Glass samples are named Agx according to their concentration in at. %.

3. Conductivity and Microstructure

Electrical conductivity measurements were performed on bulk samples using the complex impedance spectroscopy technique in the frequency range of 5 Hz–2 MHz and the temperature range of 293–363 K.[2,3]

Fresh fractures of the $Ag_x(Ge_{0.25}Se_{0.75})_{100-x}$ glasses were observed by field emission-scanning electron microscopy (FE-SEM) and electric force microscopy (EFM). FE-SEM measurements were carried out using a HITACHI S5400 instrument with an acceleration voltage of 20 kV and a magnification of 3,000–30,000. The EFM experiments were performed with a Nanoscope Dimension 3100 from Veeco Instruments operating in the lift-mode in ambient conditions. Dark regions in the diagrams represent strongly attractive zones between the tip and the sample. EFM images of the samples were recorded using different voltages from −6 and +6 V.[4,5]

The dependence of the conductivity at room temperature on the silver content for $Ag_x(Ge_{0.25}Se_{0.75})_{100-x}$ glasses was studied by Kawasaki et al.[6] and Ureña et al.[2,3] Figure 1 shows the data along with the corresponding EFM images. The EFM images show that the glasses are phase separated and, while a silver-rich phase is embedded in a silver poor phase for the glass with 5 at. % Ag, the reverse is true for the glasses with $x > 10$ at. %, with the silver-rich phase controlling the silver diffusion throughout the material. The combined investigation of both techniques helped in understanding the difference of seven to eight orders of magnitude in the conductivity of the system: the percolation of the Ag-rich phase is at the origin of the sudden jump of the conductivity.[4]

Equivalent contrasts (chemical contrasts, in this case) were also observed by FE-SEM.

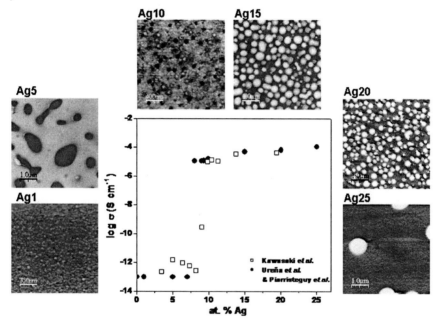

Figure 1. Variation of the conductivity at room temperature with the silver content (in at. %) for $Ag_x(Ge_{0.25}Se_{0.75})_{100-x}$ glasses (open squares: Kawasaki et al.[6]; full circles: Ureña et al.[2] and Piarristeguy et al.[3]) and EFM micrographs (applied voltage = −3 V) of the glasses with x = 1, 5, 10, 15, 20 and 25 at. % Ag (named in the figures Ag1, Ag5, Ag10, Ag15, Ag20 and Ag25, respectively).

4. Neutron Diffraction and Ab-initio Molecular Dynamics Simulation

Neutron diffraction experiments on Ag15 and Ag25 glasses were carried out on the D4 instrument at the Institut Laue Langevin (Grenoble, France).[7] Low- and room-temperature experiments were performed using a standard orange cryostat. For the experiment, the samples were placed in cylindrical vanadium containers of 8 mm outer diameter and 0.1 mm thickness. Diffraction spectra of the samples were registered at 10 and 300 K. The required ancillary measurements were also carried out. After the measurements, the raw diffraction data underwent the usual corrections performed using the CORRECT code.[8]

Ab initio molecular dynamics (MD) simulations were carried out with the Density Functional Theory (DFT) based code "Vienna Ab-initio Simulation Package" (VASP).[9,10] The Perdew–Burke–Enzerhof variation of the General Gradient Approximation to of the DFT was chosen to calculate the atomic ground state energy with VASP. The calculation runs started from a randomly distributed model with the appropriate densities for the amorphous systems (0.03695 and 0.03901 atoms/Å^3 for Ag15 and Ag25,[11] respectively). A first step of thermalization was done at 2,400 K using the compass force-field, before starting the DFT calculations. A second thermalization at 1,400 K was performed with VASP, followed by a quenching down to 400 K during 7 ps, and a third thermalization at 400 K during 4 ps. The MD time step was 2 fs.

The experimental structure factors $S(Q)$ for the samples Ag15 and Ag25 at $T = 10$ and 300 K showed similar features; only a marked decrease of the first sharp diffraction peak at 1 Å^{-1} was observed when the silver amount increases, as usual for oxides and chalcogenide glasses. Figure 2a depicts the simulated structure factor for the Ag15 glass along with that derived from neutron data. The simulation produced slightly different peak positions and widths than the experiments. Such a difference is due to the use of a soft pseudo potential for Se with a cut-off energy of 211 eV (the only one available at hand). It affected the coordination of Se by overestimating the correlation distances by about 4%. The width of the peaks in the low-Q range of the simulation data are affected by the use of a small simulation box. This is the main reason for the differences between the experimental and simulated widths of the first sharp diffraction peak.[12]

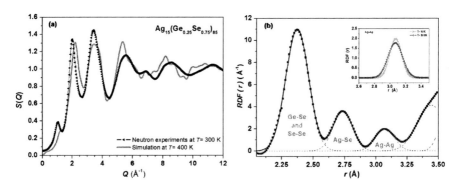

Figure 2. (a) Simulated structure factor $S(Q)$ for the Ag15 sample at $T = 400$ K compared to that derived from neutron experiments at $T = 300$ K; (b) Peak fitting of the radial distribution function RDF(r) for the Ag15 sample at $T = 10$ K. The Gaussian functions are depicted by dotted lines, the total fitted function by the solid line, and the experimental data by symbols. The inset shows the changes with temperature for the Ag–Ag correlation peak (open and full symbols for 300 and 10 K, respectively).

The radial distribution function RDF(r) for the Ag15 sample at 10 K and the Gaussian peaks corresponding to the correlations observed below 3.5 Å are shown in Figure 2b. According to the simulation, the first peak had contributions from both Ge–Se and Se–Se pairs. The next two peaks at about 2.38 and 3 Å were attributed to Ag–Se and Ag–Ag correlations, respectively.

Concerning the first peak of the RDF(r), a correlation distance for Ge–Se of 2.37–2.38 Å is observed, in agreement with literature data for a similar crystalline structure, i.e. 2.36 Å for the Ge–Se bond length in the $GeSe_2$ crystal structure.[13] The next peak at about 2.71–2.73 Å corresponds to Ag–Se correlations. A coordination distance of ~2.8 Å was found for both Ag15 and Ag25 glasses. The Ag–Se distances are in good agreement with those found in two related crystalline compounds, i.e. Ag_2Se and Ag_8GeSe_6.[14–17] The Ag15 and Ag25 glasses only retain the triangular coordination for Ag, which is in agreement with the findings of Dejus et al.[18] According to these authors this coordination might be a key for the understanding of the fast ion motion in these glasses. The third peak in the RDF(r) was attributed to Ag–Ag correlations. For both samples a correlation distance of 3.06 Å was found. It is in agreement with both the experimental data from Dejus et al. and the correlation distances that can be derived from the structure of the related crystalline Ag_2Se and Ag_8GeSe_6 phases (2.93–3.6 Å).

If we now look at the effect of the temperature on the structure of the glasses, we mainly observe the expected changes due to increased thermal vibrations with a slight increase of the lengths of Ag–Se and Ge–Se bonds. The main changes affect the peak at 3 Å attributed to the Ag–Ag correlation as shown in the inset of Figure 2b for the Ag15 sample. The peak decreases in intensity as a result of its broadening when the temperature increases. This can be explained by the increased diffusion of silver atoms at elevated temperature.

Very recently simulation results extended to Ag5 glasses showed that the silver environment depends very much on the silver content in the glassy matrix. While the glass with the lowest silver content (Ag5) showed only one correlation peak at 4.4 Å, the Ag-rich glasses possessed two peaks: a main peak at 3 Å and a small one at 4.4 Å. This result could be related to the electrical properties of the materials, namely a very low conductivity for the Ag5 sample and a high Ag^+ conductivity for the samples Ag15 and Ag25.[19]

5. Neutron Thermodiffraction

Up to this point, only static measurements were presented to investigate the Ag–Ge–Se glasses. However, the functioning of a PMC device implies a dynamic process with a voltage applied to a miniaturized component and thus some energy transfer to the glassy Ag–Ge–Se film. Therefore, a complementary investigation of the bulk glasses under dynamic conditions was thought beneficial to bring additional information to understand the phenomena underlying the functioning of PMCs. At this point the temperature was chosen as the parameter to study the glasses under dynamic conditions; more precisely, an *in situ* neutron thermodiffraction investigation of glasses from the Ag–Ge–Se system was performed.

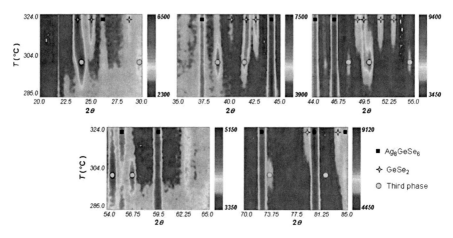

Figure 3. Neutron thermodiffractograms for the Ag25 sample during a heating ramp from 285°C up to 324°C ($\lambda = 2.5295(1)$ Å). These images reveal clearly the presence of a "third" phase.

A neutron thermodiffractometry study of the samples Ag5, Ag15 and Ag25 was carried out. The neutron diffraction experiments were performed using the D1B instrument at the Institut Laue-Langevin (Grenoble, France). The containers with the samples were placed inside a 0.2 mm thick vanadium wrap, which was located in a standard ILL furnace. The exact values of the wavelength (nominal value 2.52 Å) and the zero-angle correction were determined by means of an independent measurement using an Al_2O_3 powder sample as a standard.

Two sets of diffraction experiments were performed: the first one while the sample was heated up to 350°C and the second one under isothermal conditions at about 300°C.[20] A complete 80°-diffractogram (with steps of 0.2°) was collected in about 5 min (for the heating ramp) and 15 min (under isothermal conditions).

Figure 3 shows clearly the appearance of two main crystalline phases during the heating process for the Ag25 sample: the cubic form of the Ag_8GeSe_6 phase appeared first, followed by the $GeSe_2$ phase. Similar results were obtained for the other samples (Ag5 and Ag15). These phases were expected according to the phase diagram of the Ag–Ge–Se system as proposed by Prince.[21]

Let us now consider more closely the diffraction patterns of sample Ag25 and focus on different regions of the spectra for temperatures between 285°C and 324°C (Figure 3). While most of the peaks are the signature of the two stable crystalline phases, additional peaks are observed. For example, the peak at about $2\theta = 39°$ is the signature of a third unknown phase, that begins to crystallize at about 285°C. The three main peaks of this phase appeared at 2θ values of 24.2°, 39° and 50.2°.

(a)   **(b)**

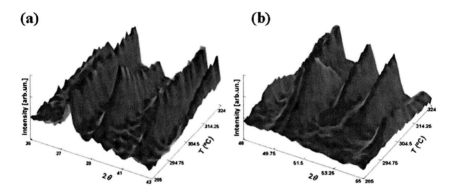

Figure 4. Neutron thermodiffractogram for the Ag25 sample during the heating ramp from 285°C up to 324°C. The graphs show specifically the peaks (a) $2\theta = 39°$, whose intensity started to decrease above 320°C and finally remained constant for the remaining part of the heating process, and (b) $2\theta = 50.2°$, which started to decrease in intensity at 300°C to disappear completely at about 315°C.

The intensity of the peaks kept increasing from 285°C, when they first appeared up to approximately 300°C. Above 315°C two different behaviors were observed. Examples of both of them are given in Figure 4 for the peaks at $2\theta = 50.2°$ and 39°. In the first case the peak completely disappeared above 315°C, while in the second case the decrease of the peak intensity stopped at this temperature and the intensity remained unchanged afterwards. Two peaks only are showing this behavior, i.e. the peaks at $2\theta = 39°$ and 54.6°.

In addition, it is clear from Figures 3 and 4 that (a) the appearance of the third phase had no impact on the crystalline Ag_8GeSe_6 phase (no change of the intensity of the peaks for example) and (b) the temperature at which the third phase started to disappear (as attested by the decrease of the peak intensity) matched the crystallization temperature of $GeSe_2$.

Another point to be noted is the fact that the two peaks of the third phase, which did not disappear during the heating ramps, were still present in the spectra recorded during the cooling process and at room temperature.

The additional phases are not present in the silver poor glass Ag5, whereas in the silver rich glasses the contribution of the phases is the stronger, the larger the Ag content.[20]

On the other hand, an annealing process at $T = 298°C$ during 5 h helped in investigating the kinetics of crystallization of sample Ag25. The cubic Ag_8GeSe_6 phase was present from the beginning of the annealing process; no change of the intensity of the related peaks was observed throughout the whole process. The contribution of the $GeSe_2$ phase was appreciable after 1 h of annealing. The new phases were also observed. An increase in the intensity of the peaks of these new phases was observed during the first 2 h of annealing; they later diminished until the end of the heat treatment. Clearly, the "lifetime" of the third phase was only a few hours, regardless of the heating treatment.

All the previous results point toward the crystallization of an unstable phase that decomposes after some time to give $GeSe_2$ and a fourth phase as attested by the presence of the two additional peaks that do not disappear upon heating, cooling or annealing.

In 2003, Ureña et al.[22] reported data obtained from several $Ag_x(Ge_{0.25}Se_{0.75})_{100-x}$ glasses by differential scanning calorimetry. A different calorimetric behavior was observed for sample Ag25 in comparison to samples with $x < 25$ at. %. A small exothermic peak between the two main peaks corresponding to the crystallization of Ag_8GeSe_6 and $GeSe_2$ was detected.

Several assumptions can be made on the composition, structure, stability of the third and the fourth. The third phase could be the ternary phase Ag_2GeSe_3. There is no agreement whether this phase exists or not. Several papers claim to prove its existence while others claim the opposite.[21,23-26]

The comparison with the analogous ternary chalcogenide glass system Ag–Ge–S could help in the interpretation of the data. A structural study of the pseudo-binary system $Ag_2S–GeS_2$ above 250°C revealed the existence of three ternary crystalline phases. In addition to the well-known Ag_8GeS_6 two other phases, i.e. Ag_2GeS_3 and $Ag_{10}Ge_3S_{11}$, were found at high temperatures.[27]

Taking into account (i) all the characteristics of the third and fourth phases derived from the neutron thermodiffraction experiments, (ii) the experimental results on the analogue system Ag–Ge–S, and (iii) the composition diagram of the Ag–Ge–Se system, an assumption on the composition of the two phases can be proposed: Ag_2GeSe_3 and $Ag_{10}Ge_3Se_{11}$. The first one has already been reported in the literature as a "transient" phase in the ternary Ag–Ge–Se system, and both of them have sulfide homologues. Ag_2GeSe_3 would appear first; being an unstable compound it would decompose at higher temperature or after few hours to give $Ag_{10}Ge_3Se_{11}$ and $GeSe_2$ according to the reaction

$$Ag_2GeSe_3 \rightarrow 0.2\ Ag_{10}Ge_3Se_{11} + 0.4\ GeSe_2.$$

Such a scheme would be in agreement with the increase in intensity of the peaks related to $GeSe_2$ when the unstable Ag_2GeSe_3 phase starts decomposing. The peaks found in the thermodiffraction pattern in addition to those related to the stable Ag_8GeSe_6 and $GeSe_2$ phases would be the signature of the new $Ag_{10}Ge_3Se_{11}$ phase.

5. Conclusion

A series of experiments was carried out on $Ag_x(Ge_{0.25}Se_{0.75})_{100-x}$ glasses with $x < 30$ at. %. The combined investigation of the electrical conductivity by complex impedance spectroscopy and the microstructure by FE-SEM and EFM helped in understanding the origin of a sudden jump of the conductivity, i.e. the existence of a percolation threshold due to a phase separation in the glasses. Neutron diffraction experiments combined with ab initio molecular dynamics simulation showed that the silver environment depended very much on the silver content; the Ag–Ag correlations are different below and above the percolation threshold.

Finally, a re-crystallization process was studied by in situ neutron thermodiffraction. This experiment evidenced the existence of a new phase, i.e. Ag_2GeSe_3 at high temperatures. Such a phase partially decomposed with time or when the temperature decreased and gave $GeSe_2$ and a new phase $Ag_{10}Ge_3Se_{11}$ that can be found in the compound at room temperature.

ACKNOWLEDGEMENTS

We thank G. J. Cuello, A. Fernández-Martínez (ILL, Grenoble F), M. Ramonda, M. Ribes (UM2, Montpellier F) for helpful advice and discussions. The work was carried out in the framework of the Project ANR-05-BLAN-0058-01.

References

1. M.N. Kozicki, M. Mitkova, M. Park, M. Balakrishnan, C. Gopalan, Superlattice Microstruct. 34, 459 (2003).
2. M.A. Ureña, A.A. Piarristeguy, M. Fontana, B. Arcondo, Solid State Ionics 176, 505 (2005).
3. A.A. Piarristeguy, J.M. Conde Garrido, M.A. Ureña, M. Fontana, B. Arcondo, J. Non-Cryst. Solids 353, 3314 (2007).
4. V. Balan, A.A. Piarristeguy, M. Ramonda, A. Pradel, M. Ribes, J. Optoelectron. Adv. Mater. 8, 2112 (2006).
5. A.A. Piarristeguy, M. Ramonda, M.A. Ureña, A. Pradel, M. Ribes, J. Non-Cryst. Solids 353, 1261 (2007).

6. M. Kawasaki, J. Kawamura, Y. Nakamura, M. Aniya, J. Non-Cryst. Solids 123, 259 (1999).
7. H.E. Fischer, G.J. Cuello, P. Palleau, D. Feltin, A.C. Barnes, Y.S. Badyal, J.M. Simonson, Appl. Phys. A: Mat. Sci. Process. 74, S160 (2002).
8. M.A. Howe, R.L. McGreevy, P. Zetterström, *Computer Code CORRECT, Correction Program for Neutron Diffraction Data* (NFL, Studsvik, 1996).
9. G. Kresse, J. Furthmüller, Phys. Rev. B 54, 11169 (1996).
10. G. Kresse, J. Furthmüller, *VASP* (Vienna University of Technology, Vienna 1999).
11. A.A. Piarristeguy, M. Mirandou, M. Fontana, B. Arcondo, J. Non-Cryst. Solids 273, 30 (2000).
12. G.J. Cuello, A.A. Piarristeguy, A. Fernández-Martínez, M. Fontana, A. Pradel, J. Non-Cryst. Solids 353, 729 (2007).
13. G. Dittmar, H. Schafer, Acta Crystallogr. B 32, 2726 (1976).
14. G.A. Wiegers, Am. Mineral 56, 1882 (1971).
15. M. Olivera, R.K. McMullan,B.J. Wuensch, Solid State Ionics 28–30, 1332 (1988).
16. W. Kuhs, R. Nitsche, K. Scheunemann, Mater. Res. Bull. 14, 241 (1979).
17. D. Carre, R. Ollitrault-Fichet, J. Flahaut, Acta Crystallogr. B 36, 245 (1980).
18. R.J. Dejus, S. Susman, K.J. Volin, D.G. Montague, D.L. Price, J. Non-Cryst. Solids 143, 162 (2007).
19. A.A. Piarristeguy et al., In preparation (2008).
20. A.A. Piarristeguy, G.J. Cuello, P. Yot, A. Pradel, M. Ribes, J. Phys.: Condens. Mat. 20, 155106 (2008).
21. A. Prince, *Ternary Alloys* (VCG, New York, 1988) p. 195.
22. M.A. Ureña, M. Fontana, B. Arcondo, M.T. Clavaguera-Mora, J. Non-Cryst Solids 320, 151 (2003).
23. R. Ollitrault-Fichet, J. Rivet, J. Flahaut, J. Less-Common Met. 114, 273 (1985).
24. A.A. Movsum-Zade, Z.Y. Salaeva, M.R. Allazov, Zh. Neorg. Khim. 32, 1705 (1987).
25. A. Velázquez-Velázquez, E. Belandría, B.J. Fernández, R.A. Godoy, G. Delgado, G.D. Acosta-Najarro, Phys. Status Solidi 220, 683 (2000).
26. A.G. Mikolaichuk, V.N. Moroz, Zh. Neorg. Khim. 32, 2312 (1987).
27. A. Nagel, K. Range, Z. Naturf. B 33, 1461 (1978).

6.3. OPTOELECTRONIC APPLICATIONS

NANOSTRUCTURED SEMICONDUCTOR MATERIALS FOR OPTOELECTRONIC APPLICATIONS

J.P. REITHMAIER[*]
Technische Physik, Institute of Nanostructure Technology and Analytics, Universität Kassel, Heinrich-Plett-St. 40, D-34132 Kassel, Germany

Abstract. This paper will give an introduction to semiconductor nanostructures, adequate fabrication technologies and examples of their applications in optoelectronics. This includes an introduction to basic properties of low dimensional electronic and photonics systems like quantum confinement, density of state function and optical transitions. An overview is given on the most relevant fabrication technologies used for optoelectronic and nanophotonic device applications. This includes bottom-up approaches like the self-organization process of quantum dots during epitaxial growth and top-down approaches based on nanolithography and dry etching. Several application examples will be discussed, where either the properties of nanostructured material or the fabrication technology itself play a dominant role for device properties. This includes quantum dot material for high-power high-brightness lasers, photonic wires for ultra-dense on-chip optical networks, and cavity enhanced single photon emitters.

Keywords: nanostructured semiconductors; nanophotonics; quantum dots; laser; amplifier; photonic crystals

1. Introduction

Semiconductor nanostructures allow the control of basic material parameters only by changing geometric factors without changing the material composition. To understand the strong impact of geometric factors in the nanometer regime one has to briefly review the bulk properties of semiconductors.

---
[*]jpreith@ina.uni-kassel.de

J.P. Reithmaier et al. (eds.), *Nanostructured Materials for Advanced Technological Applications*, 447
© Springer Science + Business Media B.V. 2009

Most of the semiconductors used in electronics and optoelectronics form diamond, like silicon (Si), or zincblende lattices, like gallium arsenide (GaAs) or indium phosphide (InP) as shown in Figure 1. During the formation of a crystal from N isolated atoms, e.g., from a liquid or gas phase, the atomic spacing decreases and the electronic valence states of each atom overlap each other forming bands with N states (Figure 2). However, the energetic distances are too small to be distinguishable. In case of the formation of a periodic lattice the electronic wavefunctions can interfere with the periodic potential fluctuation of the crystal and build a bandstructure with bands and band gap regions, where no electron can penetrate into the crystal.[1]

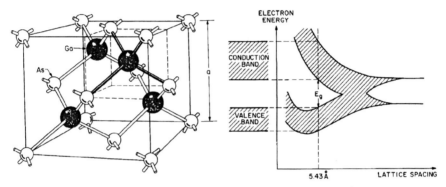

Figure 1 (left). Zincblende lattice of, e.g., GaAs or InP.[1]

Figure 2 (right). Electronic band structure formation as a function of the lattice spacing for a diamond type lattice. A band gap E_g is formed at the given lattice constant of the crystal.[1]

1.1. ELECTRONIC STATES IN BULK SEMICONDUCTOR CRYSTALS

For optoelectronic applications direct bandgap semiconductors are most appropriate because energy and momentum conservation can be easily obtained for optical transitions between two electronic states without contributions of other particles, like phonons. In Figure 3, a schematic E(k) diagram of a direct band gap material is shown, which describes the energy dispersion in k-space for the conduction band (e) and three different valence bands, i.e., heavy hole (hh), light hole (lh) and spin-orbit split bands (so). If electrons are excited to the conduction band, they occupy states near $k = 0$ and can recombine with a hole in the valence band with $\Delta k \approx 0$ by emitting a photon with an energy $E = h\nu \geq E_g$.

To understand how carriers occupy the different electronic states, one has to consider the statistical distribution function, which is in the case of electrons the Fermi–Dirac function

$$F(E) = \frac{1}{1 + \exp\left(\dfrac{E - E_F}{k_B T}\right)} \qquad (1)$$

and the density of state function D(E), which can be derived very easily in the k- or momentum space. In Figure 4, the volume of a single state in the momentum space is illustrated, which can be derived by the uncertainty relationship between space and momentum and is given by

$$\Delta x \times \Delta p_x \geq h \qquad (2)$$

where the uncertainty in space Δx can be set to the size L_x of a crystal in x-direction. For macroscopic dimensions this means a small value of the uncertainty in momentum Δp_x.

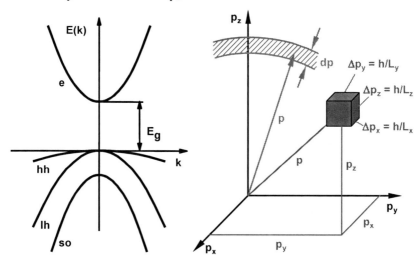

Figure 3 (left). Schematic E(k) diagram, i.e., band structure for a direct band gap semiconductor, like GaAs, with the four most important bands from top: conduction band (e), heavy hole (hh), light hole (lh) and spin-orbit split (so) valence bands.

Figure 4 (right). Sketch of the state distribution in the momentum space (p_x, p_y, p_z), which is used for the calculation of the density of state function in three dimensions. The momentum space, which a single state is occupying, is given by the uncertainty relationship $\Delta x \times \Delta p_x \geq h$ with $\Delta x = L_x$.

To get the density of state function, one has to integrate over all states within a thin spherical shell (see Figure 4) of thickness dp. Due to the homogeneous state distribution the number of states as function of p is given by

$$N(p) = 2 \cdot \int_0^p d^3p \Big/ V_p = \frac{2 \cdot \frac{4}{3}\pi p^3}{V_p} = \frac{8\pi p^3}{3h^3} \cdot L^3 , \tag{3}$$

where the volume of one state is given by $V_p = (h/L)^3$ and each state can be occupied by two electrons with different spins. The density of state function $D(p)$ is the derivative of $N(p)$ and equivalent to the number of states within the spherical shell of thickness dp. By taking into account the energy dispersion relationship given in Figure 3, which can be approximately described with a parabolic function

$$E(k) = \frac{k^2}{2m*} \tag{4}$$

where $m*$ is the effective electron mass in the crystal. With Eq. (4) the density of state function of Eq. (3) can be transferred by

$$D(E)dE = D(p)dp = \frac{dN}{dp}dp = \frac{8\pi p^2 V}{h^3}dp \tag{5}$$

and the volume $V = L^3$ to the final energy dependent density of state function

$$\rho(E) = D(E)/V = \frac{4\pi(2m)^{3/2}}{h^3}\sqrt{E} . \tag{6}$$

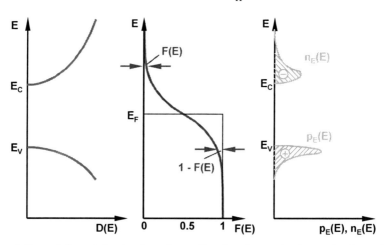

Figure 5. From left to right: density of state function for conduction and valence bands; Fermi–Dirac function for $T = 0$ K (black) and $T > 0$ (grey); carrier distribution functions for electrons and holes.

In the left diagram of Figure 5, the density of state functions for electron and valence bands are plotted. The carrier distribution can be determined by multiplying the density of state function with the Fermi–Dirac function (Figure 5, plot in the middle), which is shown on the right side of Figure 5. Due to the heavier mass, the holes are more localized near the band edge while electrons are more widely distributed in energy. Due to the shift of the carrier density maxima away from the band edges, optical transitions are more favoured at energies larger than the band gap, which leads in semiconductor lasers to a blue shift of the emission wavelength. The carriers below the carrier density maximum cannot be utilized in lasers and increase the threshold current densities. This point will be of importance once again by discussing the situation in low dimensional systems.

1.2. ELECTRONIC STATES IN LOW DIMENSIONAL SYSTEMS

In lower dimensional systems the motion of electrons is restricted and the energy states are quantized. In Figure 6, the electron wavefunctions and energy eigenstates in a quantum well system with finite potential height is illustrated.[2]

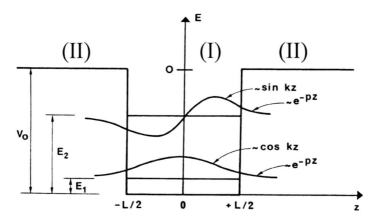

Figure 6. Eigenfunctions and quantized energy states of electrons in a finite potential well of width L and barrier height V_0.[2]

Here the motion of electrons is restricted in z direction, however, in x and y direction the electron can still increase energy and momentum, e.g., by an external field parallel to the quantum well layer as illustrated in Figure 7.

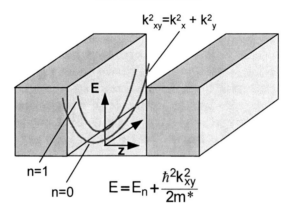

Figure 7. Dispersion functions of electrons moving parallel to the quantum film. The dispersion relationship for the n-th state is indicated.[2]

The density of state function of such a system can be derived in a very similar way as described above. Due to the restriction of motion in one dimension the states in k-space are only distributed in a plane as illustrated in Figure 8. In this case, only states within the indicated circle of radius p have to be considered.

The area of one state is $A = (h/L)^2$ and the number of states within a circle of radius p is given by

$$N(p) = 2 \times \frac{\pi p^2}{(h/L)^2} \,.$$ (7)

The density of state function is given as

$$D(p) = \frac{dN}{dp} = \frac{4\pi p A}{h^2} \,.$$ (8)

By using the same energy dispersion function given in Eq. (4) and setting $k = k_{xy}$ with $k^2_{xy} = k^2_x + k^2_y$ the density of state function is given by

$$D(E) = \sum_{i=1}^{n} \frac{m^* L^2}{\pi \hbar^2} \Theta(E - E_i)$$ (9)

by summing over all quantized energy states E_i. $\Theta(E - E_i)$ denotes the Heavyside function, which is 1 for $E = E_i$ and 0 else. This density of state function is plotted in Figure 9 and results in a step-like function with a constant density of states for each discrete energy level. In the limit of an infinite quantum well width this function converges to the 3D case as indicated by the dashed square root function in Figure 9.

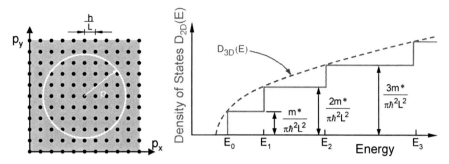

Figure 8 (left). Electron state distribution in the p space. The circle indicates the integration area up to momentum p.

Figure 9 (right). Two-dimensional density of state function (step-like function) in comparison to the density of state function of bulk material (dashed line).

By further reducing the dimensionality to one or zero, the density of state function as shown in Figures 10 and 11, respectively, degenerates to sharp peaks at the quantized energies. In the case of one-dimensional systems, the function is an inverted square root function, while it is a delta function in the case of zero dimension. In Figure 12, the density of state functions of the electronic ground state are summarized in a table for different dimensionalities.

For optical transitions in a nanostructured semiconductor, e.g., a quantum dot material, the overlap of the wavefunctions has to be significantly large. In Figure 13, the electron and hole wavefunctions are plotted as surfaces of constant probability for a pyramidally shaped quantum dot.[3] One can easily see that the electronic C_2 state has only a very weak overlap with the hole state V_2, which leads to a negligible optical transition.

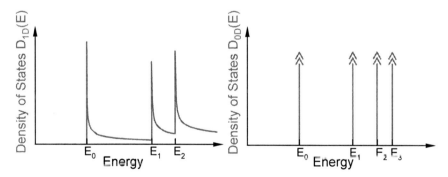

Figure 10 (left). One-dimensional density of state function.

Figure 11 (right). Zero-dimensional density of state function.

dimension	density of state D(p)	density of state D(E)
3D	$\dfrac{8\pi V}{h^3} \cdot (p)^2$	$\dfrac{V}{2\pi^2}\left(\dfrac{2m^*}{\hbar^2}\right)^{3/2} \cdot (E)^{\frac{1}{2}}$
2D	$\dfrac{4\pi A}{h^2} \cdot (p)^1$	$\dfrac{A}{2\pi}\left(\dfrac{2m^*}{\hbar^2}\right)^{1} \cdot (E)^{0}$
1D	$\dfrac{2L}{h} \cdot (p)^0$	$\dfrac{L}{\pi}\left(\dfrac{2m^*}{\hbar^2}\right)^{\frac{1}{2}} \cdot (E)^{-\frac{1}{2}}$
0D	$\delta(p-p_i)$	$\delta(E-E_i)$

Figure 12. Density of state functions as function of momentum (left side) and energy (right side) of the electronic ground state for different dimensionalities.[2]

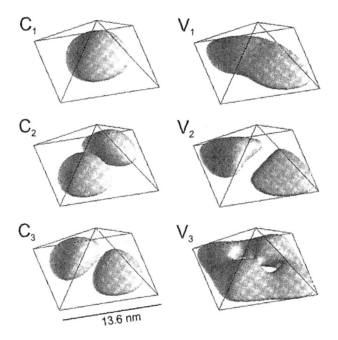

Figure 13. Electron (C_1, C_2, C_3) and hole (V_1, V_2, V_3) wavefunctions of pyramidally shaped quantum dots.[3]

1.3. PHOTONIC STATES IN PHOTONIC CRYSTALS

In optoelectronic systems one would like to have a full control over the electrons as well as the photons, which interact with each other. A possibility to control the light propagation or to confine light in a small volume is possible with so-called photonic crystal (PC) structures, which consist of a periodic modulation of the dielectric constant.

In Figure 14, several examples of photonic crystals are shown with different dimensionalities.[4] 1D PCs can be easily made by thin film deposition of materials of different refractive indices and are, e.g., used in vertical cavity surface emitting lasers (VCSELs) to build a one-dimensional optical cavity for light confinement. 2D PCs can be made by etching holes or pillars into semiconductors to form a 2D lattice, where electromagnetic waves are blocked for specific energies and directions. By introducing crystal lines or point defects one can guide or confine light, respectively, within a much reduced volume – even below the wavelength – in the material. The fabrication of 3D PCs is the most challenging approach, which exhibit, similar to electronic properties of real crystals, a 3D photonic band structure and allow a full 3D confinement of light in crystal defects.

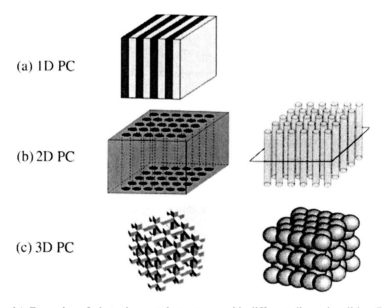

(a) 1D PC

(b) 2D PC

(c) 3D PC

Figure 14. Examples of photonic crystal structures with different dimensionalities. Details are described in the text.

In Figure 15, the light dispersion functions for different one-dimensional PC systems are shown.[5] Due to the linear relationship between energy and momentum of the photon according to the relationship

$$E = \hbar\omega = pc = \hbar kc \qquad (10)$$

one gets a linear dispersion function

$$\omega(k) = c \cdot k \qquad (11)$$

with c as speed-of-light.

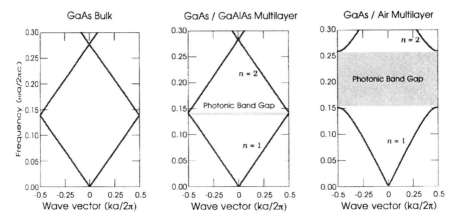

Figure 15. Photonic dispersion functions in GaAs bulk, GaAs/AlGaAs multilayers, and GaAs/air multilayers (from left to right).[5]

In bulk material with a homogeneous refractive index no restriction exists for the light propagation (Figure 15, left side). However, in periodic structures, like GaAs mulilayers, a small bandgap emerges, which forbids light propagation perpendicular to this multilayer structure (middle). By exchanging each second layer by air, the bandgap is strongly enlarged by the much higher refractive index contrast (right side).

By omitting holes (Figure 16a) or columns (Figure 16b) in a 2D-PC structure, light can be localized either in point defects or guided in line defects. Donor like states can be obtained by omitting holes (see Figure 16a), while acceptor like states by enlarging holes. Although light can be localized in a point defect the wavefunction penetrates into the surrounding of the cavity over several crystal periods as shown in Figure 17. Here, the localization is realised by an omitted column, which results in a maximum field amplitude in air.[6]

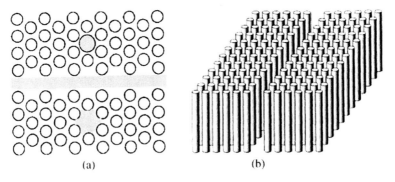

(a)   (b)

Figure 16. (a) 2D photonic crystal structure with donor and acceptor like point defects and a single line defect. (b) Schematic 3D plot of a planar PC structure with a single line defect.[4]

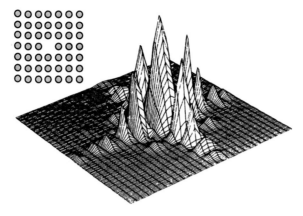

Figure 17. Intensity distribution function of a confined photonic state in a point defect PC cavity structure consisting of dielectric columns.[6]

In Figure 18, an overview is given on the influence of the dimensionality on the density of state functions for electronic and photonic systems.[7] In optoelectronic devices, the overlap of the two systems is important. For example, in semiconductor quantum dot lasers a quasi zero-dimensional electronic system is coupled with a ridge waveguide, which is equivalent to a one-dimensional photonic system.

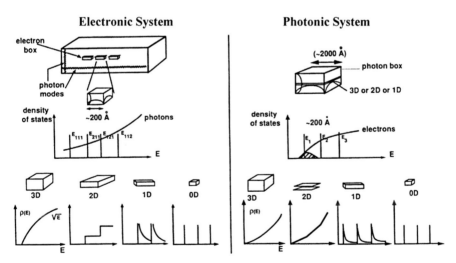

Figure 18. Overview on the influence of the dimensionality on the density of state functions in electronic and photonic systems. The middle part illustrates the spectral overlap of zero-dimensional systems with the 3D counterpart, e.g., delta-like electronic density of state functions overlaps with the quadratic density of state functions of unconfined photons.[7]

1.4. LIGHT MATTER INTERACTION

For optoelectronic devices the interactions between light and electronic
states are of fundamental interest. The basic optical transitions in a two
level system can be described by Fermi's Golden Rule

$$R_r = \frac{2\pi}{\hbar}\left|H'_{21}\right|^2 D(E)\cdot\delta(E-\hbar\omega) \tag{12}$$

where R_r is the radiative transition rate, $D(E)$ the density of state function
and H'_{21} the transition matrix element given by

$$H'_{21} \equiv \left\langle\psi_2\left|H'(\mathbf{r})\right|\psi_1\right\rangle = \int_V \psi_2^* H'(\mathbf{r})\psi_1 d^3 r \tag{13}$$

with the wavefunctions ψ_1 and ψ_2 for ground and excited state, respectively.

   The material gain of a semiconductor system depends on the density of
state function, the transition matrix element and the occupation of the
excited and ground states and is given by[8]

$$g(E_{21}) = \frac{\pi q^2 \hbar}{n\varepsilon_0 c m_0^2}\frac{1}{\hbar\omega_{21}}\left|M_T(E_{21})\right|^2 \rho_r(E_{21})(f_2 - f_1) = g_{max}(E_{21})\cdot(f_2 - f_1) \tag{14}$$

with the transition energy E_{21}, the transition matrix element $M_T(E_{21})$, the
reduced density of state $\rho_r(E_{21})$ und the Fermi–Dirac functions f_1, f_2 for
valence and conduction band, respectively.

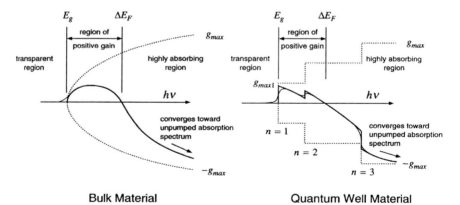

Figure 19. Material gain functions for bulk and quantum well material. The dotted lines
follow the maximum absorption ($-g_{max}$) or gain (g_{max}) possible according to the density of
state functions. The solid line show the actual gain depending on the pumping intensity.[8]

In Figure 19, the material gain functions for bulk and quantum well materials are shown. The dotted lines mark the maximum positive or negative gain according to the corresponding density of state functions. Negative gain is equivalent to optical absorption. By optical or electrical pumping the excited states can be filled and the material turns to positive gain for a certain spectral range (solid lines). Due to the step-like density of state function in a quantum well material, the carrier density at the lowest energy state is much higher than in bulk material. This allows to decrease the necessary carrier density to obtain transparent condition of a gain material and hence allows to operate QW lasers at reduced threshold current densities.

A further reduction in transparent carrier density can be achieved by using quantum dots as gain material. However, in real systems the dot size is varying leading to inhomogeneous broadening of the transition energies (see Figure 20).

The spectral gain function for such a system, considering only the fundamental dot transitions, is given by

$$g(\hbar\omega) = \frac{\pi q^2 \hbar}{n\varepsilon_0 cm_0^2} \frac{1}{\hbar\omega} \cdot \int d\varepsilon \sum_n \left| M_{T,n}(\hbar\omega) \right|^2 \frac{2}{V_0} P_n\left(\varepsilon, \varepsilon_n, \sigma_n\right)$$

$$\cdot [f_c(\varepsilon, E_{Fc}) - f_v(\varepsilon, E_{Fv})] L_n(\hbar\omega, \varepsilon) \quad (15)$$

with f_c and f_v as quasi-Fermi functions in the conduction and valence band, respectively, $L_n(\hbar\omega, \varepsilon)$ as Lorentian function, which substitutes the δ function in Eq. (12) and describes the homogeneous broadening while $P_n\left(\varepsilon, \varepsilon_n, \sigma_n\right)$ describes the inhomogeneous broadening, caused by the dot size distribution. The inhomogeneous broadening is described by a Gaussian function and given by

$$P(\varepsilon, \varepsilon_n, \sigma_n) = \frac{1}{\sigma_n \sqrt{2\pi}} \exp\left[-\frac{(\varepsilon - \varepsilon_n)^2}{2\sigma_n^2} \right] \quad (16)$$

with σ_n as full width at half maximum value for the n-th dot transition. In Figure 21, this gain function is plotted for a single transition energy and different Fermi levels, i.e., chemical potentials μ.

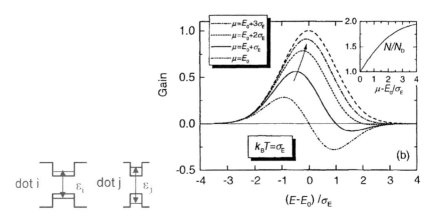

Figure 20 (left). Scheme of dot transitions in a inhomogeneously broadened dot ensemble with transition energies ε_i and ε_j.

Figure 21 (right). Spectral gain for inhomogeneously broadened dot ensemble with one confined state. μ = chemical potential; σ_E = inhomogeneous linewidth.

2. Fabrication Technologies

Due to the huge progress made in optical lithography, the fabrication of nanostructured semiconductors is – with some restrictions – also possible with standard lithographic tools, like optical steppers. However, for structure dimensions on the order of less than 50 nm, other lithographic tools are necessary in combination with advanced mask and etching techniques or self-organization effects have to be applied for forming especially high density nanostructured materials. This chapter gives a brief overview about fabrication technologies with a special emphasis on direct writing and the growth of self-assembled III/V quantum dots.

2.1. LITHOGRAPHY

In Figure 22, an overview is given on different lithographic techniques (figure from Waser et al. with addition of the imprint technology[9]). All lithographic techniques start with the structure definition by a computer aided design (CAD) tool, which allows with the help of a control programme to write the structure onto an optical mask, imprint mask or directly into the resist of a resist coated wafer. In case of optical lithography, the mask is either used for contact, proximity or projection lithography. In mass production, optical steppers are used, which project the mask on the wafer with a size reduction of typically a factor of 4.

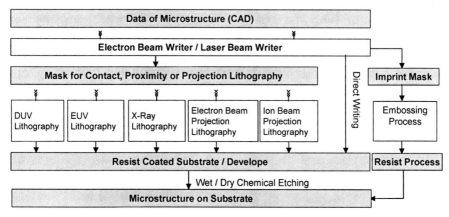

Figure 22. Overview of advanced lithographic techniques based on structure definition, mask writing, mask transfer and structure transfer into semiconductors by etching processes.[9]

With this standard technique feature sizes of 45 nm are obtained already in production and <30 nm in development. Here, deep ultra violet (DUV) light sources with wavelengths on the order of 157–250 nm are used. With extreme UV light (11–14 nm), which is in the development stage, the way might be open also to 10 nm line widths. In Figure 23, a sketch of such a tool is shown.

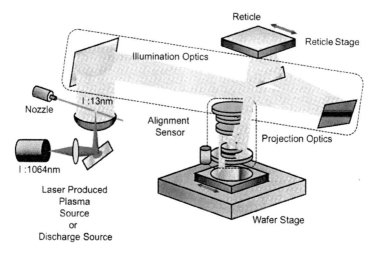

Figure 23. Optical stepper for EUV exposure based on reflection optics for 13 nm wavelength.[9]

As alternative techniques, x-ray, electron and ion beam projection technologies are in development but with an unclear future due the very expensive and challenging mask technologies related to these techniques and the limited advantages in comparison to the new types of optical lithography systems.

A more promising and cheaper lithographic technique especially for medium production volumes is nano-imprinting, where a stamp already containing the nanostructure design is used to copy it into a resist coated wafer. In Figure 24, the process flow for nano-imprinting is illustrated.[10]

For ultimate small structure sizes, direct writing with electrons or ions is necessary. In Figure 25, an research type e-beam writer system from Raith is shown, which has a beam size of <2 nm at 10 keV. A cross section of an e-beam column is shown in Figure 26.

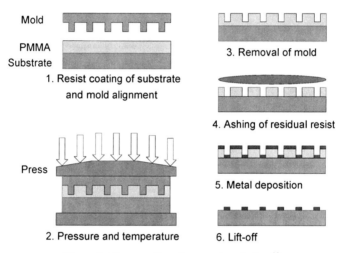

Figure 24. Process flow for nano-imprinting.[10]

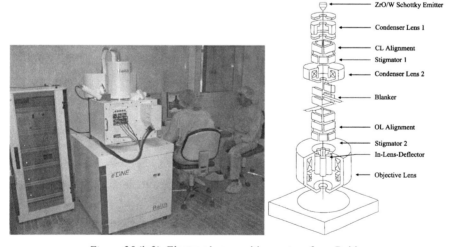

Figure 25 (left). Electron beam writing system from Raith.

Figure 26 (right). Sketch of an electron beam writer column.

2.2. PATTERN TRANSFER BY ETCHING

After mask patterning by lithography the structure has to be transferred to the semiconductor either by wet- or dry-chemical etching. In Figure 27, different properties of dry and wet etching are summarized by distinguishing between isotropic and anisotropic behavior. For crystallographic etching and damage free surface preparation wet-chemical etching is preferred, while for high aspect ratio and underetch-free structures dry etching is more favorable.

For dry etching, mainly reactive ion etching (RIE) techniques are used as illustrated in Figure 28. Here a high frequency and high voltage power supply is used to ionize injected gases.[11] The process is performed under high vacuum conditions. In III–V materials typical etch rates of 0.5 μm/min can be achieved.

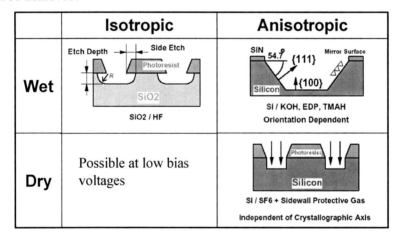

Figure 27. Isotropic and anistropic etch behavior of wet- and dry-chemical etch processes.

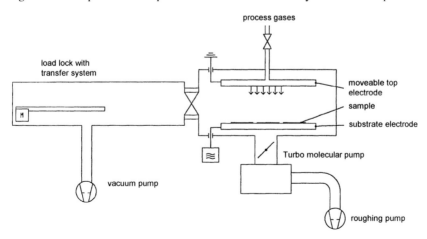

Figure 28. Schematic drawing of a reactive ion etching system with a sample transfer system.[11]

There are two major etch phenomena one has to consider in RIE, i.e. physical and chemical etching. In Figure 28, the two phenomena are shown in order of increasing gas pressure and decreasing ion energy. At high ion energy and low gas pressure physical etching is dominating, which results in an anisotropic etch behavior. At low ion energy and high gas pressure, which is equivalent to a high plasma density, the etching is more isotropic, which results in a significant undercut. By combining both effects with the right balance one can obtain vertical side-walls with high aspect ratios.

In Figures 30 and 31, two examples of nanostructured semiconductors with high aspect ratios are presented. Figure 30 shows a photonic crystal membrane structure, where the holes are dry-etched while the sacrificial layer underneath the photonic crystal is removed by wet chemical etching.[12] Figure 31 shows a grating etched in silicon with 10 nm wide ridges and an aspect ratio of 20. Here, a so-called gas chopping process in an inductively coupled plasma (ICP) RIE reactor was used with alternating sidewall passivation and etching steps.[13]

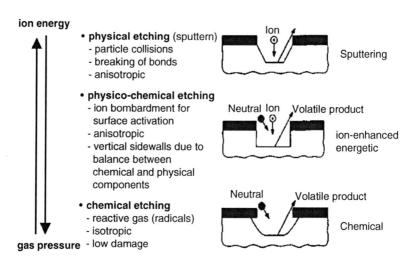

Figure 29. Basic phenomena occurring during a plasma etch process as, e.g., in RIE. The top most effect is dominating at high ion energies and low gas pressures, while the bottom process is favored at low energies and high gas pressures. By utilizing both effects, vertical sidewalls can be obtained with optimizsed etching parameters.

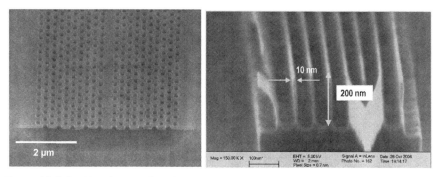

Figure 30 (left). SEM image of a PC membrane structure with a single line defect as optical waveguide.[12]

Figure 31 (right). High aspect ratio nanostructures etched into silicon by ICP-RIE.[13]

2.3. NANOSTRUCTURE FABRICATION BY SELF-ORGANISATION EFFECTS

Although dimensions in the order of 10 nm and below can be achieved by high resolution lithography and etching, the crystal structure on the atomic scale can be easily damaged, which leads to a strong degradation of the optical properties of III–V semiconductors. Since beginning of 1990s, self-organization effects during epitaxial processes were developed, which overcome this problem.

In solid state molecular beam epitaxy (MBE) elemental sources are used to form compound semiconductors.[14] In Figure 32, a cross section of an MBE from Vacuum Generators is shown, while in Figure 33 the major functional items are illustrated. The atoms and molecules are evaporated from heated cells and are deposited on a heated substrate. The atom flux is controlled by mechanical shutters, which allow the control of layer thickness on a sub-monolayer scale. Typical growth rates are in the range of 1 μm/h.

By depositing strained materials like InAs on GaAs, a stress energy driven formation of nanoscale islands occurs, which have the properties of quantum dots. In Figure 34, typical examples for quantum dots and quantum dash structures are shown, which are grown on GaAs and InP, respectively. The typical dot width of InAs on GaAs is about 20 nm in width and 5 nm in height, while InAs dashes on InP are elongated by 50–100 nm and behave more as quantum wires.[15,16]

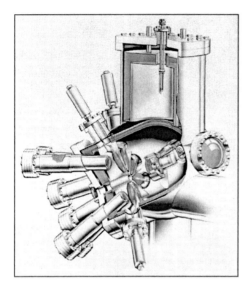 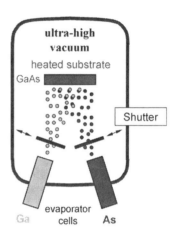

Figure 32 (left). Cross section of an MBE system (model V80H from Vacuum Generators).

Figure 33 (right). Sketch of the major functional items of an MBE system.

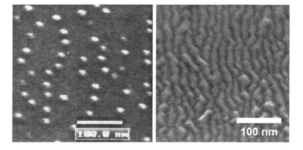

Figure 34. SEM surface images of an InAs quantum dot sample grown on GaAs (left) and an InAs quantum dash sample grown on a GaInAlAs surface lattice matched to InP (right).

3. Applications of Nanostructured Materials in Optoelectronic Devices

At the present time, nanostructured materials are used in a large variety of optoelectronic device applications. In this chapter a few examples of application of nanostructured semiconductors for new types of photonic and optoelectronic devices will be highlighted with a major focus on quantum dot material. Further reviews about QD lasers for telecom applications are given in the literature.[17–20]

3.1. HIGH-BRIGHTNESS HIGH-POWER QUANTUM DOT LASERS

As shown above, quantum dots can be formed in high crystal quality by self-organization during the epitaxial growth of layers. Due to the low-dimensional properties the gain function can be strongly modified by geometric parameters, like dot size, dot density and dot size distribution. In Figure 35, the influence of these three major parameters on the optical properties is illustrated. While the dot size, i.e., mainly the dot height, controls the splitting energy of two quantized levels, the dot uniformity controls the overlap of two transition peaks. The dot density has a direct impact on the modal gain of the structure (the modal gain is the material gain weighted by the overlap of a propagating optical mode). At low dot densities, less dots contribute to the emission and the emission intensity saturates, while more carriers are excited to higher order states.

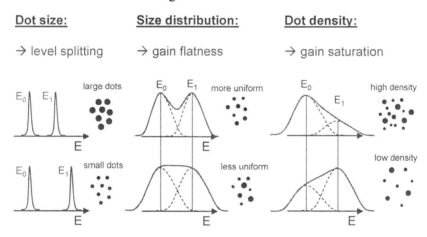

Figure 35. Illustration of the impact of three major geometrical parameters of dots layers on optical material properties.

This possibility to tailor material properties (e.g. the optical gain function) by additional geometric parameters allows to optimize device properties for specific applications. As an example, quantum dots layer can be optimised in such a way that one creates – with the right energy splitting, inhomogeneous linewidth broadening and low saturation density – a flat gain profile as illustrated in Figure 36. Is this material used in a semiconductor laser, the laser will emit at temperature T_1 at a certain emission wavelength according to the threshold condition as indicated in Figure 36. By increasing the temperature to T_2 the losses in a laser increase, which has to be compensated by higher current injection resulting in carrier filling of higher order transitions (blue shift of emission). However, at the same time the bandgap

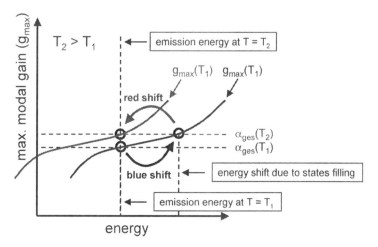

Figure 36. Illustration of the principle effects involved for the internal compensation of the emission wavelength shift in quantum dot gain material of a laser.

is reduced with temperature due to band renormalization (left curve). If the gain function is flat enough the carrier filling effect can internally compensate the usual emission red shift of the lasers to keep the emission wavelength constant.[21] The temperature dependent wavelength change is a severe problem in wavelength sensitive low-cost applications, because otherwise a rather expensive thermoelectric cooler has to be used to stabilize the operation temperature.

In this particular case, the QD lasers are developed as temperature insensitive high power light sources for un-cooled pump modules, which can be used as pump sources for Yb doped fibre lasers or amplifiers for telecom applications.[22,23] Here the absorption band of Yb is very narrow and would not allow to operate a pump laser in a normal ambient temperature range of 0–85°C.

In Figure 37, the wavelength shift of 980 nm QD lasers with different types of QD layers are shown in comparison to a typical QW laser with a similar emission wavelength. Here the wavelength shift can be reduced by more than a factor of 3.[21] The output power characteristics of such QD lasers are plotted in Figure 38 for different stripe widths and operation temperatures. The lasers were epi-side down mounted on a copper heat sink. Output powers >6 W could be obtained from a single broad area laser with 1 mm cavity length and 100 μm stripe width.[24]

Recently, also short wavelength 920 nm QD lasers were developed, which show a new record of temperature stability together with a high device performance.[25] In Figure 39, the light output characteristics and curves for the wall-plug efficiencies of lasers of three different lengths are shown. Output powers of >2.5 W/facet (i.e., nearly 5 W total power) are reached, which are limited by the maximum current of the pulse source used. A high maximum conversion efficiency (=wall-plug efficiency) of 54% could be achieved with a cavity length of 0.89 μm. In Figure 40, the wavelength shift of a device with 2.6 mm cavity length is shown, which exhibits a record low temperature coefficient of 0.08 nm/K.

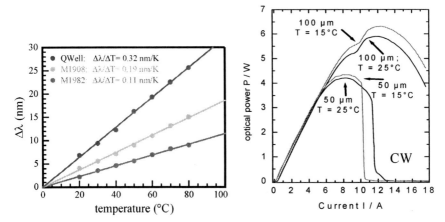

Figure 37 (left). Wavelength shift as a function of temperature for two different types of 980 nm QD lasers in comparison to a QW laser.

Figure 38 (right). High power cw output characteristics for two different temperatures and stripe widths of epi-side down mounted 980 nm broad area QD lasers with 1 mm cavity length.[24]

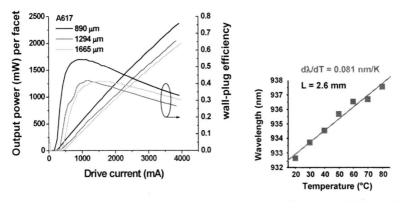

Figure 39 (left). Light output characteristics and wall plug efficiencies of 920 nm QD lasers with different cavity lengths.[25]

Figure 40 (right). Temperature dependence of the emission wavelength of a 2.6 mm long 920 nm QD laser.[25]

In many applications, the beam quality is of high importance. In the ideal case the laser beam should have a Gaussian profile, which allows the best focusing properties. The figure of merit M^2 describes how near the beam profile is to a Gaussian beam. In ideal case $M^2 = 1$, which is the diffraction limit. Values better than $M^2 = 3$ are considered already as near to diffraction limit. To combine a single emission wavelength (single lateral and longitudinal) with excellent beam properties at high output powers together with a high temperature stability, QD lasers with feedback gratings and tapered geometry were developed as shown in Figure 41. The tapered laser geometry allows a single-lobe output at high power levels. The feedback grating selects and stabilizes the longitudinal mode, and the QD gain material with its low temperature dependence and broad gain spectrum allows a temperature independent operation.[26]

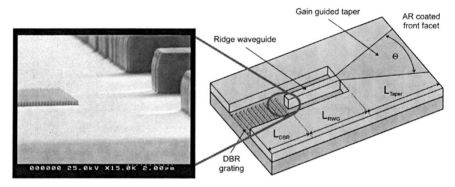

Figure 41. Sketch of tapered QD laser with surface defined feedback gratings. On the left is SEM image of the surface grating (left) nearby the ridge waveguide section (right).[26]

Due to the feedback grating, the laser emits in single mode with a side-mode suppression of 45 dB (see Figure 43) and a maximum output power of more than 1W (see Figure 42).

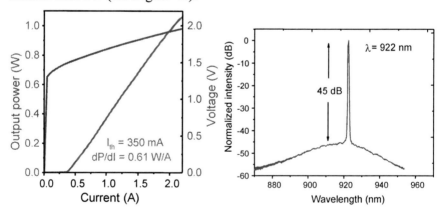

Figure 42 (left). Output power and voltage characteristics of a tapered QD laser with feedback gratings.[26]

Figure 43 (right). Single mode emission spectrum of a 920 nm high power tapered QD laser at room temperature.[26]

3.2. PHOTONIC WIRES FOR HIGHLY INTEGRATED OPTICAL NETWORKS

To miniaturize and integrate optoelectronic and passive photonic devices in the optical domain, one needs compact optical networks on a chip. There are different approaches to realize this goal. By using photonic crystals, one can achieve very compact waveguides (see Figure 30) and waveguide bends. However, the waveguide properties are strongly dependent on the wavelength due to the crystal period, and on the propagation direction due to the directionality of crystal lattices. An alternative, which propose broadband transmission over a wide wavelength range, are ridge waveguides. While standard glas fibres do only allow bend radii on the order of several millimeters due to their low index contrast, one can obtain much smaller bend radii in semiconductors. Although ridge waveguide structures are already widely used for optoelectronic integrated circuits (OEICs), the effective refractive index contrast is still limited and allows only bend radii on the order of several 100 µm, which is not enough for highly dense optical networks.

Very recently, new types of high index contrast ridge waveguides have been developed to utilize a full optical confinement of a high index semiconductor stripe with low index surrounding materials, which is either air or a dielectric.[27] Due to the single-mode behavior these waveguide structures are called photonic wires. A cross section of such a structure is shown in Figure 44 (bottom). The GaAs waveguide confines in this case the mode only partially. A significant part is guided in air or in the underlying dielectrical layer. The lateral dimensions are 500 nm, while the waveguide layer thickness is only 200 nm. The SEM picture above show a top view of a bent waveguide with a bend radius of only 1 µm. In Figure 45, the optical loss in such a waveguide is plotted as a function of the bend radius and compared to 3D simulations. Down to a bend radius of 1.6 µm the measured values are in excellent agreement with the theory and confirm very low losses of less than 0.05 dB/bend or 99% transmittance. With this low loss values more than 50 bends could be applied in a passive network before 50% of the intensity get lost.

In Figure 46, a more complex network was realised with 2 µm bend radius including 3dB splitters.[28] The optical output at the cleaved end facet show very circular beam profiles with equal intensity distributions between each channel, demonstrating excellent structure quality.

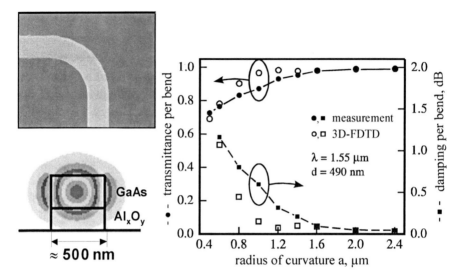

Figure 44 (left). SEM image (top view) of a photonic wire waveguide bend (top) and a vertical sketch of the layer structure and mode intensity profile.[27]

Figure 45 (right). Simulated (open dots) and measured transmittance and damping values per bend as function of the curvature radius of a photonic wire. The waveguide width is 490 nm and the measurement wavelength 1.55 μm.[27]

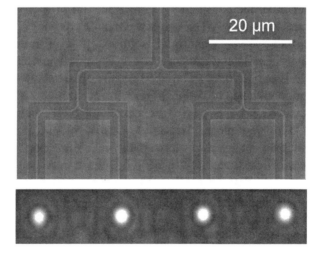

Figure 46. SEM image (top) of a passive optical network consisting of 500 nm wide photonic wires with 2 μm bend radius and 3dB splitters. The far field emission spectra recorded by an CCD camera is shown below.[28]

3.3. SINGLE PHOTON EMITTING STRUCTURES AND DEVICES BASED ON HIGH-Q CAVITIES

Ultimate small devices going to the theoretical limit of optoelectronics are single-photon emitters based on single-electron transitions. For this purpose localized electronic states in quantum dots have to be combined with high-Q optical cavities to confine in all three dimensions a single mode with a maximum spatial and spectral overlap. Two different approaches are shown in Figures 47 and 48. While in the micropillar case (Figure 47), the optical cavity is build by a Bragg resonator with bottom and top distributed Bragg reflectors (DBRs), the optical cavity in Figure 48 is built by a photonic crystal membrane cavity in lateral direction. In both cases Q values on the order of 10^5 and beyond could be demonstrated, and a single dot is localized in the maximum of the optical mode. An overview about strong-coupling effects in semiconductor cavities is given in the literature.[29]

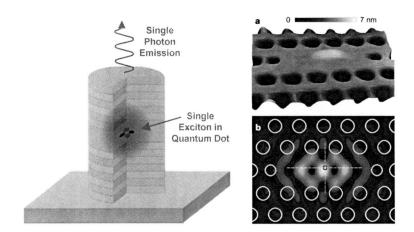

Figure 47 (left). Sketch of a single photon emitter based on a high-Q micropillar cavity. The QD is placed at the field maximum in the center.

Figure 48 (right). High-Q cavity based on a L3 PC cavity with a single QD placed near the field maximum: AFM image (top) and simulations (bottom).[30]

In Figure 49, the Q-factor as function of the micropillar diameter is plotted for two different layer structures. The MC2 structure has a few Bragg mirror layers more than MC1 and shows a clear improvement. However, at small pillar diameters the surface roughness scattering reduces the Q-value from about 180,000 down to about 20,000 at 1.5 µm.[31]

Due to the relatively high Q-value at small diameters, the photons couple strongly with the exciton localized in a single dot.[32,33] In this strong-coupling regime these two particles (exciton and photon) build a polariton as a new quasi-particle. This means that exciton and photon exchange coherently energy with each other and build a new combined quantum state.

To confirm that the emission at low excitation density is really based on single photons, autocorrelation measurements were performed. In Figure 50, the coincidence spectrum to detect two photons at the same time is plotted. At zero delay the amplitude goes down showing a high probability of single photon emission. For a conventional source one would expect similar amplitudes independent on the time delay.[34]

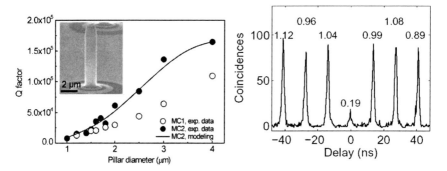

Figure 49 (left). Q-values of micropillars with different diameters and two different layer structures.[31]

Figure 50 (right). Coincidence spectrum of an autocorrelation measurement of micropillar single-photon emitters in the strong coupling regime.[34]

Although these structures clearly show single-photon emission, one need more functionality to use them as real devices. A very important part is the electrical excitation and control of the device. In Figure 51, an electrically driven high-Q micropillar structure is shown, which allow electrical excitation in forward bias as well as electrical tuning in reverse bias.[35] The spectrum in Figure 52 shows the strong cavity enhancement during temperature tuning of such a device confirming the preservation of high-Q values also after metallization.

This field of cavity quantum electrodynamics using high-Q semiconductor cavities is now approaching device level; recent results show already strong coupling effects using electrical tuning.[36] Although all these devices currently are only working at cryogenic temperatures, electrical access to the functionality and the superior quality in terms of directional coupling makes applications in the field of quantum key distribution of secure communications channels already feasible in the near future.

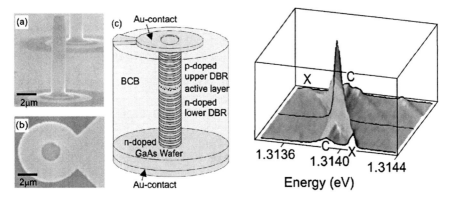

Figure 51 (left). High-Q micropillars with metal contacts for electrical injection or electrical detuning of the exciton transition.[35]

Figure 52 (right). Emission spectrum with a strong cavity enhancement at the spectral overlap of exciton (X) and cavity resonance (C).[35]

References

1. S.M. Sze, *Physics of Semiconductor Devices*, Wiley, New York (1981).
2. P. Harrison, *Quantum Wells, Wires and Dots*, Wiley, Chichester (2000).
3. D. Bimberg, M. Grundmann, N.N. Ledentsov, *Quantum Dot Heterostructures*, Wiley, Chichester (1999).
4. K. Inoue, K. Ohtaka, *Photonic Crystals: Physics, Fabrication and Applications*, Springer, Berlin (2004).
5. J.D. Joannopoulos, R.D. Meade, J.N. Winn, *Photonic Crystals: Molding the Flow of Light*, Princeton University Press, Princeton, NJ (1995).
6. K. Sakoda, *Optical Properties of Photonic Crystals*, Springer, Berlin (2004).
7. C. Weisbuch, H. Benisty, R. Houdré, J. Lumin. 85 (4), 271 (2000).
8. L.A. Coldren, S.W. Corzine, *Diode Lasers and Photonic Integrated Circuits*, Wiley, New York (1995).
9. R. Waser, *Nanoelectronics and Information Technology*, Wiley, Weinheim (2003).
10. M. Kamp, University of Würzburg, private communication.
11. G. Franz, *Oberflächentechnologie mit Niederdruckplasmen. Beschichten und Strukturieren in der Mikrotechnik*, Springer, Berlin (1998).
12. Courtesy of A. de Rossi, S. Combrie, Thales Research & Technology, 2007.
13. I.W. Rangelow, D.L. Olynick, J.A. Liddle, EIPBN Conference, Orlando, USA, 2005.
14. M. A. Herman, H. Sitter, *Molecular Beam Epitaxy - Fundamentals and Current Status*, Springer, 1989.
15. R. Krebs, F. Klopf, J.P. Reithmaier, A. Forchel, Jpn. J. Appl. Phys. 41 (2B), 1158 (2002).
16. R. Schwertberger, D. Gold, J.P. Reithmaier, A. Forchel, IEEE Phot. Technol. Lett. 14 (6), 735 (2002).
17. J.P. Reithmaier, A. Forchel, Comptes Rendus Physique 4 (6), 611 (2003).

18. J.P. Reithmaier, A. Somers, S. Deubert, R. Schwertberger, W. Kaiser, A. Forchel, M. Calligaro, P. Resneau, O. Parillaud, S. Bansropun, M. Krakowski, R. Alizon, D. Hadass, A. Bilenca, H. Dery, V. Mikhelashvili, G. Eisenstein, M. Gioannini, I. Montrosset, T.W Berg, M. van der Poel, J. Mørk, B. Tromborg, J. Phys. D 38, 2088 (2005).

19. J. P. Reithmaier, A. Somers, W. Kaiser, S. Deubert, F. Gerschütz, A. Forchel, O. Parillaud, M. Krakowski, R. Alizon, D. Hadass, A. Bilenca, H. Dery, V. Mikhelashvili, G. Eisenstein, M. Gioannini, I. Montrosset, T.W. Berg, M. van der Poel, J. Mørk, B. Tromborg, Phys. Status Solidi B 243 15, 3981 (2006).

20. J.P. Reithmaier, G. Eisenstein, A. Forchel, Proc. IEEE 95 (9), 1779 (2007).

21. F. Klopf, S. Deubert, J.P. Reithmaier, A. Forchel, Appl. Phys. Lett. 81 (2), 217 (2002).

22. F. Klopf, J.P. Reithmaier, A. Forchel, J. Cryst. Growth 227–228, 1151 (2001).

23. S. Deubert, R. Debusmann, J.P. Reithmaier, A. Forchel, Electron. Lett. 41 (20), 1125 (2005).

24. Sumpf, S. Deubert, G. Erbert, J. Fricke, J.P. Reithmaier, A. Forchel, R. Staske, G. Tränkle, Electron. Lett. 39, 1655 (2003).

25. E.M. Pavelescu, C. Gilfert, J.P. Reithmaier, A. Martín-Mínguez, I. Esquivias, Semicond. Sci. Technol. 23, 085022 (2008).

26. E.M. Pavelescu, J.P. Reithmaier, W. Kaiser, P. Weinmann, M. Kamp, A. Forchel, will appear in Phys. Status Solidi (2008).

27. Ch. Schuller, S. Höfling, A. Forchel, C. Etrich, R. Iliew, F. Lederer, T. Pertsch, J.P. Reithmaier, Electron. Lett. 42 (22), 1280 (2006).

28. Ch. Schuller, S. Höfling, A. Forchel, C. Etrich, T. Pertsch, R. Iliew, F. Lederer, J.P. Reithmaier, Appl. Phys. Lett. 91, Art. No. 221102 (2007).

29. J.P. Reithmaier, will appear in Semicond. Sci. Technol. (2008).

30. K. Hennessy, A. Badolato, M. Winger, D. Gerace, M. Atatüre, S. Gulde, S. Fält, E.L. Hu, A. Imamoglu, Nature 445, 896 (2007).

31. S. Reitzenstein, C. Hofmann, A. Gorbunov, M. Strauß, S. H. Kwon, C. Schneider, A. Löffler, S. Höfling, M. Kamp, A. Forchel, Appl. Phys. Lett. 90, 251109 (2007).

32. J.P. Reithmaier, G. Sęk, A. Löffler, C. Hofmann, S. Kuhn, S. Reitzenstein, L. Keldysh, V. Kulakovskii, T.L. Reinecke, A. Forchel, Nature 432, 197 (2004).

33. S. Reitzenstein, J.P. Reithmaier, A. Forchel, in: *Semiconductor Quantum Bits*, O. Benson, F. Henneberger (Eds.), World Scientific, Singapore (2008).

34. D. Press, S. Götzinger, S. Reitzenstein, C. Hofmann, A. Löffler, M. Kamp, A. Forchel, Y. Yamamoto, Phys. Rev. Lett. 98, 117402 (2007).

35. C. Böckler, S. Reitzenstein, C. Kistner, R. Debusmann, A. Löffler, T. Kida, S. Höfling, A. Forchel, L. Grenouillet, J. Claudon, J.M. Gerard, Appl. Phys. Lett. 92 (9), 091107 (2008).

36. C. Kistner, T. Heindel, C. Schneider, A. Rahimi-Iman, S. Reitzenstein, S. Höfling, A. Forchel, Opt. Express 16 (19), 15006 (2008).

6.4. BIOTECHNOLOGICAL APPLICATIONS

ULTRANANOCRYSTALLINE DIAMOND/AMORHOUS CARBON NANOCOMPOSITE FILMS FOR BIOTECHNOLOGICAL APPLICATIONS

W. KULISCH[*1], C. POPOV[2]
[1]*Nanotechnology and Molecular Imaging, Institute for Health and Consumer Protection, European Commission Joint Research Center, Institute for Health and Consumer Protection, Via Enrico Fermi, I-21020 Ispra (VA), Italy*
[2]*Institute of Nanostructure Technologies and Analytics, University of Kassel, Heinrich-Plett-Strasse 40, 34132 Kassel, Germany*

Abstract. Ultrananocrystalline diamond/amorphous carbon nanocomposite films have been deposited by MWCVD from CH_4/N_2 mixtures and investigated in view of their suitability for applications in modern (nano)biotechnology, e.g. as coatings for implants but also as templates for the immobilization of biomolecules e.g. for biosensors or DNA chips. First, some surface properties which are important for such applications, e.g. the roughness and the conductivity, have been established. The main emphasis, however, was laid on the development of different methods to modify the chemical nature of UNCD surfaces, and to characterize them thoroughly by a variety of techniques. Finally, the paper describes experiments to determine the biocompatibility and the bioinertness of UNCD/a-C surfaces, their affinity to unspecific interactions with biomolecules, and to develop routes for the controlled immobilization of biomolecules.

Keywords: UNCD; biosensors; surface characterization; surface modification

1. Introduction

Owing to the combination of the outstanding properties of diamond[1,2] and a relatively smooth surface, ultrananocrystalline diamond (UNCD) films have

---

*wilhelm.kulisch@jrc.it

recently attracted considerable interest for applications in such diverse fields as tribology, electronics, optics, microelectromechanical devices, and biotechnology.[3-6] UNCD films can be obtained by the standard techniques developed for the deposition of polycrystalline diamond films by rendering one or more parameters in such a way that the rate of secondary nucleation is very high,[3,7] This can be achieved by applying a bias voltage throughout the deposition, by using low temperatures or low pressures, and especially by using hydrogen poor, carbon rich gas mixtures in techniques such as microwave plasma chemical vapour deposition (MWCVD) or hot filament chemical vapour deposition (HFCVD) (for reviews see Refs. 2 and 7).

Diamond is the ideal biomaterial.[5,8] Taking into account the smooth surfaces of nano- and ultrananocrystalline diamond, quite a number of applications of NCD and UNCD films in the recently developing field of (nano)biotechnology have been proposed, among others as coatings for implants[9,10] and other devices,[11] but also as templates for the immobilization of biomolecules for biosensors, DNA chips, etc.[12-15]

Recently, we have developed UNCD/a-C nanocomposite films which consist of diamond nanocrystallites embedded in an amorphous carbon matrix, and studied very comprehensively the mechanisms of nucleation and growth, the basic film properties but also a number of application relevant properties. Details of these studies have been discussed in previous papers.[7,16-17] In this contribution we will exclusively report on aspects which regard the application of UNCD/a-C films in various fields of modern (nano)biotechnology.

The paper is organized as follows: The next section gives a brief summary of the deposition of UNCD films and their basic properties, as far as they are of importance for the remainder of this study. The first of the two main sections describes a thorough study of the surface properties of UNCD/a-C films but addresses also the question of their modification in view of a given application. The second major part then discusses those experiments carried out hitherto which involve the participation of biomolecules or larger biological entities such as cells.

2. Deposition and Properties of UNCD/a-C Films

2.1. DEPOSITION

UNCD/a-C nanocomposite films have been deposited by microwave plasma chemical vapour deposition (MWCVD) from 17% CH_4/N_2 mixtures at temperatures of 770°C and 600°C, respectively. Details of the deposition process have been described earlier.[18,19] For all experiments discussed in this contribution, (100) Si wafers have been used as substrates. In order to enhance the nucleation density, they have been pretreated ultrasonically with a suspension of diamond powder in n-pentane. Either pure nanocrystalline diamond powder with a mean grain size of 250 nm or a mixture of this powder with ultradisperse diamond (UDD, 3–5 nm) have been used for this purpose. More details of the pretreatment step and the nucleation densities achieved can be found in Ref. 20.

2.2. FILM PROPERTIES

A wide variety of techniques has been used to characterize these films very thoroughly with respect to their structure and morphology, composition, crystallinity, and bonding structure. Details of this study can be found in Refs. 7 and 17. It can be summarized as follows: The films possess a nano-composite nature; they consist of diamond nanocrystals of 3–5 nm diameter which are embedded in an amorphous carbon matrix with a grain boundary width of 1–1.5 nm. The matrix contains 20–30% sp^2 hybridized carbon and 15–20% hydrogen, which is, however, predominantly bonded to sp^3 carbon. For films grown at 770°C, the crystallite/matrix volume ratio is about unity and the density 2.75 gcm^{-3}. For films deposited at 600°C, the density is slightly lower (2.65 gcm^{-3}). Finally, besides hydrogen the films contain oxygen and nitrogen with concentrations of 0.5–1%.

3. Surface Characterization and Modification

For applications of UNCD films e.g. as coatings for implants, of course their mechanical bulk properties are of great importance. For all applications in biotechnology, however, the nature of the surface is decisive. This regards of course the chemical properties and also the possibility to modify them according to the requirements of a given application, but also surface properties such as the roughness and the surface conductivity (for biosensor applications).

3.1. SURFACE ROUGHNESS

The roughness of the surface of a given material is of utmost importance for many applications, especially in the field of biotechnology. In fact, UNCD films have been developed in order to overcome an intrinsic disadvantage of polycrystalline diamond films, the large surface roughness caused by their faceted growth.[21,22]

Figure 1 shows two AFM topography images of films deposited with quite different conditions (see the figure caption). Despite these differences, both images show the same morphology: The surfaces consist of rounded features with diameters of some hundred nanometers which are clearly composed of smaller particles. The rms roughnesses derived from these images are 11.9 and 9.9 nm, respectively. In a very comprehensive study,[20,23] the influence of a number of parameters on the morphology and surface roughness has been investigated: (i) nucleation density; (ii) deposition time; (iii) growth rate; (iv) substrate material.

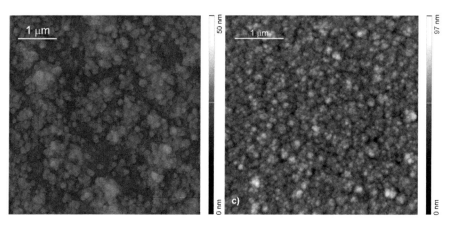

Figure 1. AFM images of UNCD/a-C nanocomposite films. Left: substrate temperature 770°C, pretreatment with pure NCD powder, nucleation density 3×10^8 cm^{-2}, thickness 4 µm. Right: $T_s = 600$°C, treatment with a UDD/NCD mixture, $d_n = 1 \times 10^{10}$ cm^{-2}, d = 350 nm.

It turned out that none of these parameters has a significant effect on morphology and roughness. For example, the latter was in all cases between 9 and 13 nm. This can be explained by the fact that, once the films are closed, their morphology is independent of the nucleation step and solely defined by the high rate of secondary nucleation, which is the decisive mechanism leading to the formation of UNCD and which is not affected by the parameters investigated in this study.[20,23]

3.2. SURFACE CONDUCTIVITY

For application in biosensors, the conductivity of a surface may play a decisive role, as it is important for the design of the sensor, especially if electrical detection methods are to be used. For this reason, the electrical properties of the UNCD/a-C films have been investigated in two sets of experiments.[18] The first used impedance spectroscopic measurements as a function of frequency and temperature; they were performed in a vertical arrangement, i.e. through the film bulk. The second was carried out in van der Pauw geometry and consisted of I/V curves and Hall measurements.

The left image of Figure 2 shows results of the impedance measure-ments. It can be seen that the conductivity increases with temperature in the range between 210 and 295 K; at room temperature, the resistivity is $\varphi = 1.3 \times 10^6$ Ωcm. From the inset of this diagram it is evident that there are three contributions to the conductivity at room temperature. For a detailed discussion of these results, the reader is referred to Refs. 18 and 24.

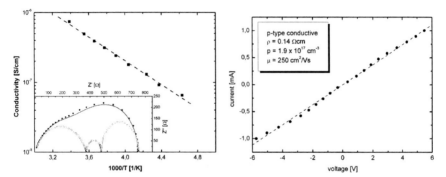

Figure 2. Left: conductivity as determined by impedance spectroscopic measurements as a function of temperature. The inset shows a Cole–Cole plot recorded at room temperature; right: I/V curve taken in van der Pauw geometry. The inset summarizes the results of I/V and Hall measurements performed in this geometry.

The right diagram in Figure 2 shows an I/V curve obtained at room temperature in van der Pauw geometry. The inset summarizes the results obtained from these I/V curves and accompanying Hall measurements. According to these results, the material is p-type conductive with a resistivity of $\varphi = 0.14$ Ωcm and a carrier concentration of 1.9×10^{17} cm^{-2}.

On a first view, there seems to be a strong contradiction between these two sets of measurements, with the resistivity differing by seven orders of magnitude. But one has to take into account that both experiments were carried out in different geometries. The first addresses the film bulk while the van der Pauw measurements regard the surface of the films. It is well known that – owing to their band structure – hydrogen terminated diamond surfaces are p-type conductive.[25,26] Obviously, this is also true for UNCD/a-C films, which are also hydrogen terminated after growth as will be shown in Sect. 3.4.2 (for a more detailed discussion of this topic we refer again to Refs. 18 and 24).

The conclusions drawn above were confirmed by an experiment the result of which is shown in Figure 3. An as-grown, H-terminated UNCD/a-C surface has been UV/O$_3$ treated (see Sect. 3.3) through a mask (a copper

SEM grid). This led to an O-termination of those parts of the surface exposed. The SEM image in Figure 3 shows a clear contrast between the squares exposed and the stripes between them which were not exposed. Since the UV/O_3 treatment does not lead to any etching of the surface (see below) this contrast can only be due to the electrical properties of the two surfaces, i.e. conductive in the case of the H-terminated untreated parts and insulating for the O-terminated exposed parts. This means that by such surface treatments not only the chemical, but also the electrical surface properties can be changed in a patterned way, which may be utilized in the design of sensors.[26]

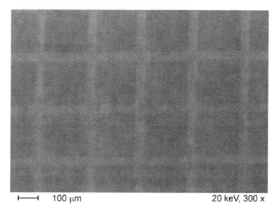

├──┤ 100 μm   20 keV, 300 x

Figure 3. SEM image of an as-grown UNCD sample which has been UV/O_3 treated through a mask.

3.3. SURFACE MODIFICATION TECHNIQUES

By means of plasma and wet chemical treatments, quite a number of different surfaces have been prepared for subsequent characterization and further experiments. The conditions of these treatments are summarized in Table 1.

i)   As-grown surfaces after the growth in a CH_4/N_2 mixture (in the following: AG)
ii) Surfaces treated additionally by a H_2 MWCVD plasma at 400°C (HP)
iii) Surfaces treated by an O_2 MWCVD plasma at room temperature (OP)
iv) Surfaces treated by a CHF_3 rf plasma at room temperature (FP)
v)   Surfaces treated chemically by aqua regia (HCl/HNO_3 3:1, CT)
vi) Surfaces treated by UV light in air (UV/O_3 treatment, UV)

TABLE 1. Conditions used for the various surface treatments discussed in this section. Also given is the surface composition after the treatments, the contact angles, and the surface roughness.

Parameter		AG	HP	OP	FP	CT	UV
Plasma		2.45 GHz	2.45 GHz	2.45 GHz	13.56 MHz	–	–
Power	(W)	800	800	200	70	–	–
Gas		17% CH$_4$/N$_2$	H$_2$	O$_2$	CHF$_3$	–	air
Flow	(sccm)	300	250	–	5	–	–
Pressure	(kPa)	2.2	6	0.2	0.066	–	Atm
Temp.	(°C)	600	400	Rt	Rt	Rt	Rt
Time	(min)	390	30	10	10	90	30
C	(%)	96.9 ± 0.6	93.3 ± 0.2	87.4 ± 0.1	81.9 ± 2.3	97.6 ± 0.5	93.8 ± 0.5
O	(%)	2.2 ± 0.4	4.1 ± 0.3	12.0 ± 0.4	4.8 ± 0.2	1.8 ± 0.1	6.2 ± 0.5
N	(%)	0.8 ± 0.1	1.2 ± 0.3	0.7 ± 0.1	0.8 ± 0.2	0.7 ± 0.1	–
Si	(%)	–	1.4 ± 0.1	–	–	–	–
F	(%)	–	–	–	12.5 ± 2.4	–	–
θ_c	(°)	86 ± 2	88 ± 2	7 ± 1	100 ± 3	67 ± 2	< 5
rms	(nm)	10.0	12.1	12.1	–	10.5	–

3.4. SURFACE CHARACTERIZATION

In order to investigate the nature of the surfaces created by these different treatments, among others XPS, TOF-SIMS, and contact angle measurements have been carried out. All measurements have been performed immediately after the treatments.

3.4.1. XPS

The composition of these various surfaces as revealed by XPS are also summarized in Table 1. It can be seen that the as-grown is extremely clean, the more if one takes into account that the bulk material contains already 0.5–1% oxygen and nitrogen each (see Sect. 2.2). The same holds for the surface treated chemically with aqua regia. Somewhat surprisingly, the HP surface contains about 4% of oxygen and also traces of Si. A more detailed investigation[27] has revealed that this silicon is present in the form of SiO$_{1.5-2}$;

it stems very probably from the quartz window through which the micro-wave plasma was coupled into the chamber used for the HP treatment. Finally, and most important, the surfaces of the OP and FP samples contain about 12% oxygen and fluorine, respectively. The oxygen content after UV/O$_3$ treatment is about 6%. This is a strong hint that these treatments led to an oxidation, respectively a fluorination, of the UNCD surfaces. More details of the bonding of the oxygen and fluorine atoms to the UNCD were revealed by the TOF-SIMS measurements discussed in the next section.

3.4.2. TOF-SIMS

Figure 4 shows normalized TOF-SIMS count rate ratios CH$_x$/C$_x$ for x = 2, 4, 6 for differently treated UNCD surfaces; these ratios can be regarded as a measure for the hydrogen termination.[27] It can be seen that the hydrogen plasma treatment leads to a small increase of these ratios with respect to the as-grown surface which is used as a reference. This means that the treat-ment caused a slight increase of surface C–H groups. On the other hand, from the magnitude of this increase it can be inferred that already the as-grown surface is to a large extent hydrogen terminated, in agreement with literature,[28] despite the hydrogen poor gas mixture used.

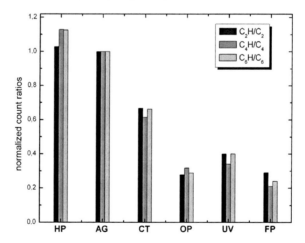

Figure 4. Normalized TOF-SIMS count rate ratios CH$_x$/C$_x$ for x = 2, 4, 6 for differently treated UNCD/a-C surfaces. Normalization has been carried out with respect to the as-grown surface.

All other treatments led to a loss of C–H surface groups, which is most pronounced for the OP and FP samples. For the UV/O$_3$ treated surface the ratios are slightly higher, in agreement with the lower oxygen content of 6.2% only. In this context it has also to be mentioned that the normalized count rates of oxygen containing groups are significantly lower for sample UV than for OP, although there are no differences in the nature of the species and their intensity ratios. Finally, for the aqua regia treated CT sample the CH_x/C_x ratios are about 0.7. Obviously, the treatment has led to a loss of surface C–H bonds; but in contrast to the other treatments it the hydrogen has not been replaced by another element.

Besides these results regarding the hydrogen termination, TOF-SIMS measurements revealed a number of further interesting information. From the filtered spectra (filtered means that not all peaks in the mass range depicted are shown but only those of interest) of an OP treated surface and of a polyacrylic acid (PAA) surface shown on the left side of Figure 5 and similar spectra in other mass regions it can be concluded that the oxygen plasma treated surface is not terminated by carboxyl (COOH) groups as in the case of PAA but rather by C–O and C–OH groups.

Similarly, a comparison of filtered spectra from the FP surface and from polytetrafluorethylene (PTFE, Teflon) (right diagram in Figure 5) leads to the conclusion that the CHF_3 treatment has not led to the deposition of a $C_xH_yF_z$ polymer but had rather caused a replacement of C–H groups of the as-grown hydrogen terminated surface by CF_x groups.[29]

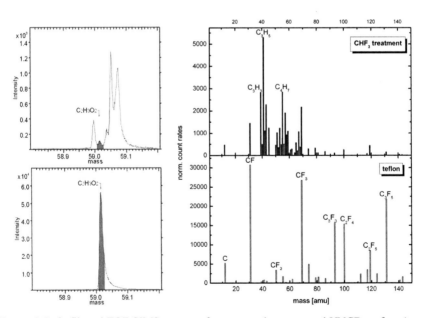

Figure 5. Left: filtered TOF-SIMS spectra of a oxygen plasma treated UNCD surface (upper panel) and a PAA film (lower panel); right: filtered TOF-SIMS spectra of a FP treated UNCD surface (upper panel) and a teflon surface (lower panel).

3.4.3. Contact Angle Measurements

The contact angles of the various surfaces investigated, as measured immediately after the treatments, are summarized in Table 1. It can be seen that for the as-grown sample θ_c is $86 \pm 2°$, which improves to $88 \pm 2°$ after hydrogen plasma treatment, in agreement with the slight improvement of the hydrogen termination revealed by the TOF-SIMS measurements shown in Figure 4. The low contact angles observed for the OP and UV treated samples, on the one hand, and the very high angle of $100 \pm 3°$ of the FP sample, on the other hand, are also in agreement with the surface analytical results presented above, indicating a successful termination of the UNCD surface with oxygen and fluorine species, respectively. Further investigations are necessary to clarify the very low contact angles for the UV sample despite the fact that according to both XPS and TOF-SIMS the surface oxygen content is lower than that of the oxygen plasma treated sample.

Interestingly, the contact angle of the aqua regia treated surface of $67 \pm 2°$ is somewhat in between those of the hydrophobic AG, HP and FP samples, on the one side, and of the hydrophilic OP and UV samples; however, this is also in agreement with the surface analytics presented above: on the one hand, the H-termination seems to have been removed to some extent; on the other hand, according to XPS it has not been replaced by another distinct element.

In order to get information on the stability of these different surfaces, their contact angle against purified water has been measured from time to time over the period of a year, during which the samples have been stored in air. The results are presented in Figure 6. The two hydrogen terminated surfaces (AG and HP) turned out to be most stable. Over the entire period, θ_c was always $\geq 80°$ (AG) and $\geq 85°$ (HP), respectively. For the fluorinated surface, the contact angle was initially very high ($100 \pm 3°$) but dropped to $95 \pm 2°$ after 399 days. However, it is not clear yet whether this change is significant.

For the oxygen terminated surfaces in contrast a strong and early degradation takes place. For the OP sample, there is a strong and early increase of θ_c in the first days up to values of $20–25°$, accompanied by a large scattering of the data. Thereafter, the values are relatively stable. TOF-SIMS measurements[27] have shown that the OP surface is prone to contaminations with amines (C_xH_yN). Very probably these and similar contaminants are responsible for the observed increase of θ_c. The increase of θ_c for the UV/O_3 treated sample was even more pronounced (not included in Figure 6): whereas directly after the treatment $\theta_c < 5°$ was observed, the values obtained after 90 days were $36 \pm 2°$.

For the aqua regia treated surface, finally, the contact angle increase from 65° to almost 80°. The reasons are not fully understood yet but it may be that the surface recovered somehow from the partial removal of the hydrogen termination caused by the aqua regia treatment.

With AG, OP and FP samples, another series of experiments has been carried out against water, formamide, and benzyl alcohol; from these measurements, also the surface energy could be determined. These experiments are described in detail in another contribution to these proceedings.[30]

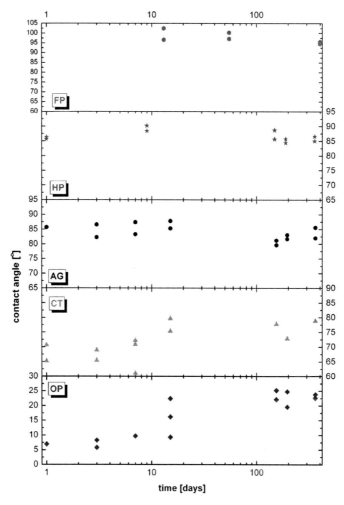

Figure 6. Development of the contact angle as a function of time for differently treated surfaces. Please note the logarithmic time scale.

3.5. GRAFTING OF AN ANTI-FOULING PEG LAYER

Any material used for bioimplants or biosensors which is exposed e.g. to biomolecules of body fluids must not interact unspecifically with the bio-molecules, i.e. they must be non-fouling. Although UNCD surfaces – as will be shown in the next section – are not prone to such unspecific inter-actions they will nevertheless absorb heavily fouling proteins such as bovine serum albumin (BSA). In order to counteract this, anti-fouling layers such as poly(ethylene glycol) PEG[31,32] can be used e.g. for those parts of a biosensor which must remain inactive. In order to investigate whether it is possible to coat UNCD surfaces with such an anti-fouling layer, first experiments have been carried out to directly graft a poly(ethylene glycol)methacrylate (PEGMA) polymer onto an UNCD film.

3.5.1. Protocol

Direct grafting of PEGMA layers on O-terminated UNCD surface was per-formed using an atomic transfer radical polymerization process (ATRP)[32,33] using α-bromoisobutyryl bromide[32] as an initiator. It consists of the follow-ing steps:

i)   UV/O$_3$ treatment for 60 min
ii) Reaction with α-bromoisobutyryl bromide/triethylamine vapour in nitrogen atmosphere for 60 min at 20°C to immobilize the initiator
iii) Rinse in ethanol
iv) Immersion for 2–8 h in a deoxygenated solution containing 10 ml water, 10 ml poly(ethylene glycol)methacrylate (PEGMA-360 Mwt) macromonomer, 41 mg of Cu(I)Cl, 9 mg Cu(II)Br$_2$ and 160 mg of 2,2'-Bipyridyl
v)   Rinse in ethanol and water

3.5.2. Results

An initial series of TOF-SIMS measurements has been performed to judge the success of the PEGMA grafting process described above. The results are presented in Figure 7. The spectrum of the as-grown surface in Figure 7a is dominated by pairs C_x/C_xH. The UV/O$_3$ treatment (Figure 7b) of this sur-face has two important effects: (i) the ratios C_xH/C_x decrease considerably for each of the pairs shown in the diagram in agreement with Figure 4 and (ii) there is a drastic relative increase of the count rates of O and OH. Both observations support the conclusions drawn above that the UV/O$_3$ treatment leads to a substitution of surface C–H bonds by C–O and C–OH species.

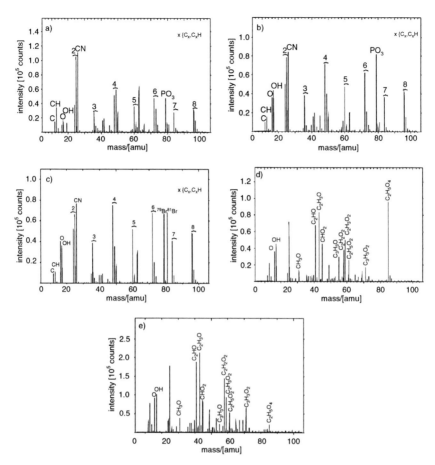

Figure 7. TOF-SIMS spectra taken during the PEGMA grafting process described in Sect. 3.5.1: (a) as-grown UNCD surface; (b) after UV/O₃ treatment; (c) after α-bromoisobutyryl bromide exposure; (d) after grafting of the PEGMA layer; (e) spectrum of a spin-casted PEG layer for comparison.

The spectrum in Figure 7c taken after the exposure of the surface to the α-bromoisobutyryl bromide initiator clearly proves the presence of bromine on the surface. After deposition of the PEGMA layer, the spectrum is dominated by $C_xH_yO_z$ species (Figure 7d). A comparison with Figure 7e, showing the spectrum of a thick spin-coated, OH-terminated PEG films reveals that the peaks and their intensity ratios of both spectra are almost identical; the only difference is the $C_2H_5O_4$ peak stemming from the methacrylate group of the PEGMA molecules which is absent in the spectrum of the film from OH-terminated PEG.

4. Interactions with Biomolecules and Biological Systems

4.1. BIOCOMPATIBILITY AND BIOINERTNESS

In order to investigate the biocompatibility of UNCD/a-C films, so-called direct cell tests were performed in which cell were brought in direct contact with the surfaces of the films for a number of days, after which the morphology of the cells was investigated microscopically and compared with that of cells on control samples which are known to be either biocompatible or cytotoxic. In the course of our investigations, four different types of cells have been used for this purpose:

i)   Osteoblast-like SaOS-2 cells[34]
ii) Human embryonal lung fibroblasts LEP$_{19}$[35]
iii) Human endothelial cells EA.hy 926[36]
iv) Mouse embryonic fibroblasts Balb/3T3

The results of these experiments can be summarized briefly as follows: Continuous, hydrogen terminated UNCD/a-C films are biocompatible; with respect to all four types of cells they are not cytotoxic and do not induce harm to the health of the cells. An example is presented in Figure 8, showing human endothelial cells EA.hy 926 in contact with an as-grown UNCD surface. It is evident that the cells show good attachment and growth; similar to control cells well-developed elongated extensions are clearly visible.

Based on tests with mouse embryonic fibroblasts, also O- and F-terminated UNCD surfaces are not cytotoxic. On the other hand, it turned out that the adhesion, growth and proliferation of cells on non-continuous, nodule-type films differ significantly from those on continuous films. An example is presented in Figure 9. The left SEM image shows such a film where the nucleation density was too low to provide a continuous film at a thickness of about 1 μm; the right image shows the same human endothelial cells as in Figure 8 in contact with such a discontinuous surface. The single nodules, but also the voids between them are clearly visible. It can also be seen that the cells are predominantly rounded and had not developed elongated extensions. Furthermore, they are reduced in size and show irregular cell borders. Similar results for discontinuous films were also obtained with mouse embryonic fibroblasts. From literature it is well-known that adhesion, growth and proliferation of cells are strongly influenced by the micro/nanostructure of the surface.[41,42]

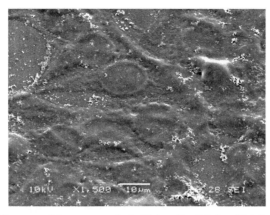

Figure 8. SEM image of human endothelial cells EA.hy 926 in contact with a UNCD/a-C surface after 5 days of growth.

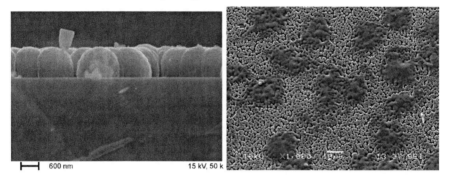

Figure 9. Left: SEM image of a not closed, nodule-type of film. Right: SEM image of human endothelial cells EA.hy 926 in contact with such a UNCD/a-C film.

In another experiment, a so-called simulated body fluid test (SBF),[34] as-grown UNCD surfaces have been brought into contact with a fluid the composition of which resembles the composition of blood plasma. As even after ten days no hydroxyapatite has developed on the surface of the films, it could be concluded that the UNCD films were bioinert. For more details of this test we refer to Ref. 34.

4.2. UNSPECIFIC INTERACTIONS WITH BIOMOLECULES

4.2.1. Bovine Serum Albumin

A first series of experiments to investigate the affinity of different UNCD surfaces to non-specific interactions with biomolecules was carried out by exposing AG, HP, CT and OP samples to bovine serum albumin (BSA).

To this end, the samples were immersed for 60 min in 30 ml of phosphate buffered saline (GIBCO) at 20°C containing 50 μg/ml of bovine serum albumin protein (Sigma-Aldrich). After immersion the samples were rinsed copiously in ultrapure analytical grade water and dried in flowing nitrogen. Thereafter, they have been analysed by XPS and TOF-SIMS. The TOF-SIMS measurements revealed that all four surfaces investigated were covered by a BSA film approaching a monolayer.[27] Unfortunately, these layers were too thick to use substrate related peaks as a reference to quantify the coverage.

Table 2 shows in the upper row the composition of a thick (i.e. thicker than the information depth of XPS) BSA film. The table summarizes also the surface composition of the four samples of this investigation. It can be seen that in all cases the composition deviates from that of the thick BSA film which means that the BSA coverage is in all cases limited. In order to get further information on the degree of BSA attachment we present in Table 2 also the compositional differences between the four surfaces before (Table 1) and after BSA exposure. It can be seen from Table 2 that changes were smallest for the AG surface, indicating that this H-terminated surface is not very prone to BSA attachment. Interestingly, for the HP surface the changes are larger, despite the slightly improved hydrogen termination (see Sect. 3.4.2). Probably, this is caused by the slight SiO$_x$ contamination of this surface revealed by XPS (Sect. 3.4.1). Similarly, the easy contamination of OP treated surfaces with amines etc. may be a reason for a comparably high coverage with BSA. Most importantly, the highest changes, and thus the highest BSA coverage were found for the aqua regia treated CT sample. This may be a consequence of the partial removal of the H-termination described above, which was not replaced by another functionality.

TABLE 2. Surface composition and compositional changes after exposure of differently treated UNCD surfaces to BSA. In the first row, the composition of a thick BSA film is given.

	C	O	N	ΔC	ΔO	ΔN
BSA	66.2	17.2	15.3	–	–	–
HP	88.4	7.2	3.3	−4.9	0.6	2.1
AG	95.3	2	0.8	−1.6	−0.2	1.8
CT	88.4	6.1	5.2	−9.2	4.3	4.5
OP	82	13.4	3.7	−5.4	1.4	3

In another series of experiments, scanning force spectroscopy has been applied between a Si AFM tip functionalized with BSA, and an as-grown UNCD surface and – for comparison purposes – a glass surface. During the measurements, 120 force/distance curves have been recorded on different positions of the samples. In the case of the glass surface, in 38% of the tests a clear interaction between tip and surface was observed, leading to curves of type (a) in Figure 10. In the case of UNCD, none of the curves showed any interaction (curve (b) in Figure 10).

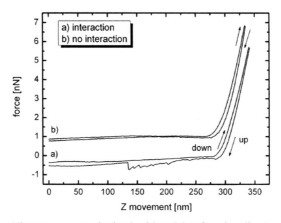

Figure 10. Force/distance curves obtained with a BSA functionalized AFM tip. (a) Interaction between tip and molecules on the surface; (b) no interaction.

These experiments of BSA exposure can therefore be summarized as follows:

i) UNCD surfaces are much less prone to unspecific interactions with the heavily fouling BSA as e.g. glass.
ii) Nevertheless, there are such interactions, the degree of which depends on the nature of the surface (termination, contamination, etc.).

4.2.2. RNA and Antibodies

Another series of experiments addressed the affinity of UNCD/a-C surfaces to unspecific interactions with RNA molecules. To this end, double stranded RNA with a length of 118 ± 4 nm, but also proteins which can bind to this RNA (20 nm diameter) have been deposited on UNCD surfaces. For control purposes, also mica samples have been used. After careful cleaning, these surfaces have been investigated by AFM and scanning force microscopy (for a more detailed description of this experiment see Ref. 35). The results of these studies are presented in Figure 11.

In the AFM image on the left side of the figure, showing a mica surface exposed to dsRNA and proteins, both the elongated strands and the rounded proteins are clearly visible. They are completely absent in the AFM image of the corresponding UNCD/a-C sample, which means that its surface is much less prone to unspecific interactions with RNA and antibodies as e.g. mica.

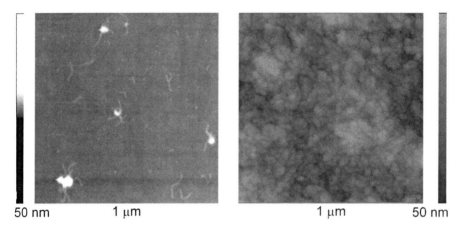

50 nm   1 μm   1 μm   50 nm

Figure 11. AFM images of a mica (left) and a UNCD/a-C surface (right) which have been exposed to dsRNA and proteins.

4.3. IMMOBILIZATION OF BIOMOLECULES

During the last years two major processes have been developed to functionalize the surfaces of PCD, NCD and UNCD films for the immobilization of biomolecules, both starting from hydrogen terminated surfaces. In the approach developed by the group of Hamers,[12,37–39] ω-unsaturated long-chain molecules (e.g. 10-amino-dec-1-ene) are covalently bonded by a photochemical process to the diamond surface, while the process of Nebel et al. uses electrochemical reduction of diazonium salts to form nitrophenyl-linker molecules on boron-doped diamond films.[13,40]

In a first experiment to immobilize RNA molecules on UNCD/a-C surfaces, a modification of the Hamers-process was employed: Instead of a ω-unsaturated long-chain molecule, a unsaturated cyclic compound was used to provide the surface with amine functionalities. The following protocol, which is schematically shown in Figure 12, was applied[36]:

i)   Photochemical attachment of 1-amino-3-cyclopentene hydrochloride ($C_5H_{10}ClN$) by exposure to UV light (244 nm) for 195 min

ii) Deprotection of the amino groups in 10% NH4OH

iii) Attachment of sulfosuccinimidyl 4-(N-maleimi-domethyl)cyclohexane-1-carboxylate (SSMCC) which serves as a linker molecule

iv) Functionalization with a thiolated DNA oligonucleotide (5' GATCCC CGGTACCGAGCTCGAATTCGCCC- SH 3')

v)   Attachment of a 374 base pair long double RNA strand with a 30 nucleotide single stranded overhang at the 5' end, which can form Watson-Crick base pairs with the thiolated DNA oligonucleotide

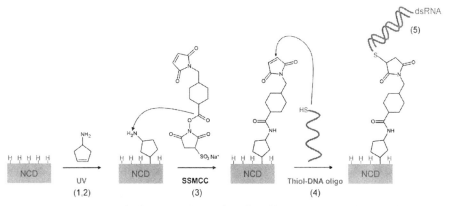

Figure 12. Photochemical approach to functionalize UNCD surfaces with RNA (schematical).

After the process, the samples were investigated by AFM. Figure 13 shows a phase image of a UNCD/a-C surface functionalized according to the scheme in Figure 12. In the image, clearly a number of small objects can be seen (some of them are marked by an arrow). As discussed in detail in Ref. 36, they can be identified as RNA molecules which are freely accessible on the UNCD surface. This identification is confirmed by the fact that these object are found neither on UNCD surface exposed to RNA but not functionalized with amine groups nor surfaces which have been functionalized but not exposed to RNA.

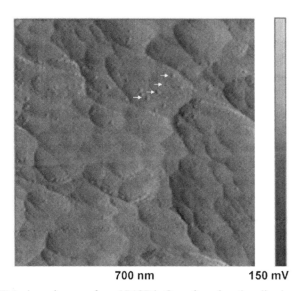

700 nm   150 mV

Figure 13. AFM phase image of an UNCD/a-C surface functionalized according to the scheme shown in Figure 12.

5. Summary

Ultrananocrystalline diamond/amorphous carbon composite films have been investigated in view of possible applications in biotechnology. Emphasize has been laid on the properties of the surface of the films and their modification. The surfaces are rather smooth but not perfect with a rms roughness of about 10 nm. The hydrogen terminated surface is p-type conductive but can be rendered insulating by replacing the H-termination by an oxygen one. By means of plasma or photochemical processes, differently terminated surfaces can be achieved. H- and F-terminated surfaces are highly hydrophobic, oxygen terminated ones highly hydrophilic. While the former are extremely stable, this is not the case for the latter. Oxygen terminated surfaces can be used to graft a non-fouling poly(ethylene glycol) layer.

Concerning the bioproperties of UNCD/a-C films, the following results have been obtained:

i) UNCD/a-C surfaces are biocompatible.

ii) UNCD/a-C surfaces are bioinert.

iii) UNCD/a-C surfaces are not prone to unspecific interactions with heavily fouling proteins such as bovine serum albumin.

iv) UNCD/a-C surfaces are not prone to unspecific interactions with RNA and antibodies.

v) UNCD/a-C surfaces can be functionalized by a photochemical approach with RNA.

ACKNOWLEDGEMENTS

We would like to thank our coworkers at our home institutes, the JRC in Ispra and the INA in Kassel, and our colleagues from all over the world who contributed to this work: At the JRC: G. Ceccone (XPS), L. Cereotti (cell tests), D. Gilliland (TOF-SIMS), A. Ruiz (SEM), L. Sirghi (AFM), F. Rossi (discussion); at the INA: S. Bliznakov (deposition), S. Boycheva (functionalization), J.P. Reithmaier (discussion); throughout the world: M.D. Apostolova and N. Milinovik (Bulgarian Academy of Sciences, cell tests), A. Bock (PhysTech GmbH, Moosburg, electrical measurements), T. Dostalova (Charles University Prague, cell tests), C. Hammann and N. Anspach (University of Kassel, RNA and protein interaction), E. Lefterova (Bulgarian Academy of Sciences, electrical measurements), Marco Morra and Clara Casinelli (Nobil Bio Ricerche, Villafranca d'Asti, cell tests), J. Strnad (Lasak Ltd., Prague, SBF tests).

References

1. K.E. Spear and J.P. Dismukes, *Synthetic Diamond: Emerging CVD Science and Technology* (Wiley, New York, 1994).
2. W. Kulisch, *Deposition of Superhard Diamond-Like Materials* (Springer Tracts on Modern Physics, Heidelberg, 1999).
3. D.M. Gruen, Annu. Rev. Mater. Sci. 29, 211 (1999).
4. D.M. Gruen, Mater. Res. Bull. 26, 771 (2001).
5. J.A. Carlisle and O. Auciello, Electrochem. Soc. Interface 12, 38 (2003).
6. J.E. Butler, Electrochem. Soc. Interface 22, 12 (2003)
7. W. Kulisch and C. Popov, Phys. Status Solidi A 203, 203 (2006).
8. C.E. Nebel, B. Rezek, D. Shin, H. Uetsuka, and N. Yang, J. Phys. D 40, 6443 (2007).
9. M.J. Papo, S.A. Catlegde, Y.K. Vohra, and C. Machado, J. Mater. Sci. Mater. Med. 15, 773 (2004).
10. S.A. Catledge, M.D. Fries, Y.K. Vohra, W.R. Lacefield, J.E. Lemons, S.Woodard, and R. Venugopalan, J. Nanosci. Nanotechnol. 2, 1 (2002).
11. X. Xiao, J. Wang, C. Liu, J.A. Carlisle, M. Mech, R. Greenberg, D. Guven, R. Freda, M.S. Humayun, J. Weiland, and O. Auciello, J. Biomed. Mater. Res. B 77B, 273 (2006).
12. W. Yang, O. Auciello, J.E. Butler, W. Cai, J.A. Carlisle, J.E. Gerbi, D.M. Gruen, T. Knickerbocker, T.L. Lasseter, J.N. Russell Jr., L.M. Smith, and R.J. Hamers, Nature Mater. 1, 294 (2002).
13. J. Wang, M.A. Firestone, O. Auciello, and J.A. Carlisle, Langmuir 20, 11450 (2004).
14. A. Härtl, E. Schmich, J.A. Garrido, J. Hernando, S.C.R. Catharino, S. Walter, P. Feulner, A. Kromka, D. Steinmüller, and M. Stutzmann, Nature Mater. 3, 736 (2004).
15. P. Christiaens, V. Vermeeren, S. Wenmackers, M. Daenen, K. Haenen, M. Nesladek, M. vandeVen, M. Ameloot, L. Michiels, and P. Wagner, Biosens. Bioelectron. 22, 170 (2006).
16. C. Popov, W. Kulisch, P.N. Gibson, G. Ceccone, and M. Jelinek, Diamond Relat. Mater. 13, 1371 (2004).
17. C. Popov, W. Kulisch, S. Boycheva, K. Yamamoto, G. Ceccone, and Y. Koga, Diamond Relat. Mater. 13, 2071 (2004).
18. W. Kulisch and C. Popov, in: L. M. Krause and J.T. Walter (Eds.) *New Research on Nanocomposites* (Nova Publishers, New York, 2009).
19. C. Popov, M. Novotny, M. Jelinek, S. Boycheva, V. Vorlicek, M. Trchova, and W. Kulisch, Thin Solid Films 506–507, 297 (2006).
20. W. Kulisch, C. Popov, H. Rauscher, L. Sirghi, T. Sasaki, S. Blisznakov, and F. Rossi, Diamond Relat. Mater. 17, 1116 (2008).
21. P. Koidl and C.-P. Klages, Diamond Relat. Mater. 1, 1065 (1992).
22. S.P. Hong, H. Yoshikawa, K. Wazumi, and Y. Koga, Diamond Relat. Mater. 11, 887 (2002).
23. W. Kulisch and C. Popov, Paper submitted to Phys. Status Solidi A (2008).
24. W. Kulisch, C. Popov, S. Bliznakov, E. Lefterova, A. Riuz, J.P. Reithmaier, and F. Rossi, Paper submitted to Diamond Relat. Mater. (2008)
25. J. Ristein, Appl. Phys. A 82, 377 (2006).

26. A. Härtl, S. Nowy, J. Hernando, J.A. Garrido, J. Hernando, and M. Stutzmann, Sensors 5, 496 (2005).
27. W. Kulisch, C. Popov, S. Bliznakov, G. Ceccone, D. Gilliland, L. Sirghi, and F. Rossi, Thin Solid Films 515, 8407 (2007).
28. D. Steinmüller-Nethl, F.R. Kloss, M. Najam-Ul-Haq, M. Rainer, K. Larsson, C. Linsmeier, G. Köhler, C. Fehrer, G. Lepperdinger, X. Liu, N. Memmel, E. Bertel, C.W. Huck, R. Gassner, and G. Bonn, Biomaterials 27, 4556 (2006).
29. C. Popov, W. Kulisch, S. Bliznakov, G. Ceccone, D. Gilliland, L. Sirghi, and F. Rossi, Diamond Relat. Mater. 17, 1229 (2008).
30. H. Vasilchina, C. Popov, S. Ulrich, J. Ye, F. Danneil, M. Stüber, and A. Welle, This proceedings, p. 479.
31. B. Zdyrko, V. Klep, and I. Luzinov, Langmuir 19, 10179 (2003).
32. S. Sun, J. Liu, and M.L. Lee, Anal. Chem. 80, 856 (2008).
33. J. Liu, B. Zdyrko, V. Klep, and I. Luzinov, Langmuir 20, 6710 (2004).
34. C. Popov, W. Kulisch, M. Jelinek, A. Bock, and J. Strnad, Thin Solid Films 494, 92 (2006).
35. C. Popov, W. Kulisch, J.P. Reithmaier, T. Dostalova, M. Jelinek, N. Anspach, and C. Hammann, Diamond Relat. Mater. 16, 735 (2007).
36. C. Popov, S. Bliznakov, S. Boycheva, N. Malinovik, M.D. Apostolova, N. Anspach, C. Hammann, W. Nellen, J.P. Reithmaier, and W. Kulisch, Diamond Relat. Mater. 17, 882 (2008).
37. T. Strother, T. Knickerbocker, Schwartz, J.N. Russel Jr., J.E. Butler, L.M. Smith, and R.J. Hamers, Langmuir 18, 1968 (2002).
38. T. Knickerbocker, W. Strother, M.P. Schwartz, J.N. Russel Jr., J.E. Butler, L.M. Smith, and R.J. Hamers, Langmuir 19, 1938 (2003).
39. W. Yang, S.E. Baker, J.E. Butler, C.S. Lee, J.N. Russell Jr., L. Shang, B. Sun, and R.J. Hamers, Chem. Mater. 17, 938 (2005).
40. C.E. Nebel, B. Rezek, D. Chin, H. Uetsuka, and N. Jang, J. Phys. D 40, 6443 (2007).
41. P.K. Chu, J.Y. Chen, L.P. Wang, and N. Huang, Mat. Sci. Eng. R 36, 143 (2002).
42. A.I. Teixeira, G.A. Abrams, P.J. Bertics, C.J. Murphy, and P.F. Nealey, J. Cell Sci. 116, 1881 (2003).

WETTING BEHAVIOUR AND PROTEIN ADSORPTION TESTS ON ULTRANANOCRYSTALLINE DIAMOND AND AMORPHOUS HYDROGENATED CARBON THIN FILMS

H. VASILCHINA[1*], C. POPOV[2], S. ULRICH[1], J. YE[1],
F. DANNEIL[1], M. STÜBER[1], A. WELLE[3]
[1]*Research Center Karlsruhe, Forschungszentrum Karlsruhe, Institut für Materialforschung I, Hermann-von-Helmholtz-Platz 1, 76344 Eggenstein-Leopoldshafen, Germany*
[2]*University of Kassel, Institute of Nanostructure Technologies and Analytics (INA), 34132 Kassel, Germany*
[3]*Forschungszentrum Karlsruhe, Institut für Biologische Grenzflächen, Hermann-von-Helmholtz-Platz 1, 76344 Eggenstein-Leopoldshafen, Germany*

Abstract. Hydrogenated amorphous carbon (a-C:H) and ultra-nanocrystalline diamond/amorphous carbon (UNCD/a-C:H) composite films are considered as excellent candidates for use as biocompatible coatings on biomedical implants. The aim of this work is a comparative study of the wetting behaviour and protein adsorption on a-C:H and UNCD/a-C:H films prepared by microwave plasma chemical vapour deposition (MWCVD) under different process conditions. All films were characterized by a variety of methods: their morphology by scanning electron microscopy (SEM), the topography by atomic force microscopy (AFM), the chemical bonding structure by Raman spectroscopy. SEM revealed that the a-C:H films possess a column-like morphology, while the UNCD/a-C:H surfaces are composed of structures with diameters of several hundreds nanometers, which themselves have a sub-structure. All films under investigation were rather smooth with rms roughnesses down to 2 nm for a-C:H and to 12 nm for UNCD/a-C:H, as shown by AFM topography analyses. The wettability of the coatings was investigated by contact angle measurements, from which the surface energy was determined. An inverted Enzyme-Linked ImmunoSorbent Assay (ELISA) was performed in order to investigate the protein adsorption on the surfaces of the materials. The proteins studied were albumin and fibrinogen which

*hristina.vasilchina@imf.fzk.de

J.P. Reithmaier et al. (eds.), *Nanostructured Materials for Advanced Technological Applications,* 501
© Springer Science + Business Media B.V. 2009

are known to affect the blood compatibility. The albumin/fibrinogen ratio, which can be used to evaluate the tendency of thrombus formation, was calculated and compared for all samples under investigation.

Keywords: UNCD/a-C:H; a-C:H; wettability; protein adsorption; biofunctionality

1. Introduction

Amorphous carbon and nanocrystalline diamond films exhibit beneficial chemical and physical properties such as high chemical inertness and stability, low friction, high wear resistance, hardness, etc. These unique properties make the films attractive as potential wear-resistant coatings for biomedical applications (mechanical heart valves, artificial blood vessels).[1,2] It is the surface of a biomaterial which first contacts the living tissue when the biomaterial is placed in the body. Therefore, the initial response of the living tissue to the biomaterial must depend on its surface properties.[3] The first event to occur when a foreign material comes into contact with blood is plasma protein adsorption.[4] Subsequently platelets interact with the adsorbed protein layer.[5] In order to evaluate the haemocompatibility of a surface, in the sense of reducing thrombus formation, at least two parameters are usually studied. The first parameter is the adsorption of two proteins, albumin and fibrinogen. It has been found that the albumin adsorption on surfaces inhibits thrombus formation. The fibrinogen, among many other coagulation stimulating proteins, takes part in blood coagulation, facilitates adhesion and aggregation of platelets, and is important in the processes of both haemostasis and thrombosis.[6,7] The second parameter is the amount and the shape of the platelets adherent to the surface. The present work is focused on the evaluation of the first parameter, i.e. the protein adsorption on a-C:H and UNCD/a-C:H coatings. The aim of the work is a comparative study of the properties – basic and relevant to the biomedical applications - of both types of carbon-based thin films.

2. Experimental

Hydrogenated amorphous carbon and ultrananocrystalline dimond/amorphous carbon nanocomposite films have been deposited by microwave plasma chemical vapor deposition (MWCVD) with different deposition equipment under different process conditions: a-C:H coatings were deposited from gaseous mixtures of acetylene and argon in a PE-CVD hybrid coating set-up[8] using a microwave plasma of 2.45 GHz. The gas phase composition $C_2H_2/(C_2H_2 + Ar)$ was varied (values of 0.07; 0.17; 0.33 were used), while the microwave power applied was 700 W for a first series of experiments and 1,200 W for a second one. The total gas pressure in all experiments was 6×10^{-3} mbar. The coatings were deposited onto double-side polished (100) silicon substrates (thickness 350 μm). The films were grown at room temperature without applying a bias voltage; the deposition time was varied from 35 to 120 min in dependence on the acetylene flow rate. Prior to the film deposition the substrates were cleaned for 10 min by argon-plasma etching with a microwave power of 1,000 W and an additional r.f. substrate bias of −500 V.

UNCD/a-C:H nanocomposite films have been deposited from a mixture of 17% CH_4 in N_2 at a pressure of 23 mbar and a temperature of 600°C using the set-up and process described elsewhere.[9] (100) oriented silicon wafers were used as substrates (thickness 380 μm); the deposition time was 390 min at a constant microwave input power of 800 W. The substrates were pretreated prior the deposition in a suspension of diamond powder in n-pentane in order to promote the diamond nucleation.

The substrates used for the protein tests were double-side coated both, with a-C:H and UNCD/a-C:H films. All films have been characterized by a variety of methods with respect to their morphology (SEM), topography (AFM) and bonding structure (Raman spectroscopy). The wettability of the coatings was characterised by contact angle measurements applying a sessile drop method using two polar liquids (water and formamide) and one non-polar liquid (benzyl alcohol), whose surface tensions are well known. From the measured contact angles the surface energy of the layers as well as its polar and disperse components were calculated. An inverted Enzyme-Linked ImmunoSorbent Assay (ELISA) (a biomedical test with standard human plasma) was performed to investigate the protein (albumin and fibrinogen) adsorption on the surface of the a-C:H and UNCD/a-C:H films. HSA (2.5 mg/ml) (human serum albumin) and HSF (2.5 mg/ml) (human serum fibrinogen) solutions in phosphate buffer (pH = 7.4) were used.

3. Results and Discussion

3.1. BASIC FILM PROPERTIES

3.1.1 Scanning Electron Microscopy

Figure 1 shows SEM micrographs of the cross-section of a-C:H and UNCD/a-C:H films, taken with a JEOL JSM 840 microscope. A columnar microstructure and relatively smooth surfaces are observed for all hydrogenated amorphous carbon samples. The deposited films are densely packed with a uniform thickness of about 1.6 μm. (Figure 1, left). The UNCD/a-C:H layers are composed of structures with diameters of several hundreds nanometers. The thickness of the films is 1.15 μm (Figure 1, right).

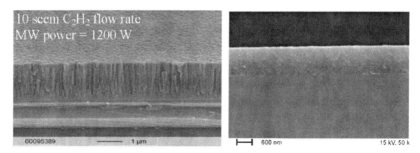

Figure 1. SEM micrographs of a-C:H (left) and UNCD/a-C:H (right) films.

3.1.2. Atomic Force Microscopy

The topography of the layers was studied by AFM (Digital Instrument Dimension 3100) operated in the contact mode. Figure 2 shows typical AFM images of hydrogenated amorphous carbon and ultrananocrystalline diamond/amorphous carbon composite films.

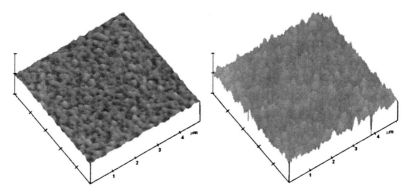

Figure 2. AFM images of a a-C:H film deposited with 10 sccm C_2H_2 flow rate and a MW power of 1,200 W (left) and a UNCD/a-C:H film (right). The area is 5×5 μm^2, the height scale comprises 100 nm. The roughness was as follows: a-C:H : Ra = 2.3 nm and rms = 2.8 nm; UNCD/a-C:H : Ra = 9.1 nm and rms = 15.8 nm.

The surfaces of all samples exhibited essentially a similar topography as shown in Figure 2. The rms roughness and average roughness (Ra) determined from the AFM images show that all the films are rather smooth, which is required for many biomedical applications. As can be seen from the surface analysis, the UNCD/a-C:H coatings have, however, a significantly higher roughness than the a-C:H thin films.

3.1.3. Raman Spectroscopy

The bonding structure of the coatings was evaluated by micro-Raman spectroscopy at room temperature, using a Renishaw–1000 system equipped with an argon ion laser (excitation wavelength 514.5 nm) and a HeCd laser (325 nm). Figures 3 and 4 show the visible and UV-Raman spectra of the films, respectively.

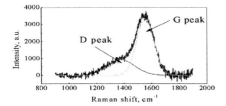 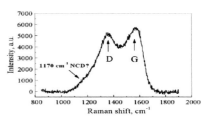

Figure 3. Raman spectra of a-C:H (left) and UNCD/a-C:H (right) films at 514 nm excitation.

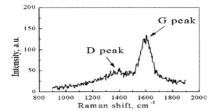 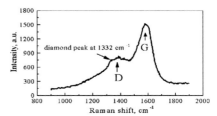

Figure 4. UV Raman spectra of a-C:H (left) and UNCD/a-C:H (right) films at 325 nm excitation.

All spectra are dominated by two features: the G peak at ca. 1,550 cm^{-1} and the D mode around 1,350 cm^{-1}. The G peak is due to the bond stretching of all pairs of sp^2 atoms in both, rings and chains. The D peak is due to the breathing modes of sp^2 atoms in rings. In the visible spectrum of the UNCD/a-C:H films a shoulder at about 1,170 cm^{-1} is observed, which could be attributed to the presence of diamond nanocrystallites, which was observed as well by XRD.[10] The diamond peak at 1,332 cm^{-1}, displayed in Figure 4 (right) as a sharp edge, appears only at the higher excitation energy of the UV-Raman.

3.2. GENERAL BIOPROPERTIES

3.2.1. Surface Energy and Wettability

In many biomedical applications the surface wetting and adsorption properties are critical. The most important characteristic of the surface of a
solid (single crystal, composite, coating) is the surface energy γ.[11] Its value,
which is defined by the amount of uncompensated bonds of the surface
atoms, impacts the adsorption, the adhesion, and the catalytic and
tribological properties of the surface. The protein adsorption on the surface
of biomaterial is also related to the surface energy.[12]

Wettability examinations were performed with the sessile drop method
using a DataPhysics contact angle goniometer. The test liquids used are
listed in Table 1 together with their relative surface tension components.

TABLE 1. Surface tension γ_l and surface tension components of the liquids used in the
contact angle measurements – γ_l^d: dispersive component; γ_l^p: polar component.

Liquid	γ_l (mN/m)	γ_l^d (mN/m)	γ_l^p (mN/m)
Water	72.1	19.9	52.2
Formamide	56.9	23.5	33.4
Benzyl alcohol	39.0	30.3	8.7

The contact angle measurements were carried out with at least five
drops of each liquid for each sample in order to get reasonable statistical
averages. The surface energy of the samples were calculated according to
the Owens, Wendt–Kaelble method.[13] Figure 5 shows the dispersive and
polar components of the surface energies of the films together with the
contact angle of water.

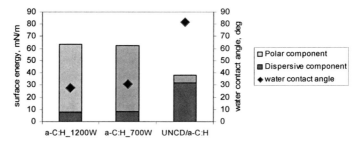

Figure 5. Water contact angles and surface energies of different coatings, together with their
polar and dispersive components.

The a-C:H films show a larger polar component of their surface energies,
while the dispersive component predominates for the UNCD/a-C:H films.
The water contact angle decreased as the polar component in the surface

energy increased. The polar components attract the electric dipoles of water molecules, which minimizes the interfacial energy between the surface and the water and lowers the water contact angle.[14]

All a-C:H coatings (deposited at different process conditions, e.g. MW power and acetylene flow rate) showed similar values of the surface energy and its polar and dispersive components as the coatings shown in Figure 5. Further detailed investigations of various UNCD/a-C:H composite films are however needed for a general conclusion about the surface energy results for these materials.

3.2.2. Protein Adsorption

The effect of protein adsorption on the material surface, as a possible indicator of blood compatibility, was studied by Inverted Enzyme-Linked ImmunoSorbent Assay (ELISA), a biomedical test with standard human plasma. The proteins studied were albumin and fibrinogen. The ratio of albumin to fibrinogen is important when assessing the adhesion of blood platelets to artificial surfaces; the higher the ratio, the lower the number of adhering platelets, and the lower the tendency of thrombus formation.[16] The ratios were calculated from the individual levels of the two proteins adsorbed to the surfaces, and the results are shown in Figure 6.

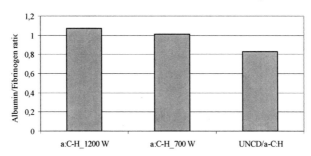

Figure 6. Albumin/fibrinogen adsorption ratio on a-C:H and UNCD/a-C:H coatings.

Higher values of albumin adsorption were observed on the a-C:H samples, in comparison to the UNCD/a-C:H films. This is probably due to the different polar and dispersive components of the surface energy of the films. The higher fibrinogen adsorption on UNCD films should be attributed to the strong interaction between fibrinogen molecules and the hydrophobic film surface. The higher albumin adsorption on a-C:H layers is caused by the larger polar component of their surface energy. Albumin with its hydrophilic nature is attracted to the hydrophilic surfaces through electrostatic interactions. Ma et al.[15] also reported higher albumin to fibrinogen ratios for surfaces with higher surface energies. The ratios determined in the present work are close to those in literature. Dion et al. reported a value of the protein ratio of 1.24 for DLC coatings produced by CVD.[16] Jones et al.[17] have observed an albumin to fibrinogen ratio of 1.12 for DLC coatings, about 0.8 for titanium, 0.7 for TiC and about 0.6 for TiN.

4. Conclusions

A comparative study of the properties – basic and relevant to the biomedical applications – of a-C:H and UNCD/a-C:H thin films deposited by MWCVD have been performed. The different morphology, topography and bonding structure of the amorphous and nanocrystalline carbon coatings impact their wetting behaviour and the protein adsorption on their surfaces. The a-C:H coatings revealed higher surface energies with larger polar components than the UNCD/a-C:H films. The ratio albumin to fibrinogen adsorption was higher for the more hydrophilic a-C:H coatings as compared to the ratio obtained for the UNCD/a-C:H nanocomposite films. The results are comparable with literature data and are indicative of the ability of the investigated coatings to prevent thrombus formation.

ACKNOWLEDGMENTS

The autors gratefully acknowledge the financial support of the "Dr. Anton Kalojanoff" Foundation.

References

1. R. Hauert, Diamond Relat. Mater. 12, 583 (2003).
2. J.A. Carlisle and O. Auciello, Electrochem. Soc. Interf. 12, 28 (2003).
3. F.Z. Cui and D.J. Li, Surf. Coat. Technol. 131, 481 (2000).
4. J.D. Andrade, in: J.D. Andrade (Ed.), *Surface and Interfacial Aspects of Biomedical Polymers* (Plenum Press, New York, 1985).
5. K.K. Chittur, Biomaterials 19, 301 (1998).
6. E. Lee and S. Kimm, ASAIO J. 3, 355 (1980).
7. S. Slack and T. Horbet, J. Biomater. Sci. Polym. 2, 227 (1991).
8. L. Niederberger et al., Surf. Coat. Technol. 174–175, 708 (2003).
9. C. Popov, M. Novotny, et al., Thin Solid Films 506–507, 297 (2006) .
10. C. Popov, W. Kulisch, P.N. Gibson, G. Ceccone, and M. Jelinek, Diamond Relat. Mater. 13, 1371 (2004).
11. D.H. Kaelblc, Polymer 18, 475 (1997).
12. V. Adamson, *Chemistry of Diamond Surface* (Naukova Dumka, Kiev, 1990), p. 200.
13. D.K. Owens and R.C. Wendt, J. Appl. Polym. Sci. 13, 1741 (1969).
14. R.K. Heon-Woong Choi, Diamond Relat. Mater 16, 1732 (2007).
15. W.Y. Ma et al., Biomaterials 28, 1620 (2007).
16. I. Dion et al., Biomed. Mater. Eng. 3, 51 (1993).
17. M.L. Jones et al., J. Biomed. Mater. Res. 52, 413 (2000).

TANTALUM PENTOXIDE AS A MATERIAL FOR BIOSENSORS: DEPOSITION, PROPERTIES AND APPLICATIONS

W. KULISCH[*], D. GILLILAND, G. CECCONE,
L. SIRGHI, H. RAUSCHER, P.N. GIBSON, M. ZÜRN,
F. BRETAGNOL, F. ROSSI
*Nanotechnology and Molecular Imaging, Institute
for Health and Consumer Protection,
European Commission Joint Research Centre, Institute for
Health and Consumer Protection, Via Enrico Fermi, I-21027
Ispra (VA), Italy*

Abstract. Tantalum pentoxide (Ta_2O_5) thin films have been deposited at room temperature by single and dual ion beam deposition (IBS and DIBS, respectively) for applications as a waveguide in biosensors. X-ray photo-electron spectroscopy, Fourier transform infrared spectroscopy and X-ray diffraction have been employed, among others, to establish composition, bonding structure and crystallinity of the films. The optical properties were established by variable angle ellipsometry, UV/VIS spectroscopy and direct measurements of the optical losses. Provided the oxygen partial pressure during deposition is high enough, stoichiometric Ta_2O_5 films are obtained, which are amorphous and extremely smooth. Their refractive index is in the range of 2.05–2.2, while the optical losses are below 3 dB/cm. Finally, in view of applications in biosensors experiments are presented to functionalize the surfaces of these films with amine and epoxy groups.

Keywords: tantalum pentoxide; dual ion beam deposition; waveguide; biosensor

1. Introduction

Tantalum pentoxide (Ta_2O_5) is a material with outstanding electrical, optical and chemical properties,[1,2] among which the high refractive index, the high transmission in a large part of the visible spectrum, and the chemical stability

---

[*]wilhelm.kulisch@jrc.it

J.P. Reithmaier et al. (eds.), *Nanostructured Materials for Advanced Technological Applications,*   509
© Springer Science + Business Media B.V. 2009

make it an interesting candidate for application in label-free biosensors.[3-7] Ta_2O_5 films can be deposited by physical as well as by chemical vapour deposition methods (for a review see Ref. 1). But as applications in the field of biosensors require more and more deposition on temperature sensitive substrates such as thermoplasts, deposition methods are required which can be applied at temperatures below 100°C, which excludes CVD techniques.[8] Furthermore, substrates such as Topas® and Zeonor® are also sensitive against ion bombardment and exposure to UV light[8] which excludes also most of the plasma based PVD methods. Reactive ion beam deposition can be carried out at room temperature without exposing the substrate to energetic particles. In DIBS, a second ion beam is used to densify the films or to adjust the stoichiometry. Exposure of the substrate to ions can be avoided by using the growing film as a shield.

In this paper we report the deposition of Ta_2O_5 films by reactive IBS and DIBS. The oxygen partial pressure p_{O2} and the energy of the substrate ion beam E_s were the main parameters varied. A variety of techniques has been employed to establish the composition, bonding structure and crystallinity of these films. Specific attention was paid to their optical properties (refractive index n, extinction coefficient k, optical losses) as a function of p_{O2} and E_s. Finally, first experiments are reported to functionalize the surface of Ta_2O_5 films with amine and epoxy groups.

2. Experimental

Tantalum pentoxide films have been deposited by single (IBS) and dual ion beam deposition (DIBS) using the set-up schematically shown in Figure 1. Details and also a description of the deposition procedure have been reported in a previous paper.[9] Two identical Kaufman-type ion guns (3 cm from CSC, Alexandria, Virginia) were used for the target and the substrate ion beams. Oxygen was introduced directly into the deposition chamber to allow a reactive deposition process. The parameters used throughout this investigation are summarized in Table 1. (100) silicon pieces, standard cover glasses, quartz and Zeonor® 1420 R chips were used as substrates.

The Ta_2O_5 films were characterized with respect to their morphology and structure by scanning electron microscopy (SEM) and atomic force microscopy (AFM); the composition and also the bonding environment were investigated by X-ray photoelectron spectroscopy (XPS). Further insight into the bonding structure was brought by Fourier transform infrared spectroscopy (FTIR). X-ray diffraction was used to investigate the crystalline properties. A more detailed description of these characterization methods applied can be found in Ref. 9. The optical properties of the films were studied by variable angle ellipsometry at 532 nm and by UV–VIS spectroscopy in the wavelength range from 200 to 1,000 nm. Finally, the optical losses were directly measured by coupling light of a He/Ne laser via an optical grating into the waveguide as described in detail in Sect. 4.

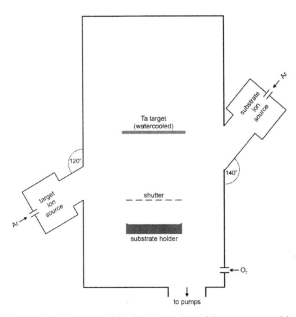

Figure 1. Schematical diagram of the ion beam deposition set-up used in this study.

TABLE 1. Deposition parameters used throughout this study. The indexes t and s stand for target and substrate, respectively.

Parameter		IBS	Reactive IBS	Reactive DIBS
Temperature		rt	rt	rt
E_t	(eV)	1,000	1,000	1,000
J_t	(mA/cm^{-2})	2.8 – 5.5	3.9	3.9
E_s	(eV)	–	–	0–200
J_s	(mA/cm^{-2})	–	–	2.2
$p_{Ar}(t)$	(10^{-2} Pa)	4.2	4.2	4.2
P_{O2}	(10^{-2} Pa)	–	0–4	2.7
$p_{Ar}(s)$	(10^{-2} Pa)	–	–	2.1
P_{total}	(10^{-2} Pa)	4.2	4–8	9

3. Basic Film Properties

3.1. METALLIC FILMS

A number of experiments have been carried out without oxygen addition to deposit metallic Ta films in order to learn about the possibilities and limitations of the set-up shown in Figure 1. The only parameter varied was the current density of the target ion beam. The results obtained – as far as they are of relevance for the remainder of this paper – can be summarized as follows:

1. The Ta films were free of any contamination within in the limits of XPS (ca. 1%).
2. According to SEM, the films were smooth and structure-less.
3. XRD measurements showed that the films were nanocrystalline with extremely small crystallite sizes (2–4 nm).
4. The growth rates obtainable with the set-up were about 6–8 nm/min.

3.2. IBS Ta_2O_5 FILMS

The formation of Ta_2O_5 films has been investigated by adding oxygen to the gas phase of the deposition chamber. The composition of the resulting films was examined by XPS (Figure 2). The data shown have been taken after 3 min of 3 keV Ar^+ sputtering; for the film deposited without oxygen addition data after 6 min sputtering are shown also. The tantalum content of the metallic film apparent from the first measurement is ca. 70% only; this value increases to 80% after 6 min of sputtering. This result can be explained by oxidation of the Ta surface as soon as it is exposed to air; this surface oxygen is knocked-on into the film by the Ar^+ ions. This effect has been

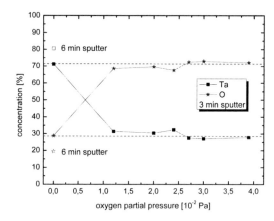

Figure 2. Composition of IBS films as a function of the oxygen partial pressure. The dashed horizontal lines mark the stoichiometry.

observed previously for other film systems.[10] All films deposited in the presence of oxygen in the deposition chamber are close to stoichiometry. However, the latter is only reached for oxygen partial pressures $p_{O2} \geq 2.5 \times 10^{-2}$ Pa.

In order to investigate whether the deviations from stoichiometry for lower p_{O2} are significant, XPS core level spectra have been recorded. Figure 3 shows Ta 4f spectra for a slightly understoichiometric and a stoichiometric film, recorded after 3 min of Ar^+ sputtering (3 keV). The spectra consist of doublets (Ta $4f_{7/2}$ and Ta $4f_{5/2}$) with a spacing of 1.9 eV. They are dominated by a doublet on the high energy side which can be assigned to Ta^{5+}, i.e. Ta in Ta_2O_5 (please note: due to calibration problems at the time of the measurements, the absolute peak positions of these spectra are not reliable. This does, however, not hold for the energy differences). According to literature,[11,12] the energy difference between Ta^{5+} and Ta^0 (i.e. metallic Ta) is 4.5–4.7 eV. As such high differences are not observed in Figure 3, it can be concluded that no metallic Ta is present in the films. Nevertheless it is also evident that both spectra possess peaks which can be assigned to various Ta suboxides. These peaks are much more pronounced for the understoichiometric film. From literature it is well known that Ta_2O_5 films are damaged by the ion bombardment used for sputter cleaning, which causes a (partial) reduction of the Ta.[12,13] As both films have been subjected to the same ion bombardment, it must be concluded that the suboxides observed for the understoichiometric film in Figure 3 can not be due to the ion damage alone but must be partly intrinsic to the film. On the other hand, from the inset in the right diagram of Figure 3, showing a spectrum of a stoichiometric film obtained without sputtering it is clear that the film consists of Ta_2O_5 only, without any suboxide present.

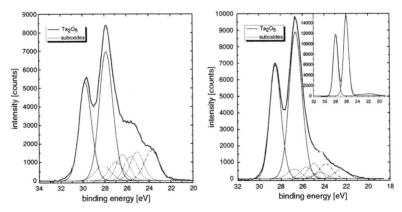

Figure 3. XPS Ta 4f spectra of a slightly understoichiometric (left) and a stoichiometric film (right) obtained after 3 min of Ar^+ sputtering. The inset in the right diagram shows a spectrum of the same surface without sputtering.

Figure 4 shows FTIR spectra of two films, one of them stoichiometric and the other slightly understoichiometric. No peaks outside the range from 1,100 to 400 cm⁻¹ depicted in the figure have been observed for any film of this study which means that e.g. no O–H or C–H groups are present in the films. The spectra are dominated by a peak at about 636 cm⁻¹ which is typical for amorphous Ta_2O_5 films,[14–18] while for crystalline films the peak shifts to 510–480 cm⁻¹. It can be assigned to O≡3Ta or Ta–O–Ta stretching vibrations. The shoulder at about 920 cm⁻¹ in the spectrum of the understoichiometric film (indicated by the arrow) can, according to Refs. 19 and 20, be ascribed to TaO_2, i.e. understoichiometric tantalum oxide units, in agreement with the XPS results presented above.

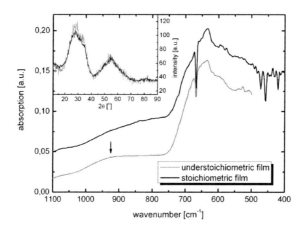

Figure 4. FTIR spectra of a stoichiometric and a slightly understoichiometric film. The inset shows two grazing angle ($\varphi = 1°$) XRD scans for these two types of film.

The inset in Figure 4 shows XRD measurements, performed under a grazing angle of 1°, of a slightly understoichiometric and a stoichiometric film, from which it is evident that these films are amorphous, in agreement with the conclusion drawn above from the FTIR spectra.

Figure 5a shows an AFM topography image of a stoichiometric IBS film deposited on silicon. It is extremely smooth, with a rms surface roughness of 0.14 nm only. Similarly smooth surfaces have been observed for all films; irrespective of the deposition conditions (oxygen partial pressure, IBS or DIBS) the roughness was always below 0.2 nm. On Zeonor 1420 substrates the roughness was 0.27 nm, but here the roughness of the uncoated surface was 0.2 nm, i.e. higher than that of all films deposited on Si.

In order to check whether the films possess any stress, a 340 nm thick film was deposited on a silicon substrate containing an area of micromachined cantilevers.[21] From the SEM image in Figure 5b it can be seen that the cantilever are slightly bended downward, indicating a compressive stress. From the film thickness and the dimensions of the cantilever a stress of -0.4 ± 0.2 MPa could be derived, which should be no impediment for the planned application

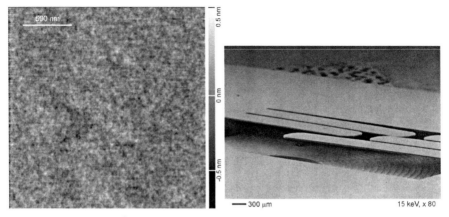

Figure 5. (a) $2 \times 2 \; \mu m^2$ AFM topography image of a stoichiometric Ta_2O_5 film (right); (b) SEM image of a 340 nm thick Ta_2O_5 film on a silicon cantilever area.

3.3. DIBS Ta_2O_5 FILMS

In the DIBS experiments, the growing films were subjected to an Ar^+ ion bombardment of 2.2 mA/cm^{-2} (Table 1) in order to achieve a densifying effect. The parameter varied was the ion energy (0–200 eV). This range has been chosen to provide enough energy for subplantation effects but low enough to avoid damage of the growing film. It turned out, however, that this additional ion bombardment had no pronounced effect on the growth and the properties of the films. This concerns on the one hand the growth rates, which were unaffected within the range of error of the ellipsometric measurements; on the other hand, AFM measurements showed no roughening of the surface (e.g. rms = 0.14 nm for 0 V and rms = 0.18 nm for 200 V). Also, FTIR spectra showed no differences indicating any influence of the substrate ion beam on the bonding environment. XPS measurements to confirm these results are currently under way.

4. Optical Properties

The decisive properties of Ta_2O_5 films for applications in biosensors are the refractive index and the transmission in the visible part of the spectrum. Consequently, emphasis has been laid to establish these two properties as a function of the two main parameters of this studies, the oxygen partial pressure p_{O2} for IBS and the energy of the substrate ion beam E_s for DIBS experiments. Variable angle ellipsometry at 532 nm has been used to determine the refractive index n and the extinction coefficient k as a function of p_{O2} and E_s. The results are presented in Figure 6.

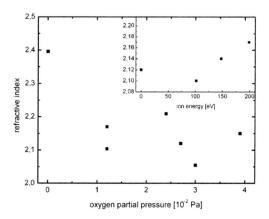

Figure 6. Refractive index n as a function of the oxygen partial pressure during IBS experiments. The inset shows n in dependence of the substrate ion energy for DIBS experiments.

From the main diagram in Figure 6 it can be seen that for all films grown under reactive conditions, i.e. with oxygen addition, the refractive index is between 2.05 and 2.2. However, there is no clear trend of n with p_{O2}. The reasons for these observations, and also for the scattering of the data is not clear at the present time.

On the other hand, it is of great importance to note that for all films the best fits to the ellipsometric data were obtained by setting the extinction coefficient $k = 0$. From the resolution of the procedure applied for the evaluation of these measurements this means that $k \leq 10^{-3}$ which is a first indication that the absorption of the films is low.

The data in the inset of Figure 6 shows n as a function of the energy of the substrate ion beam. The data seem to indicate a slight increase with increasing energy, hinting at a slight densification, in agreement with Lee et al.,[23] according to whom n is determined by the packing density but k by the stoichiometry. Since again in all cases the best fits to the ellipsometric data were obtained by setting $k = 0$, it can be concluded that the stoichiometry of the films is not affected by the ion bombardment of the growing film, in contrast to Cevro and Carter who reported DIBS experiments in which for high ion currents the extinction coefficient increased rapidly due to a loss of stoichiometry, but the ion energy in those experiments was 300 eV.

Figure 7 presents UV/VIS measurements in the range from 190 to 1,000 nm on a variety of substrates. The spectra shown in the left diagram stem from standard cover glasses. The absorption edge of ca. 300 nm is clearly determined by the substrate. But it is also clear that for the two under-stoichiometric films the transmission does not reach that of the substrate in the entire wavelength region. In contrast, the two stoichiometric films are free of absorption at least for $\lambda \geq 500$ nm. This also holds for the spectrum of a film deposited on a Zeonor substrate shown in the right diagram of the figure, and also for that on a quartz substrate presented in the inset. From the latter measurement, an absorption edge of 260 nm can be derived for the IBS Ta_2O_5 films.

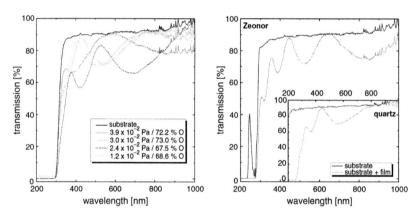

Figure 7. Left: UV–VIS spectra of a series of IBS samples on glass substrates. Parameter is the oxygen partial pressure. Right: UV/VIS spectrum of a stoichiometric Ta_2O_5 film on a Zeonor substrate. The inset shows a spectrum of a stoichiometric film on quartz.

From the above results it can be seen that the refractive index of IBS Ta_2O_5 films is on the order of 2.1 ± 0.1 and that the extinction coefficient is well below 10^{-3}. However, the UV/VIS and ellipsometric measurements discussed above have been performed in a vertical geometry, i.e. the sensored length is the film thickness (ca. 150 nm). For applications as waveguides in biosensors, the path the light has to travel is on the order of centimetres.

In order to get at least an estimation of the optical losses of the IBS Ta_2O_5 waveguides, the experimental set-up shown in Figure 8 has been used. The light of a He/Ne laser (633 nm) was directed via an adjustable mirror on a optical grating on a Zeonor 1420 substrate which was covered by a ca. 155 nm thick Ta_2O_5 film. This grid (ca. 0.5 mm wide) had a pitch of 330 nm and a grating depth of 23 nm. AFM measurements have shown that this grid was not damaged in any form by the deposition of the films.[22]

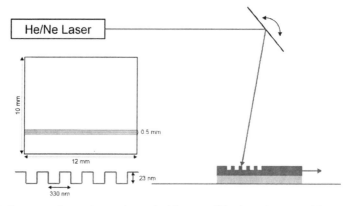

Figure 8. Set-up used to estimate the optical losses of Ta_2O_5 IBS waveguides. Also shown are the dimensions of the Zeonor chip and of the optical grating.

Figure 9 shows a picture of a Zeonor chip during such an experiment. The spot in the background indicates the position of the grid where the laser beam is coupled into the waveguide. The coupling angle of 17° was close to the value calculated from the refractive index, the thickness of the film and the optical data of the Zeonor substrate. The spot in front indicates the point where the light leaves the waveguide. Most important is the observation that the path of the light within the waveguide is not visible which means that at neither of the two interfaces scattering takes place, in agreement with the smoothness of the film evident from the AFM measurements presented above. It has to be pointed out that the optical losses within the waveguide can not be determined quantitatively from this simple experiment but they can be estimated to be below 3 dB/cm which is sufficient for the applications envisioned.

Figure 9. Picture of a Zeonor chip bearing an optical grating and coated with a 155 nm thick Ta_2O_5 film, which is illuminated with a laser beam according to the scheme in Figure 8.

5. Surface Characterization and Modification

For applications in optical biosensors, not only outstanding optical properties, but also knowledge of the chemical nature of the surface of the Ta_2O_5
films, methods to clean them and to provide a suitable substrate for further
chemical treatments, and finally routes to functionalize them according to a
given application are required. To this end, first experiments have been carried
out to apply different surface treatment methods, to characterize as-grown
and treated surfaces, and to functionalize them for future immobilization of
biomolecules.

5.1. SURFACE TREATMENTS

Besides as-grown Ta_2O_5 films (AG in the following), three different surfaces
have been investigated in the present study, which have been subjected to
the following treatments:

1. O_2 microwave plasma treatment (240 Pa, 200 W, 10 min) (OP)
2.   UV/O_3 treatment in air (room temperature, 30 min) (UV)
3.   Treatment in piranha solution (H_2O_2/H_2SO_4 1:3, 80°C, 30 min) (PI)

Thereafter, the surfaces have been investigated by contact angle measurements, XPS and TOF-SIMS.

5.2. NATURE OF THE SURFACES

The contact angles measured immediately after deposition/treatment were
as follows: $86 \pm 2°$ for the as-grown sample, $26 \pm 1°$ for the oxygen plasma
treated film, and below 5° for the UV and PI surfaces. These values are in
good agreement with data reported by De Palma et al.[25]

The surface compositions of all four samples as revealed by XPS are
presented in Figure 10. It can be seen that all four surfaces are heavily con-
taminated by organic species, despite the fact that they were introduced into
the XPS chamber immediately after the treatments. In all four cases, the
surface carbon content is between 15% and 25%, but there is no apparent
correlation with the nature of the treatment or with the corresponding contact angles. Again, these observations are in agreement with the results of
De Palma et al.,[25] who observed even higher carbon concentrations on the
surfaces and ascribed them to a rapid re-contamination after the treatments.

These results have been confirmed by TOF-SIMS analysis of these surfaces (not shown graphically). They proved the existence of an abundance
of organic species; although there were differences between the four surfaces
it was not possible to establish a clear correlation with the pretreatments

performed and with the contact angles measured. However, a closer look at those peaks stemming from tantalum containing species provided at least some preliminary insight. The left diagram in Figure 11 shows the normalized count rate ratios of the type Ta_xO_{y+1}/Ta_xO_y, the right diagram ratios of the type Ta_xO_yH/Ta_xO_y. Normalization has been performed in both cases by setting the values of the as-grown surface to unity. Two trends become visible from Figure 11:

- The ratios Ta_xO_{y+1}/Ta_xO_y, increase after any of the treatments.
- The ratio Ta_xO_yH/Ta_xO_y decrease after any of the treatments.

This means, on the one hand, that all three treatments lead to a more complete oxidation of the surface as compared to the as-grown film. On the other hand, the concentration of tantalum surface groups containing hydrogen decreases. However, from the diagrams in Figure 11 no "ranking" of the three treatments investigated can be inferred.

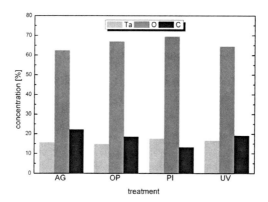

Figure 10. Surface composition of an as-grown Ta_2O_5 film and after various surface treatments according to XPS.

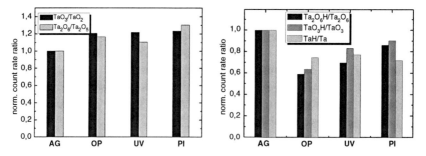

Figure 11. Normalized TOF-SIMS count ratios of the types Ta_xO_{y+1}/Ta_xO_y (left) and Ta_xO_yH/Ta_xO_y (right).

5.3. SURFACE FUNCTIONALIZATION

For a typical use of tantalum pentoxide films in biosensors, biomolecules have to be immobilized on their surfaces. The most common routes reported in literature rely either on pure adsorption[4,7] or, more reliably, on covalent bonding via silanization.[3-5] In order to test the suitability of our IBS Ta_2O_5 surfaces for this purpose, two different approaches have been chosen to functionalize them for possible applications in biosensors: (i) creation of a self assembled monolayer bearing NH_2 groups via a silanization process; and (ii) deposition of a polymer layer bearing epoxy groups. In both cases, the characterization of the resulting surfaces has not yet been completed, but there are preliminary results indicating that both approaches have been successful.

5.3.1. Silanization

Starting from a freshly grown Ta_2O_5 surface, the silanization process consisted of the following steps:

1. UV/O_3 treatment for 60 min (see Sect. 5.1)
2. 18 h immersion in 1% (vol) 3-aminopropyl trimethoxisilane (APTMS) in dry ethanol
3. Rinse in ethanol
4. Baking on a hot plate for 60 min at 80°C

Figure 12 shows filtered TOF-SIMS spectra of a Ta_2O_5 surface after the UV/O_3 cleaning step and after the silanization process. It is evident that the concentration of amine containing groups of the form C_xH_yN has increased

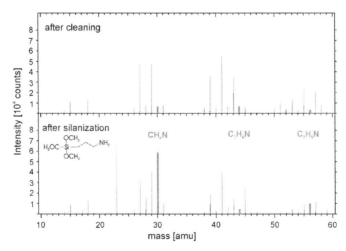

Figure 12. Filtered TOF-SIMS spectra of a Ta_2O_5 surface after UV/O_3 treatment and after silanization with APTMS. The APTMS molecule is shown in the lower panel of the figure.

drastically. Similar filtered spectra also prove the presence of Si containing groups on the surface (not shown graphically). In the ideal case, the APTMS molecules attach covalently with their silane end at OH groups on the Ta_2O_5 surface, while the amine groups at the other end of the molecule form a new surface ready for the attachment of biomolecules. Whether this ideal case has been realized in our experiments will be revealed by future angle resolved XPS measurements.

5.3.2. Deposition of PGMA

Another experiment aimed at the deposition of a thin layer of poly(glycidyl methacrylate) (PGMA) which provides epoxy groups at the surface, but also in its interior which can be used to attach further molecules. The following process has been used for the deposition of PGMA layers:

1. UV/O_3 treatment for 60 min
2. Spin casting of a solution of PGMA in methyl ethyl ketone (MEK)
3. Heat treated on a hot plate at 100°C for 60 min to bind the polymer to the Ta_2O_5 via surface OH groups
4. Rinsing away excess polymer with MEK to leave an active surface rich in epoxy groups

Figure 13 shows filtered TOF-SIMS spectra of a Ta_2O_5 surface after UV/O_3 cleaning and after deposition of the PGMA layer. It is evident that for the latter an abundance of $C_xH_yO_z$ species is present on the surface, which indicates a successful deposition of PGMA. The epoxy groups of the polymer are not completely used for the covalent attachment of PGMA to the Ta_2O_5 surface; many of them remain situated in the loops and tails of the polymer which can be used for further attachment processes.[26] Fist experiments to use this surface to graft a non-fouling poly(ethylene glycol) layer by atom transfer radical polymerization yield very promising results as will be discussed in a forthcoming paper.[27]

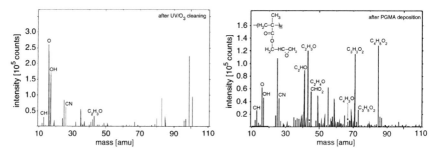

Figure 13. Filtered TOF-SIMS spectra before and after deposition of a PGMA polymer layer. Also shown is the structure of the PGMA molecules.

6. Summary

Ta_2O_5 thin films have been deposited at room temperature by reactive (dual) ion beam sputtering from a tantalum target for applications as a waveguide in optical biosensors. It has been shown that, provided the oxygen partial pressure is sufficiently high, stoichiometric, amorphous, very smooth Ta_2O_5 films can be obtained. The refractive index is on the order of 2.1 ± 0.1, while from an experiment in which a laser beam was coupled into the waveguiding film via an optical grating, the losses could be estimated to be below 3 dB/cm. First experiments to functionalize the surfaces of these films with NH_2 and epoxy groups delivered promising preliminary results.

ACKNOWLEDGEMENTS

This work was partially funded by the European Commission through the Integrated Project CAREMAN No. IP 017333 of the Sixth Framework program. We like to thank A. Nicol from Bayer Technology Services for providing the Zeonor chips with the optical grating.

References

1. C. Chaneliere, J.L. Autran, R.A.B. Devine, and D. Balland, Mater. Sci. Eng. R 22, 269 (1998).
2. Z. Geretovszky, T. Szörenyi, J.P. Stoquert, and I.W. Boyd, Thin Solid Films 453–454, 245 (2004).
3. C. Boozer, Q. Yu, S. Chen, C.-Y. Lee, J. Homola, S.S. Yee, and S. Jiang, Sensor. Actuator. B 90, 22 (2003).
4. K. Schmitt, B. Schirmer, C. Hoffmann, A. Brandenburg, and P. Meyrueis, Biosensor. Bioelectron. 22, 2591 (2007).
5. R. Polzius, T. Schneider, F.F. Bier, U. Bilitewski, and W. Koschinski, Sensor. Actuator. B 11, 503 (1996).
6. H.B. Lu, J. Homola, C.T. Campbell, G.G. Nenninger, S.S. Yee, and B.D. Ratner, Sensor. Actuator. B 74, 91 (2001).
7. J. Dostalek, J. Ctyroky, J. Homola, E. Brynda, M. Skalsky, P. Nekvindova, J. Spirkova, J. Skvor, and J. Schröfel, Sensor. Actuator. B 76, 8 (2001).
8. P. Munzert, U. Schulz, and N. Kaiser, Surf. Coat. Technol. 174–175, 1048 (2003).
9. W. Kulisch, D. Gilliland, G. Ceccone, H. Rauscher, L. Sirghi, P. Colpo, and Francois Rossi, J. Vac. Sci. Technol. A. 26 991 (2008).
10. W. Kulisch, P. Colpo, P.N. Gibson, G. Ceccone, D.V. Shtansky, E.A. Levashov, and F. Rossi, Surf. Coat. Technol. 188–189, 735 (2004).
11. H. Demiryont, J.R. Sites, and K. Geib, Appl. Opt. 24, 490 (1985).
12. J.Y. Zhang, V. Dusastre, and I.W. Boyd, Mater Sci. Semiconduct. Process. 4, 313 (2001).
13. S.M. Rossnagel and J.R. Sites, J. Vac. Sci. Technol. A 2, 376 (1984).

14. H. Grüger, C. Kunath, E. Kurth, S. Sorge, W. Pufe, and T. Pechstein, Thin Solid Films 447–448, 509 (2004).
15. A.P. Huang, S.L. Xu, M.K. Zhu, B. Wang, H. Yan, and T. Liu, Appl. Phys. Lett. 83, 3278 (2003).
16. H. Ono and K. Koyanagi, Appl. Phys. Lett. 77, 1431 (2000).
17. R.A.B. Devine, Appl. Phys. Lett. 68, 1924 (1996).
18. J.Y. Zhang and I.W. Boyd, Appl. Phys. Lett. 77, 3574 (2000).
19. J.Y. Zhang, I.W. Boyd, V. Dusastre, and D.E. Williams, J. Phys. D 32, L91 (1999).
20. W. Weltner and D. McLeod, J. Chem. Phys. 42, 822 (1965).
21. A. Klett, R. Freudenstein, F.M. Plass, and W. Kulisch, Surf. Coat. Technol. 116–119, 86 (1998).
22. W. Kulisch et al., Paper to published in Sensor. Actuator. B (2008).
23. C.C. Lee, J.C. Hsu, and D.H. Wong, Appl. Surf. Sci. 171, 151 (2001).
24. M. Cevro and G. Carter, Opt. Eng. 34, 596 (1995).
25. R. De Palma, W. Laureyn, F. Frederix, K. Bonroy, J.-J. Pireaux, G. Borghs, and G. Maes, Langmuir 23, 443 (2007).
26. B. Zdyrko, V. Klep, and I. Luzinov, Langmuir 19, 10179 (2003).
27. D. Gilliland and W. Kulisch, Unpublished data (2008).

NOVEL NANOSTRUCTURED MATERIALS ACCELERATING OSTEOGENESIS

B. TRAJKOVSKI[1], J. KARADJOV[2], B. SHIVACHEV[3],
A. DIMITROVA[4], S. STAVREV[2], M.D. APOSTOLOVA[*][1]
[1]*Medical and Biological Research Lab, Roumen Tzanev
Institute of Molecular Biology, Bulgarian Academy of
Sciences, Acad. G. Bonchev, St., bl. 21, 1113, Sofia, Bulgaria*
[2]*Institute for Space Research, Bulgarian Academy of Sciences,
Sofia, Bulgaria*
[3]*Central Laboratory of Mineralogy and Crystallography,
Bulgarian Academy of Sciences, Sofia, Bulgaria*
[4]*Department of Biology and Pathophysiology, Medical
University Pleven, Bulgaria*

Abstract. Osteoporosis is a disease in which the mineral density of bone is reduced, its microarchitecture disrupted, and the expression profile of non-collagenous proteins altered. Normal fracture treatment is a complicated, multistage process, involving different cellular events and regulated by local and systematic factors. The objective of this study was to develop bio-active nanostructured materials, in both acellular and autologous cell-seeded forms to enhance bone fracture fixation and healing thought creating highly porous structures which will promote osteogenesis. The approach to develop highly effective hydrogels for counteracting the effects of osteoporosis followed different ways: (a) synthetic and biological polymer chemistry, by using a thermally gelling polymer and nanodiamonds (ultradisperse diamond UDD) (b) experiments with cell models (endothelial progenitor cells EPC). Our investigations demonstrated that *in vitro* EPC transformation to osteo-blasts was enhanced in the presence of a osteoprogenitor medium and UDD. These results provided initial evidence that synthetic nanomaterials may exhibit certain properties that are comparable to natural one, and nano-material architecture may serve as a superior scaffolding for promoting EPC transformation and biomineralization.

[*]margo@obzor.bio21.bas.bg

J.P. Reithmaier et al. (eds.), *Nanostructured Materials for Advanced Technological Applications*, 525
© Springer Science + Business Media B.V. 2009

Keywords: nanodiamonds; angiogenesis; osteogenesis; hydrogels; EPC

1. Introduction

Osteoporosis is a disease in which the mineral density of bone is reduced, its microarchitecture disrupted, and the expression profile of non-collagenous proteins altered. This predisposes bones to fracture, particularly the hip, spine and wrist, and is a major cause of disability, severe back pain and deformity. It is responsible for significant morbidity in Europe, being the direct cause of 3.79 million fractures in 2000 alone.[1]

Normal fracture treatment is a complicated, multistage process, involving different cellular events and regulated by local and systematic factors. It also relays on the structure of the bones. The bone tissue of vertebrate skeletons is comprised of several cell types: osteoblasts (mononuclear bone-forming cells which descend from osteoprogenitor cells), osteocytes (participating in the formation of bone, matrix maintenance and calcium homeostasis), and osteoclasts (the cells responsible for bone resorption).

Recent data in understanding the regulatory factors controlling fracture healing have suggested that some chemical compounds can be used to stimulate bone growth, initiate and enhance the cascade of events involved in callus formation and maturation. Pharmacological methods of osteoporosis prevention and treatment include systematic treatment, where the main point is to increase the bone weight by decreasing osteoclastic bone resorption, using estrogen, bisphosphonates, calcitonin, calcium plus, cholecalciferol, calcitriol and selective estrogen receptor modulators. There is therefore a clinical need to develop new implants, which could be easily applied to fractures and/or osteoporotic bones to promote wound healing and regeneration, while avoiding the surgical and post-surgical complications.

During the last years, the several different studies were undertaken to developed new materials to stimulate bone healing and regeneration.[2,3] Such implants have to ensure suitable conditions for cell adhesion, proliferation and differentiation, and should be biocompatible and biodegradable. One promising approach is to culture cells in three-dimensional (3D) scaffolds or "intelligent" polymers, and to develop tissues of a practical size scale.

Thermosensitive polymers such as those of poly(ethylene glycol) and poly(lactic-co-glycolic acid) (PEG/PLGA) have been extensively studied as possible injectable, site-specific drug delivery depots.[4–8] One example of a biodegradable, water soluble, amphiphilic triblock copolymer that has received much attention is PLGA–PEG–PLGA. Its aqueous formulation is a free flowing solution below room temperature, but it forms a hydrogel at body temperature. Applications in the controlled release of bioactive agents such as insulin,[9] testosterone,[5] glucagon-like peptide-1 (GLP-1)[10] and ganciclovir-loaded PLGA microspheres[4] from this hydrogel system have been documented.

In this study, the suitability of an injectable, bioresorbable PLGA–PEG–PLGA–UDD polymer gel, in the autologous cell-seeded form, was investigated concerning its potential *in vitro* biocompatibility.

2. Materials and Methods

2.1. NANODIAMOND PRODUCTION

The shock-wave synthesis was used to produce ultradispersed diamonds (UDD) by explosive conversion of a trinitrotoluene/hexogene mixture with a negative oxygen balance.[11,12] A water-cooled combustion chamber with 3 m^3 volume was used. A mixture of diamond blends (DB) containing 85% UDD was obtained. The UDD was purified by oxidative removal of the non-diamond carbon using a mixture of K_2CrO_4 and H_2SO_2 according to Tsoncheva et al.[13] The UDD was characterized by X-ray diffraction (XRD), Fourier transform infrared spectroscopy (FTIR) and electron microscopy (EM), and showed an average grain size of 6 nm.

2.2. SCAFFOLDS

Injectable biodegradable temperature-responsive poly(dl-lactide-co-gly-colide-b-ethylene glycol-b-dl-lactide-co-glycolide) (PLGA–PEG–PLGA) triblock copolymers with a DL-lactide/glycolide molar ratio of 6:1 were synthesized from monomers of DL-lactide, glycolide and polyethylene glycol as described in Ref. 6. The resulting copolymers were soluble in water, being fluid at room temperature but forming hydrogels at body temperature. The hydrogels ± UDD (here and in the following: with and without UDD) were prepared under sterile conditions in 24-well plates and in culture dishes. Prior to cell seeding they were immersed overnight in serum-free culture medium at 37°C at 5% CO_2. The measurement of the gelation temperature and the *in vitro* erosion was done following the procedures described by Yong et al.[14] and Qiao et al.,[6] respectively.

2.3. ISOLATION OF HUMAN ENDOTHELIAL PROGENITOR CELLS

EPCs were isolated and cultured with minor modifications of protocols described previously.[15,16] In brief, mononuclear cells were harvested from human peripheral blood buffy coats using Ficoll (Sigma-Aldrich, Steinbach,

Germany) gradient centrifugation, and cultured in endothelial cell growth medium-2 (EGM-2 kit; CC-3162, Lonza, Belgium), 5% fetal calf serum (FCS; Lonza, Belgium), and 1% penicillin/streptomycin, on collagen-coated (BD Europe, Heidelberg, Germany) well plates; 5×10^6 cells per well were seeded on a 24 culture-well plate. The colonies of EPCs appeared after 3–4 weeks in culture. Two days after colony formation the EPCs were trypsinized to single cells, passed through a 70-μm filter, and placed onto the hydrogels at 25 cells/cm^2 in osteoprogenitor medium [improved modified Eagle's medium (IMEM, Life Technologies/Invitrogen) supplemented with 10% fetal bovine serum (FBS; Lonza, Belgium), 0.1 mM 2-mercaptoethanol, 2 mM Glutamax, 2 mM BMP-2 and 0.2 mM ascorbic acid]. The cultures were fed with osteoprogenitor medium every 2–3 days and allowed to differentiate for 30 days to form mature bone nodules.

The transformation of the endothelial phenotypes of EPCs into a long-term osteoblast culture was assessed as described previously.[17] The selection of primers and the reverse-transcriptase polymerase chain reaction (RT-PCR) was performed following Woo et al.,[18] while the immunofluorescence was performed as described by Apostolova and Cherian[19] using mouse monoclonal anti-osteocalcin antibodies (OCG4, clone ab13421, abcam, USA), kindly provided by Tania Christova (University of Toronto, Canada).

2.4. STATISTICAL ANALYSES

The Data were evaluated by analysis of variance (ANOVA) with experiments included as a block term (GenStat 4.2, VSN International Ltd.). Differences between the treatments were assessed by means of the t-statistic, where the standard error of the difference (s.e.d.) was based on the residual mean square (m.s.) obtained from ANOVA.

3. Results and Discussion

3.1. GELATION TEMPERATURE OF PLGA–PEG–PLGA COPOLYMER WITH UDD

The gelation temperature of copolymers with a DL-lactide/glycolide molar ratio of 6:1 is shown in Figure 1A. The gelation temperature changed significantly ($p < 0.5$) with the addition of increasing concentrations of UDD up to 5% (w/v) at the low temperature of gel formation, compared to PLGA–PEG–PLGA alone.

Figure 1B shows the effect of UDD addition on the erosion of the copolymers. The PLGA–PEG–PLGA copolymer displayed a significant amount of erosion from day 15 to day 30. The UDD supplementation significantly (p < 0.05) decreased the erosion of the PLGA–PEG–PLGA within the same time period. To investigate the time-dependent gel properties, the degradation of PLGA–PEG–PLGA ± UDD hydrogels was also studied as an open system using SEM (Figure 1C). The final degradation products of PEG–PLGA–PEG triblock copolymers were PEG, glycolic acid, lactic acid, and UDD.

SEM analyses revealed that the triblock copolymers possess an interconnected porous network which was physically associated upon gelation. The average pore diameter varies from 2 to 30 μm. Following incubation in cell culture media (pH = 7.4) for 21 days, the structure of the copolymers was disrupted but visible nodes containing UDD still remained (Figure 1C).

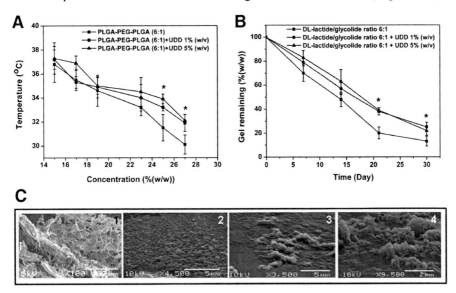

Figure 1. Effect of the molar ratio of DL-lactide/glycolide and of UDD on the gelation temperature of the copolymers (A) and on the erosion at 37°C and 5% CO_2 (B). (C): Representative SEM images of PLGA–PEG–PLGA hydrogels ± UDD following incubation in cell culture media (pH 7.2–7.4): **C1**: PLGA–PEG–PLGA, Day 0; **C2**: PLGA–PEG–PLGA, Day 21; **C3**: PLGA–PEG–PLGA + 1% (v/w) UDD, Day 21; **C4**: PLGA–PEG–PLGA + 5% (v/w) UDD, Day 21. * p < 0.05 compared to PLGA–PEG–PLGA, n = 5.

3.2. IN VITRO BIOCOMPATIBILITY STUDIES

The PLGA–PEG–PLGA ± UDD hydrogels were screened for their *in vitro* properties to enhance the transformation of EPC to osteoblasts. The adhesion, spreading and proliferation of human EPC was studied by fluorescent microscopy and RT-PCR analysis. The EPCs were characterized as described by Fuchs et al.[15]

Following 7, 14, and 21 days of EPC culturing on the PLGA–PEG–PLGA ± UDD hydrogels in the osteoprogenitor medium, the cells were collected, and the expression of genes associated with the osteoblast phenotype was examined. Figure 2A shows the mRNA expression of genes encoding *Runx-2, Osteocalcin, and Collagen I. Runx-2* is critical for the function of osteoblast as a major regulator of the osteoblast phenotype. It expression pattern was examined in the cells grown on both types of hydrogels; it was found that cells cultured on PLGA–PEG–PLGA have a maximum expression on day 7 and 14, in contrast to the PLGA–PEG–PLGA + UDD hydrogel, where the expression reached a maximum at day 7 and stayed unchanged until day 21. Repeated experiments showed that the Runx-2 mRNA level was consistently higher in the cells grown on PLGA–PEG–PLGA + UDD. There are *Runx-2* binding sites in the specific osteoblast genes encoding osteocalcin, bone sialoprotein, and osteopontin, as well as type I collagen[20]; since we have detected a specific expression pattern it was of interest to study other genes, which are under control of *Runx-2*.

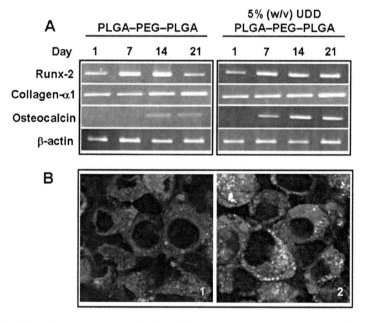

Figure 2. (A) RT-PCR gene expression in EPC grown in the osteoprogenitor medium on the PLGA–PEG–PLGA ± UDD hydrogels. β-actin was the selected housekeeping gene. Similar results were obtained from three independent experiments. The figure shows representative results. (B) Immunofluorescent micrographs of EPC grown for 30 days in osteoprogenitor medium on PLGA–PEG–PLGA ± UDD hydrogels and stained for *Osteocalcin* (×1,000).

Type I *collagen* is the most abundant extracellular protein of bone, constituting approximately 95% of the matrix.[21] Although type I collagen is not unique to the bone, it is very important for proper bone growth and development. The changes of collagen expression at the gene level were minimal, with a trend for slightly increased expression on days 14 and 21 in cells grown on both hydrogels in three separate experiments (Figure 2A).

Osteocalcin is the most abundant noncollagenous protein of bone and is found exclusively in osteoblasts.[22–24] It transcripts were detected at day 7, with a greater expression in cells grown on PLGA–PEG–PLGA + UDD hydrogels versus those grown on PLGA–PEG–PLGA, where the mRNA level was observed on day 14 (Figure 2A).

To further examine the effect of the PLGA–PEG–PLGA ± UDD hydrogels on the osteoblastic phenotype we tested the Osteocalcin protein localization by immunofluorescence. Fluorescent cells were detected predominantly on the PLGA–PEG–PLGA + 5% (v/w) UDD hydrogel, providing visual evidence for the presence of mature osteoblasts on day 30 (Figure 2B).

4. Conclusions

Temperature-responsive PLGA–PEG–PLGA hydrogels with different contents of UDD were synthesized and found to have advantages over PLGA–PEG–PLGA alone. It was found that EPCs adhered to PLGA–PEG–PLGA ± UDD hydrogels and grew over much of the surface area. Moreover, the hydrogels supported osteogenesis at the conditions chosen. The results demonstrate that PLGA–PEG–PLGA + UDD hydrogels substantially enhance the expression of the osteoblast phenotype as measured by the expression of genes associated with the osteoblast phenotype. Overall, the findings in this study substantiate the potential usage of PLGA–PEG–PLGA ± UDD hydrogels for the treatment of bone injuries.

ACKNOWLEDGEMENTS

We are grateful to the National Science Fund of Bulgaria (Grant TKX-1704) for their financial support.

References

1. J. A. Kanis and O. Johnell, Osteoporos. Int. 16, 229 (2005).
2. T. A. Ahmed, E. V. Dare, and M. Hincke, Tissue Eng. Part B Rev. 14, 199 (2008).
3. Cheng, X. et al., Ulus. Travma. Acil. Cerrahi. Derg. 14, 87 (2008).

4. S. Duvvuri, K. G. Janoria, and A. K. Mitra, J. Control Release 108, 282 (2005).
5. S. Chen and J. Singh, Int. J. Pharm. 295, 183 (2005).
6. M. Qiao, D. Chen, X. Ma, and Y. Liu, Int. J. Pharm. 294, 103 (2005).
7. S. Chen, R. Pieper, D. C. Webster, and J. Singh, Int. J. Pharm. 288, 207 (2005).
8. C. Pratoomsoot et al., Biomaterials 29, 272 (2008).
9. S. Choi and S. W. Kim, Pharm. Res. 20, 2008 (2003).
10. S. Choi, M. Baudys, and S. W. Kim, Pharm. Res. 21, 827 (2004).
11. U.S. Patent 5353708, 1994.
12. BG Patent 49267 A. 1991.
13. T. Tsoncheva et al., J. Colloid Interface Sci. 300, 183 (2006).
14. C. S. Yong et al., Int. J. Pharm. 226, 195 (2001).
15. S. Fuchs, A. Motta, C. Migliaresi, and C. J. Kirkpatrick, Biomaterials 27, 5399 (2006).
16. S. Fuchs, M. I. Hermanns, and C. J. Kirkpatrick, Cell Tissue Res. 326, 79 (2006).
17. N. L. Woll, J. D. Heaney, and S. K. Bronson, Stem Cells Dev. 15, 865 (2006).
18. K. M. Woo et al., Biomaterials 28, 335 (2007).
19. M. D. Apostolova and M. G. Cherian, J. Cell Physiol. 183, 247 (2000).
20. P. Ducy, R. Zhang, V. Geoffroy, A. L. Ridall, and G. Karsenty, Cell 89, 747 (1997).
21. S. C. Marks and P. R. Odgren, in: J. P. Bilezikian, L. G. Raisz, and G. A. Rodan (Eds.), *Principles of Bone Biology* (Academic, San Diego, CA, 2002), pp. 3–15.
22. C. Desbois, D. A. Hogue, and G. Karsenty, J. Biol. Chem. 269, 1183 (1994).
23. P. Ducy et al., Nature 382, 448 (1996).
24. J. Glowacki, C. Rey, M. J. Glimcher, K. A. Cox, and J. Lian, J. Cell Biochem. 45, 292 (1991).

EFFECTS OF FUNCTIONAL GROUPS, BIOSIGNAL MOLECULES AND NANOTOPOGRAPHY ON CELLULAR PROLIFERATION

H.T. ŞAŞMAZEL[*], S. MANOLACHE,
M. GÜMÜŞDERELİOĞLU
*Atılım University, Department of Materials Engineering,
Incek, Gölbaşı, Kizilcasar Mahallesi 06836, Ankara, Turkey*

Abstract. The aim of this study is the development of novel cell support materials for fibroblast cell cultivation by using low-pressure plasma assisted treatment. Poly(ε-caprolactone) (PCL) membranes were prepared by solvent-casting technique. The plasma assisted treatment was focused on generating a nano-topography and obtaining COOH functionalities on the surface of the membranes. The immobilization of biomolecules onto the PCL membranes was realized after the plasma treatment. The membranes prepared were characterized by various methods before and after the bio-modification. L929 mouse fibroblasts were used for cell culture evaluation. The prominent roles of surface nano-topography and carboxylic groups generated by the plasma in obtaining better cell growth on PCL surfaces are highlighted in this study.

Keywords: poly(ε-caprolactone); low pressure plasma; biomolecules immobilization; cell proliferation

1. Introduction

Synthetic polymers need selective modifications in order to introduce specific functional groups (e.g. amines, carboxyls) to the surface for the binding of biomolecules and enhancing cell growth.[1] Surface modification of polymers can be achieved by wet (acid, alkali), dry (plasma) and radiation treatments (ultraviolet radiation, laser) without affecting the bulk properties.[2-4] Plasma treatments have many advantages compared to wet-chemical methods.[5]

---

[*]htsasmazel@atilim.edu.tr+

J.P. Reithmaier et al. (eds.), *Nanostructured Materials for Advanced Technological Applications,*   533
© Springer Science + Business Media B.V. 2009

In the present study, utilizing these advantages, we introduced a new plasma assisted treatment of PCL membranes in order to immobilize insulin or heparin biomolecules through PEO (polyoxyethylene bis (amine)). The biologically modified PCL membranes were tested with L929 mouse fibroblasts in cell culture experiments.

2. Experimental Section

2.1. PCL MEMBRANES

PCL (poly-ε-caprolactone) membranes were prepared by solvent-casting technique and cut in the form of discs, with diameters of 12.5 mm, before plasma treatment.

2.2. COLD PLASMA TREATMENT USING VACUUM ENVIRONMENT

Plasma assisted treatment was carried out in a cylindrical, capacitively coupled RF plasma reactor equipped with a 40 kHz power supply in three steps: H_2O/O_2 plasma treatment; in situ or ex situ gas/solid reaction (oxalyl chloride vapors; pressure 65 Pa; time 30 min); and hydrolysis (open laboratory condition; 2 h) for final –COOH functionalities.

3. Characterization Studies

Experimental parameters such as the flow rates of water and oxygen, power, pressure and time of the plasma modifications were optimized according to a design of experiments (DoE) program (Design Expert 7, Stat-Ease, Inc., Minneapolis, MN). COOH and OH functionalities on the modified surfaces were detected quantitatively by using fluorescent labeling technique and an UVX 300G sensor. Structural chemical information of untreated and plasma treated PCL membranes were acquired using pyrolysis gas chromatography/ mass spectroscopy (GC/MS) analysis. Electron spectroscopy for chemical analysis (ESCA) analysis was used to evaluate the relative surface atomic compositions and the carbon and oxygen linkages located in non-equivalent atomic positions. Atomic force microscopy (AFM) analysis (Molecular Imaging PicoSPM instrument; contact mode; silicon and hydrazine (HZD) functionalized cantilevers) were carried out on H_2O/O_2 plasma treated, oxalyl chloride functionalized and untreated PCL samples in order to observe the micro- and/or nano-topography of surfaces.

4.   Biological Modification of PCL Membranes

The biomodification of PCL membranes with insulin and heparin biosignal molecules in the presence of a PEO spacer was realized after water/O_2 plasma treatment. The success of the immobilization process was checked qualitatively with ESCA. In addition, the amount of immobilized biomolecules was determined by using fluorescent labeling techniques.

5.   L929 Cell Culture

Anchorage-dependent L929 mouse fibroblasts [HUKUK (Cell Culture Collection) 92123004] were cultured using Dulbecco's Modified Eagle's Medium supplemented with 10% (v/v) fetal bovine serum in 24-well polystyrene tissue-culture dishes containing PCL membranes. Untreated and plasma-oxalyl chloride treated discs were placed under UV light for 30 min in order to sterilize them. Biomolecule-immobilized discs were sterilized by UV light for 10 min. An MTT assay was used in order to determine the proliferation of cells in the course of a 6-day cell culture period.

6.   Results and Discussion

In our previous study,[6] the glass transition temperature T_g, the melting temperature T_m and the values of melting enthalpy ΔH_m of PCL membranes were determined by differential scanning calorimetry (DSC) analysis at $-59.20°C$, $62.24°C$ and 68.00 Jg^{-1}, respectively. The FTIR spectrum of a PCL membrane exhibited the characteristic peaks of PCL structures in the range of 400–4,000 cm^{-1}.

6.1.   PLASMA TREATMENT

After evaluation with DoE, the best conditions for plasma treatment of further samples were selected as follows: flow rate = 5 sccm; pressure = 53 Pa; power = 35 W; time = 3.5 min. For these contiditions, the carboxyl density is high without decreasing the OH density too much; they are extremely close to some of the computed optimal conditions, and the values are easier to be controlled during the experiments.

The amount of carbon dioxide released by pyrolysis from the samples increases for oxalyl chloride functionalized samples proving the efficient attachment of carboxylic functionalities to the PCL surfaces. Water/O_2

plasma treated samples show a sensible increase of the CO_2 amount, which proves a negligible decarboxylation and the generation of carbonyl and/or carboxyl groups during the plasma treatment. ESCA results showed that decarboxylation of PCL membranes during the plasma procedure is negligible. However, free carboxyl and ester functionalities are overlapping in high resolution C 1s peaks, so labeling was performed for identifying/quantifying free carboxyls. High resolution AFM images (1×1 µm) revealed that nano patterns were more affected than micro patterns by the plasma treatments. AFM images recorded with HZD functionalized tips showed an increased size of the features on the surface, which suggests a higher density of carboxyls on these nanotopographical elements (Figure 1).

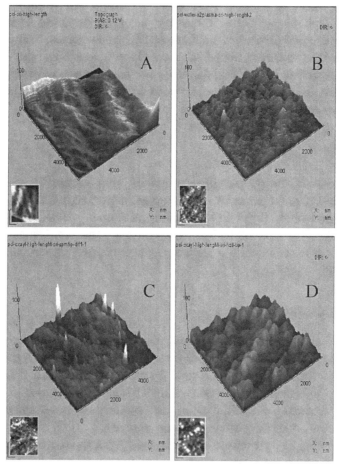

Figure 1. AFM images of PCL membranes: A – untreated; B – water/oxygen plasma treated; C – water/oxygen plasma treated and oxalyl chloride functionalized; D – same as C, recorded with a hydrazine functionalized tip.

6.2.   BIOMOLECULE IMMOBILIZATION

ESCA results proved the immobilization of biosignal molecules qualitatively. By using fluorescent labeling techniques, the average amounts of immobilized insulin and heparin on PCL surfaces were determined as 219.23 and 271.20 nmol/cm^2, respectively.

6.3.   L929 CELL CULTIVATION

According to the results of cell proliferation studies, heparin immobilized PCL samples were the most suitable materials for L929 cell growth (Figure 2).

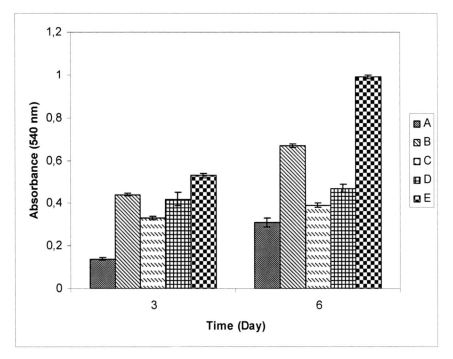

Figure 2. L929 fibroblast cell growth on PCL membranes. A: Unmodified; B: Low pressure water/O$_2$ plasma treated and then oxalyl chloride functionalized; C: PEO grafted; D: Insulin immobilized; E: Heparin immobilized.

7.   Conclusion

The present study showed that low pressure water/O$_2$ plasma assisted treatment method works well for the immobilization of biomolecules onto PCL membranes; these improved PCL membranes can be used as artificial tissue substituents. Our suggestion for related future studies is an investigation of the effects of the nano-topography created by the plasma and the bio-molecules on the cell growth in detail.

References

1. Y. Ikada, Biomaterials 15, 705 (1994).
2. M. Gümüşderelioğlu, H. Türkoğlu, Biomaterials 23, 3927 (2002).
3. H. Türkoğlu Şaşmazel, S. Manolache, M. Gümüşderelioğlu, J. Biomater. Sci.: Poly. Edition (2009), in press.
4. D. Falconnet, G. Csucs, H.M. Grandin, M. Textor, Biomaterials 27, 3044 (2006).
5. Denes, F.S., Manolache, S., Prog. Poly. Sci. 29 (2004).
6. H. Türkoğlu Şaşmazel, M. Gümüşderelioğlu, A. Gürpınar, M.A. Onur, Bio-Med. Mater. Eng. 18, 119 (2008).

SUBJECT INDEX

CONTRIBUTORS

Margarita Apostolova
Medical and Biological Research Lab
Roumen Tzanev Institute of Molecular
Biology
Bulgarian Academy of Sciences
Acad. G.Bonchev St., bl. 21
1113 Sofia
Bulgaria
margo@obzor.bio21.bas.bg

Zulfia Bakaeva
Heat Physics Department
Uzbekistan Academy of Sciences
700130 Tashkent
Uzbekistan
zulonok@yandex.ru

Ina Berezovska
O.O. Chuyko Institute of Surface
Chemistry
Ukrainian National Academy of
Sciences
General Naumov St. 17
03164 Kyiv
Ukraine
berinna2003@rambler.ru

Victor Boev
Institute of Electrochemistry and
Energy Systems
Bulgarian Academy of Sciences
Acad. G. Bonchev St., bl. 10
1113 Sofia
Bulgaria
v_boev@yahoo.com

Sylvia Boycheva
Department of Thermal and Nuclear
Engineering
Technical University of Sofia, 8 Kl.
Ohridsky Blvd. 1000 Sofia
Bulgaria
sylvia_boycheva@yahoo.com

Serap Dalgic
Department of Physics
Faculty of Art and Sciences
Trakya University
Güllapoglu Campus
22030 Edirne
Turkey
dserap@yahoo.com

Doriana Dimova-Malinovska
Central Laboratory of Solar Energy
and New Energy Sources
Bulgarian Academy of Sciences
72 Tzarigradsko chaussee Blvd.
1784 Sofia
Bulgaria
doriana@phys.bas.bg

Andriy Dmytruk
Institute of Physics of National
Academy of Sciences of Ukraine
Prosp. Nauky 46
03028 Kyiv
Ukraine
admytruk@gmail.com

Ema Fidancesvka
Faculty of Technology and Metallurgy
Ruger Boskovic 16
University St. Cyril and Methodius
1000 Skopje
FYR Macedonia
Republic of Macedonia
emilijaf@tmf.ukim.edu.mk

Elena Galeamov
Center of Optoelectronics
Institute of Applied Physics
Academy of Sciences of Moldova
Academiei St. 5
Chisinau MD-2028
Republic of Moldova
egaleamov@gmail.com

Plamen Ilchev
Central Laboratory of Photoprocesses
"Acad. J. Malinowski"
Bulgarian Academy of Sciences
Acad. G. Bonchev St., bl. 109
1113 Sofia
Bulgaria
p_ilchev2001@yahoo.com

Vanya Ilchev
Institute of Electrochemistry and
Energy Systems
Bulgarian Academy of Sciences
Acad. G. Bonchev St., bl.
1113 Sofia
Bulgaria
vania_ilcheva@yahoo.com

Daria Ilieva
Department of Physics
University of Chemical Technology
and Metallurgy
Metallurgy 8
Kl. Ohridski Blvd.
1756 Sofia
Bulgaria
darjailieva@abv.bg

Pal Jóvári
Research Institute for Solid State
Physics and Optics
H-1525 Budapest
POB 49
Hungary
jovari@sunserv.kfki.hu

Ivan Kaban
Institute of Physics
Chemnitz University of Technology
D-09107 Chemnitz
Germany
ivan.kaban@physik.tu-chemnitz.de

Kseniia Katok
O.O. Chuyko Institute of Surface
Chemistry
Ukrainian National Academy of
Sciences
17 Naumov St.
03164 Kyiv-164
Ukraine
smpl@ukr.net

Taras Kavetskyy
Institute of Materials
Scientific Research Company "Carat"
Stryjska 202
79031 Lviv
Ukraine
kavetskyy@yahoo.com

Justyna Keczkowska
Kielce University of Technology
25-314 Kielce
Poland
j.keczkowska@tu.kielce.pl

Bernt Ketterer
Forschungszentrum Karlsruhe, IMF I,
Hermann-von-Helmholtz-Platz 1,
D-76344 Eggenstein-Leopoldshafen,
Germany
Bernt.Ketterer@imf.fzk.de

Renat Khaydarov
Institute of Nuclear Physics
Uzbekistan Academy of Sciences
702132 Tashkent
Uzbekistan
renat2@gmail.com

Ofeliya Kostadinova
Institute of Chemical Engineering and
High Temperature Chemical Processes
FORTH/ICE-HT
GR-26 504 Patras
Greece
ofeliya@iceht.forth.gr

Alexander Kukhta
B.I. Stepanov Institute of Physics
National Academy of Sciences of
Belarus
Nezalezhnastsi Ave. 68
220072 Minsk
Belarus
kukhta@imaph.bas-net.by

Wilhelm Kulisch
Nanotechnology and Molecular
Imaging
Institute for Health and Consumer
Protection
European Commission Joint Research
Centre
Institute for Health and Consumer
Protection, Via Enrico Fermi
I-21020 Ispra (VA)
Italy
wilhelm.kulisch@jrc.it

Fotis Kyriazis
Institute of Chemical Engineering and
High Temperature Chemical Processes
FORTH/ICE-HT
GR-26 504 Patras
Greece

Dominic Lencer
I.Physikalisches Institut (IA), RWTH
Aachen University, 52056 Aachen,
Germany
lencer@physik.rwth-aachen.de

Charlie Main
University of Dundee
Division of Electronic Engineering
and Physics
Dundee DD1 4HN
UK
c.main@dundee.ac.uk

Joe Marshall
University Wales Swansea
UK
joe.marshall@killay9.freeserve.co.uk

Wolfgang Maser
Instituto de Carboquímica (CSIC),
Department of Nanotechnology,
C/Miguel Luesma Castán 4, E-50018
Zaragoza, Spain
wmaser@icb.csic.es

Boris Monchev
Institute of Electrochemistry and
Energy Systems
Bulgarian Academy of Sciences
Acad. G. Bonchev St., bl. 10
1113 Sofia
Bulgaria
boris_monchev@yahoo.com

Per Morgen
Institut for fysik og kemi
SDU
Campusvej 55
DK-5230 Odense M
Denmark
per@fysik.sdu.dk

George Mousdis
NHRF-National Hellenic Research
Foundation
Theoretical and Physical Chemistry
Institute-TPCI
48 Vass. Constantinou Ave.
Athens 11635
Greece
gmousdis@eie.gr

Christo Nichev
Central Laboratory of Solar Energy
and New Energy Sources
Bulgarian Academy of Sciences
Bulv. Tzarigradsko chaussee 72
1784 Sofia
Bulgaria
nitschew@yahoo.de

Adkham Paiziev
Positron Physics Laboratory
Arifov Institute of Electronics
Uzbek Academy of Science
100125 Tashkent
Uzbekistan

Perica Paunovic
University St. Cyril and Methodius
1000 Skopje
Faculty of Technology and Metallurgy
FYR Macedonia
paunovic@tmf.ukim.edu.mk

Plamen Petkov
Institute of Electrochemistry and
Energy Systems
University of Chemical Technology
and Metallurgy
Department of Physics
8 Kl. Ohridski Blvd.
1756 Sofia
Bulgaria
p.petkov@uctm.edu

Tamara Petkova
Institute of Electrochemistry and
Energy Systems
Bulgarian Academy of Sciences
Acad. G. Bonchev St., bl. 10
1113 Sofia
Bulgaria
tpetkova@bas.bg

Cyril Popov
Institute of Nanostructure
Technologies and Analytics
University of Kassel
Heinrich-Plett-St. 40
34132 Kassel
Germany
popov@ina.uni-kassel.de

Jean-Claude Pivin
Centre de Spectrométrie Nucléaire
et de Spectrométrie de Masse
CNRS-IN2P3
bâtioment 108
91405 Orsay Campus
France
pivin@csnsm.in2p3.fr

Annie Pradel
Institut Charles Gerhardt Montpellier
UMR 5253 UM2 CNRS ENSCM
UM1Equipe Physicochimie des
Matériaux Désordonnés et Poreux
CC 1503, Université Montpellier 2
F-34095 Montpellier Cedex 5
France
apradel@lpmc.univ-montp2.fr

Johann Peter Reithmaier
Technische Physik
Institute of Nanostructure
Technologies and Analytics
University of Kassel
Heinrich-Plett-St. 40
D-34132 Kassel
Germany
jpreith@physik.uni-kassel.de
jpreith@ina.uni-kassel.de

Hilal Turkoglu Şaşmazel
Atılım University
Department of Materials Engineering
Incek
Gölbaşı
Kizilcasar Mahallesi
06836 Incek-Ankara
Turkey
htsasmazel@atilim.edu.tr

Peter Sharlandjiev
Central Laboratory of Optical Storage
and Processing of Information
Bulgarian Academy of Sciences
Acad. G. Bonchev St., bl. 101
Sofia PS-1113
P.O. Box 95
Bulgaria
pete@optics.bas.bg

Toma Stankulov
Institute of Electrochemistry and
Energy Systems
Bulgarian Academy of Sciences
Acad. G. Bonchev St., bl. 10
1113 Sofia
Bulgaria
tstankulov@gmail.com

Andrey Stepanov
Kazan Physical-Technical Institute
Russian Academy of Sciences
Sibirsky trakt 10/7
420029 Kazan
Russia
anstep@kfti.knc.ru

Dan Tonchev
Department of Electrical
and Computer Engineering
University of Saskatchewan
Saskatoon, SK, S7N 5A9
Canada
dan.tonchev@usask.ca

Marina Ţurcan
Center of Optoelectronics
Institute of Applied Physics
Academy of Sciences of Moldova
Academiei St. 5, Chisinau
MD-2028
Republic of Moldova
tmaryna@yahoo.com

Sven Ulrich
Research Center Karlsruhe
Forschungszentrum Karlsruhe
GmbH
Institute for Materials Research I
Hermann-von-Helmholtz-Platz 1
76344 Eggenstein-Leopoldshafen
Germany
Sven.Ulrich@imf.fzk.de

Hristina Vasilchina
Research Center Karlsruhe,
Forschungszentrum Karlsruhe,
Institute for Materials Research I
Hermann-von-Helmholtz-Platz 1
76344 Eggenstein-Leopoldshafen
Germany
hrisi_v@yahoo.com

Miklos Veres
Research Institute for Solid State
Physics and Optics of the Hungarian
Academy of Sciences
H-1525 Budapest
Hungary
vm@szfki.hu

Spyros Yannopoulos
Foundation for Research and
Technology Hellas
Institute of Chemical Engineering and
High Temperature Chemical Processes
FORTH/ICE-HT
P.O. Box 1414
GR-26 504 Patras
Greece
sny@iceht.forth.gr

Arsham Yeremyan
Institute of Radiophysics and
Electronics of NAS of Armenia
1 Brs. Alikhanian St.
378410 Ashtarak
Armenia
arsham@irphe.am

Printed in the United States
141934LV00001BA/18/P